CRUDE AWAKENING

The Modern Truths and Future of Oil

Geoffrey Zachary

CONTENTS

Title Page
Part I: The History of Crude Oil ... 2
Chapter 1: Origins of Oil ... 3
Chapter 2: Early Uses of Oil ... 7
Chapter 3: The Birth of the Modern Oil Industry ... 12
Chapter 4: Rockefeller and Standard Oil ... 16
Chapter 5: Oil and the Industrial Revolution ... 21
Chapter 6: Oil and the Automobile ... 26
Chapter 7: The Birth of OPEC ... 31
Chapter 8: Oil Shocks of the 1970s ... 36
Chapter 9: The North Sea and New Frontiers ... 40
Chapter 10: The Gulf War and Oil ... 45
Part II: The Economics of Oil ... 50
Chapter 11: Oil Markets and Pricing ... 51
Chapter 12: Oil Futures and Speculation ... 57
Chapter 13: The Cost of Production ... 62
Chapter 14: Oil and National Economies ... 67
Chapter 15: The Petrodollar System ... 73
Chapter 16: Subsidies and Taxation ... 78
Chapter 17: Booms and Busts ... 84
Chapter 18: Oil and Global Trade ... 89

Chapter 19: The Economics of Renewable Energy vs. Oil	94
Chapter 20: Oil and Inflation	100
Part III: The Geopolitics of Oil	105
Chapter 21: Middle East Oil Dynamics	106
Chapter 22: Russia and Oil Power	112
Chapter 23: Oil in Africa	119
Chapter 24: US Oil Independence	124
Chapter 25: Oil Pipelines and Routes	130
Chapter 26: China's Oil Strategy	137
Chapter 27: Oil Embargoes and Sanctions	143
Chapter 28: Oil and War	149
Chapter 29: The Arctic Oil Rush	155
Chapter 30: The Future of OPEC	162
Part IV: Environmental Impact of Oil	168
Chapter 31: Oil Spills and Disasters	169
Chapter 32: Oil and Climate Change	175
Chapter 33: Carbon Footprint of Oil	181
Chapter 34: Oil and Air Pollution	188
Chapter 35: Water Contamination	195
Chapter 36: Oil and Biodiversity	202
Chapter 37: Mitigating Environmental Damage	209
Chapter 38: The Role of Environmental Regulations	216
Chapter 39: Renewable Energy Transition	223
Chapter 40: Corporate Responsibility	231
Part V: Technological Innovations in Oil	238
Chapter 41: Advanced Drilling Techniques	239
Chapter 42: Fracking and Shale Oil	247

Chapter 43: Deep water Drilling	254
Chapter 44: Enhanced Oil Recovery	262
Chapter 45: Digital Oilfields	270
Chapter 46: Oil Refining Innovations	278
Chapter 47: Transportation and Storage	285
Chapter 48: Safety Technologies	293
Chapter 49: Energy Efficiency in Oil Production	301
Chapter 50: Future Technologies	309
Part VI: The Social Impact of Oil	317
Chapter 51: Oil and Indigenous Communities	318
Chapter 52: Economic Inequality	325
Chapter 53: Public Health	332
Chapter 54: Job Creation and Loss	339
Chapter 55: Urbanization and Oil	347
Chapter 56: Cultural Changes	354
Chapter 57: Oil and Education	361
Chapter 58: Migration and Oil	369
Chapter 59: Human Rights and Oil	376
Chapter 60: Corporate Social Responsibility	383
Part VII: The Future of Oil	391
Chapter 61: Peak Oil Theory	392
Chapter 62: Global Oil Demand	399
Chapter 63: Alternative Energy Sources	407
Chapter 64: Electric Vehicles	415
Chapter 65: Energy Storage Technologies	422
Chapter 66: Climate Policies	430
Chapter 67: Investments in Renewable Energy	438
Chapter 68: Oil and Geopolitical Stability	447

Chapter 69: Sustainable Oil Practices	454
Chapter 70: Public Perception of Oil	463
Part VIII: Case Studies	470
Chapter 71: The Rise and Fall of Venezuela	471
Chapter 72: Norway's Oil Fund	478
Chapter 73: Saudi Arabia's Vision 2030	484
Chapter 74: Nigeria and Oil	492
Chapter 75: The US Shale Revolution	500
Chapter 76: Russia's Oil Strategy	508
Chapter 77: Canada's Oil Sands	516
Chapter 78: The North Sea Oil Boom	524
Chapter 79: Brazil's Offshore Oil	532
Chapter 80: Oil in the Gulf of Mexico	540
Part IX: Policy and Regulation	548
Chapter 81: International Oil Agreements	549
Chapter 82: National Oil Policies	557
Chapter 83: Environmental Regulations	566
Chapter 84: Taxation and Royalties	574
Chapter 85: Subsidies and Incentives	582
Chapter 86: Corporate Governance	590
Chapter 87: Anti-Corruption Measures	599
Chapter 88: Transparency Initiatives	607
Chapter 89: Climate Change Legislation	616
Chapter 90: Energy Security Policies	624
Part X: Innovations and Future Directions	633
Chapter 91: Biofuels and Oil	634
Chapter 92: Hydrogen Economy	642
Chapter 93: Carbon Capture and Storage	650

Chapter 94: Synthetic Fuels	658
Chapter 95: Advanced Refining Techniques	666
Chapter 96: Digital Transformation in Oil	674
Chapter 97: Innovations in Drilling Technology	682
Chapter 98: Oil and Artificial Intelligence	690
Chapter 99: Future of Oil Transportation	698
Chapter 100: The Role of Oil in a Sustainable Future	707

GEOFFREY ZACHARY

Crude Awakening

The Modern Truths and Future of Oil

PART I: THE HISTORY OF CRUDE OIL

CHAPTER 1: ORIGINS OF OIL

Crude Awakening: The Modern Truths and Future of Oil

The Dawn of Black Gold

As the sun dipped below the horizon, casting long shadows over the rugged terrain, a hidden world beneath our feet began to tell its ancient story. This is the story of crude oil, a resource that has shaped human civilization in profound ways, often compared to the lifeblood of modern industry. But how did this mysterious substance come to be, and what secrets does it hold from its inception millions of years ago?

The Geological Alchemy

Imagine a world vastly different from today—a prehistoric Earth where lush forests and vast seas dominated the landscape. Over 300 million years ago, during the Carboniferous period, the planet was a verdant paradise teeming with ferns, algae, and primitive trees. As these plants died, they fell into swampy waters, accumulating in thick, oxygen-poor layers that prevented complete decay.

As millennia passed, these organic layers were buried under sediment, compressing into a dense, carbon-rich material known as kerogen. The Earth's crust, dynamic and ever-moving, subjected these deposits to intense heat and pressure, initiating a slow but transformative process.

Deep within the Earth's mantle, the kerogen underwent a chemical metamorphosis, breaking down into liquid hydrocarbons—crude oil.

Unearthing the Past: A Journey Through Time

Fast forward to the early 19th century when human curiosity and technological advancement converged to uncover this buried treasure. The first commercial oil well, drilled by Edwin Drake in 1859 in Titusville, Pennsylvania, marked the beginning of a new era. But this was just the tip of the iceberg. Beneath the Earth's surface lay vast reserves of oil, each drop a testament to a time long forgotten.

Voices from the Past: The Human Connection

Consider the story of Maria Ramos, a geologist working in the arid deserts of Saudi Arabia. Her job is to piece together the Earth's history from rock samples, using clues to locate potential oil deposits. Maria's work is a blend of science and intuition, a modern-day treasure hunt. She recalls her first significant find—a reservoir holding millions of barrels of oil, a moment that connected her to the ancient Earth in a profound way.

"Every time we discover oil, it's like shaking hands with the past," Maria says. "It's humbling to think about the countless years and natural processes that went into creating this resource."

Contextual Insights: The Science Behind the Formation

To truly grasp the formation of crude oil, it's essential to understand the geochemical processes at play. The transformation from organic matter to oil is dependent on several factors:

1. Type of Organic Material: Different types of organic matter produce different types of oil. Marine organisms,

rich in lipids, tend to form lighter, sweeter oils, while terrestrial plants, abundant in cellulose and lignin, yield heavier, more sulphur-rich crude.

2. Temperature and Pressure: The "oil window" is the specific range of temperatures (approximately 60-120°C) and pressures that facilitate the conversion of kerogen to oil. Below this window, kerogen remains largely unchanged; above it, oil further breaks down into natural gas.

3. Time: The formation of oil is not a quick process. It requires millions of years of steady geological activity to convert organic material into usable crude.

Expert Insights: The Modern Implications

Dr. James Thompson, a petroleum engineer, explains how understanding the origins of oil is crucial for modern exploration. "By studying the conditions that led to oil formation, we can better predict where untapped reserves might be located. This not only aids in exploration but also in developing more efficient extraction technologies."

Thought-Provoking Questions: Reflecting on Our Reliance

As we delve deeper into the origins of oil, it's essential to ask ourselves: How sustainable is our dependence on this ancient resource? What alternatives exist that could lessen our environmental impact? And perhaps most crucially, how can we balance the demand for energy with the need to preserve our planet?

Storytelling Techniques: A Narrative Approach

Imagine a time-lapse journey, starting in the lush prehistoric swamps, traveling through the layers of sediment, and witnessing the transformation of organic material into crude oil. This narrative not only illustrates the scientific process but also connects readers emotionally to the Earth's dynamic history.

Actionable Insights: Looking to the Future

Understanding the origins of oil provides a foundation for discussing its future. As we navigate the complexities of energy production and environmental stewardship, it's clear that innovation is key. Investing in renewable energy sources, improving energy efficiency, and supporting sustainable practices are vital steps toward a balanced energy future.

Conclusion: The Legacy of Crude Oil

Crude oil's journey from ancient organic matter to a cornerstone of modern industry is a testament to the Earth's remarkable processes and humanity's ingenuity. As we continue to harness this resource, we must do so with an awareness of its origins and a commitment to a more sustainable future. This crude awakening is not just about understanding the past, but about shaping the path forward.

CHAPTER 2: EARLY USES OF OIL

Crude Awakening: The Modern Truths and Future of Oil

A Journey to the Beginning

As dawn breaks over the vast deserts of the Middle East and the ancient valleys of Mesopotamia, imagine the scene thousands of years ago. Tribes and early civilizations are beginning to discover the unique properties of a mysterious black substance seeping from the earth. This substance, known as bitumen or pitch, would become an integral part of their lives, long before the dawn of the modern oil industry. Let's journey back to explore how ancient civilizations utilized these natural oil seeps and the significant roles they played in shaping human history.

The Gift of the Earth

Natural oil seeps, where crude oil escapes from underground reservoirs to the surface, were common in many regions. Ancient peoples, ever resourceful, quickly recognized the value of this naturally occurring resource. Bitumen, a thick, sticky form of crude oil, was one of the first forms of petroleum that ancient civilizations used extensively.

Mesopotamian Marvels

One of the earliest and most significant utilizations of

oil seeps was in Mesopotamia, the cradle of civilization. The Sumerians, Akkadians, and Babylonians were among the first to harness the properties of bitumen. They used it primarily as a building material, leveraging its waterproofing qualities. The famous ziggurat of Ur, a massive step pyramid from around 2100 BCE, was constructed using bitumen to bond bricks and make the structure watertight. This ancient use of oil was crucial in ensuring the durability and stability of their monumental architecture.

Egyptian Ingenuity

In ancient Egypt, bitumen had a sacred purpose. It played a vital role in the mummification process. The Egyptians believed that preserving the body was essential for the soul's journey in the afterlife. Bitumen was used to embalm the bodies of the deceased, acting as a preservative that slowed decay and helped maintain the body's integrity for millennia. This practice not only highlights the early practical uses of oil but also its cultural and spiritual significance.

The Roads of Rome

Moving forward in history, the Romans also found innovative uses for natural oil seeps. They utilized bitumen to construct and maintain their extensive network of roads, which were vital for the expansion and control of their empire. Bitumen's adhesive and waterproofing properties made it an ideal material for binding stones and creating durable road surfaces that could withstand the wear and tear of time and travel. These ancient roads, some of which still exist today, were a testament to the practical applications of natural oil seeps.

Human Stories: Voices from the Past

Imagine the life of an ancient Mesopotamian builder, tasked with constructing a grand temple. He stands by the Euphrates River, mixing bitumen with mud bricks under the scorching sun. This simple yet profound act of utilizing the earth's resources would not only provide shelter and infrastructure but also leave an enduring legacy.

Consider also the perspective of an Egyptian priest; overseeing the embalming of a Pharaoh. The priest's careful application of bitumen during mummification rituals symbolized a deep connection to the divine, a belief in the power of preservation and eternal life.

Contextual Insights: The Science Behind the Substance

Bitumen is a naturally occurring form of crude oil, composed of complex hydrocarbons. Its sticky, viscous nature makes it an excellent adhesive and waterproofing agent. Modern science reveals that bitumen's ability to preserve organic material is due to its antimicrobial properties, which inhibit the growth of bacteria and fungi, preventing decay. These qualities made it invaluable to ancient peoples who needed durable building materials and methods to preserve their dead.

Expert Insights: The Archaeological Perspective

Dr. Linda Shepherd, an archaeologist specializing in ancient materials, emphasizes the importance of bitumen in early civilizations. "The use of bitumen was revolutionary," she says. "It provided a reliable, versatile material that ancient engineers and craftsmen could depend on. From the waterproofing of structures to the preservation of bodies, its impact on early human development cannot be overstated."

Thought-Provoking Questions: Reflecting on Ancient Practices

Reflecting on these ancient uses of oil prompts several questions: How did these early discoveries influence the development of human societies? What can we learn from their ingenuity and adaptability? And, as we face modern challenges, can we draw inspiration from their ability to utilize natural resources sustainably?

Storytelling Techniques: Bringing History to Life

To truly appreciate the early uses of oil, imagine a series of vivid scenes: the bustling construction sites of ancient Mesopotamia, the sacred embalming chambers of Egypt, and the meticulously built Roman roads stretching across Europe. These scenes connect us to the human stories behind the ancient use of oil, illustrating its profound impact on the development of civilization.

Actionable Insights: Lessons for Today

The innovative uses of oil by ancient civilizations offer valuable lessons for today. They demonstrate the importance of resourcefulness and sustainability. Modern society can learn from this historical ingenuity by finding new ways to use existing resources more efficiently and developing alternative materials and technologies that respect the environment.

Conclusion: The Legacy of Early Oil Uses

The early uses of natural oil seeps by ancient civilizations remind us of the resource's enduring significance. From construction and preservation to transportation and ritual, oil has played a crucial role in human development for millennia. As we continue to navigate the complexities of our energy future, remembering these ancient practices can inspire us to seek innovative, sustainable solutions that honour both our past and our planet. This crude awakening is not just a journey through history but a call

GEOFFREY ZACHARY

to embrace the wisdom of our ancestors as we forge a path forward.

CHAPTER 3: THE BIRTH OF THE MODERN OIL INDUSTRY

Crude Awakening: The Modern Truths and Future of Oil

The Dawn of a New Era

In the mid-19th century, a revolutionary discovery beneath the quiet hills of Pennsylvania sparked the birth of an industry that would transform the world. The modern oil industry, with its origins in a small American town, has become the backbone of the global economy, shaping industries, influencing geopolitics, and impacting daily life. But how did it all begin? Let's delve into the story of the first oil wells and the birth of the oil industry.

A Man with a Vision

In the summer of 1859, Edwin Drake, a former railroad conductor, stood at the edge of a small creek in Titusville, Pennsylvania, overseeing a motley crew of drillers and labourers. Drake, hired by the Seneca Oil Company, was on a mission to find a reliable source of oil. At that time, oil was primarily collected from natural seeps and used for medicinal purposes and as lamp fuel. The challenge was to find a method to extract oil in large quantities.

Drake's initial attempts met with scepticism and failure. The local townspeople mocked him, calling his endeavour "Drake's Folly." However, Drake's persistence and ingenuity led him to adapt a technique used in drilling salt wells. He decided to drill deeper into the earth, using a steam engine to power the drill.

The Birth of the Oil Well

On August 27, 1859, after drilling to a depth of 69 feet, Drake's crew struck oil. Black gold gushed from the ground, marking the birth of the first commercial oil well. This event was a watershed moment, heralding the dawn of the modern oil industry. News of Drake's success spread quickly, attracting entrepreneurs, speculators, and workers to Titusville, transforming it into a bustling hub of activity.

Human Stories: Pioneers of the Oil Rush

Among the early oil pioneers was "Colonel" Edwin Drake, who despite his pivotal role, did not profit significantly from his discovery. Drake spent his later years in relative obscurity, suffering financial hardships. His story, a mix of triumph and tragedy, reflects the volatile nature of the nascent oil industry.

Another notable figure was John D. Rockefeller, who entered the oil industry in the 1860s. Recognizing the potential of this new resource, Rockefeller founded Standard Oil Company in 1870. Through strategic acquisitions and innovations in refining, transportation, and marketing, Rockefeller built a vast empire that would dominate the industry for decades. His story illustrates the rapid growth and consolidation that characterized the early oil industry.

Contextual Insights: The Technological Breakthroughs

Drake's success was not just a stroke of luck; it was built on a series of technological advancements. The use of a steam engine to power the drill was a significant innovation, allowing for deeper and more efficient drilling. Additionally, the development of the derrick, a framework over the drill hole, improved the stability and safety of drilling operations. These innovations set the stage for the rapid expansion of the oil industry.

Expert Insights: The Economic and Social Impact

Dr. Sarah Lawson, an economic historian, emphasizes the transformative impact of the early oil industry. "The discovery of commercial oil wells in the 19th century was a turning point. It provided a new, efficient source of energy that fuelled industrial growth, powered transportation, and spurred economic development. The rise of the oil industry also had profound social implications, creating new jobs, fostering urbanization, and shaping modern society."

Thought-Provoking Questions: Reflecting on the Birth of an Industry

The birth of the modern oil industry raises several important questions: How did the early pioneers' vision and innovation shape the future of energy? What can we learn from their successes and failures? And as we face contemporary energy challenges, how can we draw inspiration from their resilience and ingenuity?

Storytelling Techniques: Bringing History to Life

Picture the bustling streets of Titusville in the 1860s, filled with aspiring oilmen, merchants, and labourers. The air is thick with the smell of crude oil, and the sound of drills echoes through the town. Visualize the transformation of a quiet, rural area into a vibrant epicentre of economic activity. These vivid scenes connect readers to the human experiences behind the birth of the oil industry.

Actionable Insights: Lessons for the Future

The early days of the oil industry offer valuable lessons for today. Innovation, persistence, and adaptability were key to the success of early oil pioneers. Modern industries can learn from these principles by fostering a culture of innovation, investing in research and development, and being adaptable to changing circumstances. Additionally, the environmental impacts of early oil extraction underscore the importance of sustainable practices in today's energy production.

Conclusion: A Legacy of Innovation

The birth of the modern oil industry was a momentous event that reshaped the world. From Edwin Drake's first well to the rise of industry giants like John D. Rockefeller, the early pioneers of the oil industry demonstrated vision, ingenuity, and determination. As we navigate the complexities of our current energy landscape, remembering the origins of the oil industry can inspire us to pursue innovative, sustainable solutions. This crude awakening is a testament to human potential and a call to shape a future that honour s both our history and our planet.

CHAPTER 4: ROCKEFELLER AND STANDARD OIL

Crude Awakening: The Modern Truths and Future of Oil

The Rise of an Industrial Titan

In the late 19th century, one man emerged as a central figure in the nascent oil industry, transforming it into a global powerhouse. John D. Rockefeller, through his company Standard Oil, not only revolutionized the production, refining, and distribution of oil but also left an indelible mark on the business world. His legacy is a complex tapestry of innovation, monopoly, philanthropy, and controversy. This chapter delves into Rockefeller's pivotal role in shaping the modern oil industry.

The Humble Beginnings

John D. Rockefeller was born in 1839 in Richford, New York, to a modest family. His early life was marked by frugality and a strong work ethic, traits instilled by his mother. These characteristics would later define his business philosophy. Rockefeller's entry into the oil industry began in 1863 when he invested in a refinery in Cleveland, Ohio. Recognizing the burgeoning demand for kerosene and other petroleum products, he saw an opportunity to streamline the chaotic and competitive oil market.

The Formation of Standard Oil

In 1870, Rockefeller co-founded the Standard Oil Company with the aim of consolidating and controlling the oil industry. His strategy was both innovative and ruthless. He employed vertical integration, controlling every aspect of the oil business—from extraction to refining, transportation, and distribution. This approach not only reduced costs but also allowed Standard Oil to dominate the market.

Rockefeller's business acumen extended to his use of modern technologies and efficient management practices. He introduced pipelines to transport oil, reducing dependency on railroads and lowering transportation costs. Standard Oil's refineries employed the latest techniques, ensuring higher yields and better quality products. By negotiating favourable rates with railroads, often through secretive rebates and preferential treatment, Rockefeller further strengthened his competitive edge.

The Monopoly and Its Impact

By the late 1880s, Standard Oil controlled approximately 90% of the oil refining capacity in the United States. This dominance led to significant economies of scale and immense profits. However, it also attracted scrutiny and criticism. Rockefeller's aggressive tactics, such as buying out competitors and securing exclusive deals, were seen as monopolistic and anti-competitive.

The public and government response to Standard Oil's monopoly was mixed. While many admired Rockefeller's business success and contributions to the economy, others decried his methods as unethical. The growing concern over Standard Oil's dominance led to legal challenges, culminating in the landmark antitrust case of 1911.

The U.S. Supreme Court ruled that Standard Oil must be dismantled, resulting in the creation of 34 separate companies, including Exxon, Mobil, and Chevron.

Human Stories: The Rockefeller Legacy

Rockefeller's story is not just one of business triumph but also of personal contradiction. Despite his ruthless business tactics, he was known for his philanthropy. Over his lifetime, Rockefeller donated over $500 million to various causes, including education, public health, and scientific research. Institutions such as the University of Chicago and the Rockefeller Foundation owe their existence to his generosity.

One poignant story is that of Ida Tarbell, a journalist whose investigative reporting on Standard Oil played a key role in its eventual breakup. Tarbell's father had been driven out of business by Rockefeller's aggressive tactics, fuelling her determination to expose the company's practices. Her work highlights the personal toll of Rockefeller's business empire and underscores the complex interplay between industry and individual lives.

Contextual Insights: The Industrial Revolution and Beyond

Rockefeller's rise coincided with the broader context of the Industrial Revolution, a period marked by rapid technological advancement and economic growth. The oil industry was a crucial component of this transformation, providing energy for factories, lighting for homes, and fuel for transportation. Rockefeller's ability to harness and control this resource was instrumental in shaping modern industrial society.

Expert Insights: Economic and Legal Perspectives

Dr. Michael Watson, an economist specializing in industrial organization, notes, "Rockefeller's impact on the oil industry is unparalleled. His use of vertical integration and economies of scale set the blueprint for modern corporations. However, his monopolistic practices also highlighted the need for regulatory oversight to ensure fair competition."

Legal expert Professor Anne Clark adds, "The Standard Oil case was a turning point in antitrust law. It established the principle that monopolistic practices, even if they lead to efficiency gains, are harmful to competition and consumer welfare. This case laid the foundation for future antitrust regulations."

Thought-Provoking Questions: Reflecting on Rockefeller's Legacy

Rockefeller's legacy prompts several critical questions: How do we balance the drive for business efficiency with the need for fair competition? What lessons can modern corporations learn from Standard Oil's rise and fall? And how can we ensure that the benefits of industrial and technological advancements are distributed equitably?

Storytelling Techniques: A Narrative Approach

To bring Rockefeller's story to life, imagine the bustling streets of Cleveland in the 1870s, where Rockefeller's refinery operations began. Picture the intricate network of pipelines snaking across the country, a testament to his vision and ingenuity. These vivid scenes help readers connect with the historical narrative and appreciate the human and societal dimensions of Rockefeller's impact.

Actionable Insights: Lessons for Today's Industry

Rockefeller's story offers valuable lessons for contemporary business leaders and policymakers. It underscores the importance of innovation, strategic vision, and efficiency in building successful enterprises.

At the same time, it highlights the need for ethical practices and regulatory frameworks to ensure that market dominance does not stifle competition or harm public interests.

Conclusion: The Dual Legacy of John D. Rockefeller

John D. Rockefeller's role in shaping the modern oil industry is a testament to both the potential and pitfalls of industrial capitalism. His pioneering efforts in establishing Standard Oil revolutionized the industry, setting standards for efficiency and innovation. However, his monopolistic practices also led to significant legal and ethical challenges, prompting a re-evaluation of the balance between business interests and public good.

As we navigate the complexities of the contemporary energy landscape, Rockefeller's legacy serves as a reminder of the enduring impact of visionary leadership, the importance of competitive markets, and the necessity of balancing economic growth with societal welfare. This crude awakening invites us to reflect on the lessons of the past as we chart a sustainable and equitable future.

CHAPTER 5: OIL AND THE INDUSTRIAL REVOLUTION

Crude Awakening: The Modern Truths and Future of Oil

The Wheels of Progress

As the 19th century dawned, the world stood on the brink of monumental change. The Industrial Revolution, characterized by the shift from agrarian economies to industrial powerhouses, was well underway. Steam engines roared to life, factories multiplied, and railways stretched across continents. Yet, it was the discovery and utilization of oil that would truly turbocharge this era, transforming economies and societies in ways that were previously unimaginable.

The First Sparks

Before oil became the lifeblood of the industrial world, coal was the primary energy source driving the Industrial Revolution. Coal powered steam engines, locomotives, and factories, catalysing unprecedented industrial growth. However, the limitations of coal, including its heavy smoke and labour-intensive extraction, soon became apparent. Enter oil—lighter, more energy-dense, and easier to transport.

Illuminating the Night: Kerosene and Lighting

In the early days, oil's most significant contribution to the Industrial Revolution was through kerosene. By the mid-19th century, kerosene lamps had become a popular means of illumination, replacing whale oil lamps and candles. Kerosene was cheaper and more efficient, bringing light to homes and factories and extending working hours. This simple innovation had profound effects on productivity and quality of life.

Imagine a bustling 19th-century factory, where workers once confined to daylight hours could now continue their tasks late into the night. The increased production capacity fuelled economic growth, and cities began to buzz with activity around the clock.

The Rise of Internal Combustion

The real game-changer, however, came with the advent of the internal combustion engine. Invented in the latter part of the 19th century, these engines ran on gasoline, a refined product of crude oil. Unlike steam engines, which were large and cumbersome, internal combustion engines were compact, efficient, and versatile.

Henry Ford's Model T, introduced in 1908, epitomized this revolution. Affordable and mass-produced, the Model T brought personal transportation to the masses. Roads teemed with automobiles, reducing travel times, and connecting distant regions like never before. The internal combustion engine also powered agricultural machinery, transforming farming practices and boosting food production to support growing urban populations.

Human Stories: Life Transformed by Oil

Consider the life of a factory worker in the late 1800s. Before the widespread use of oil, his day was dictated by the sun. With the advent of kerosene lamps, his factory could operate longer hours, increasing his earnings but also intensifying his workload. As internal combustion engines began to replace steam-powered machinery, his job became more efficient but also demanded new skills.

Or picture a farmer who, with the help of gasoline-powered tractors, could now plough fields in a fraction of the time, increasing his productivity and enabling him to cultivate larger areas. This mechanization transformed rural life, contributing to the migration of labour to urban centres and fuelling the growth of cities.

Contextual Insights: The Economic Boom

Oil not only powered machines but also lubricated the gears of economic progress. The availability of cheap and efficient energy sources facilitated mass production, leading to economies of scale. Factories could produce goods faster and more cheaply, making products accessible to a broader population and driving consumerism.

The infrastructure of the modern world—roads, bridges, and ports—expanded rapidly, supported by oil-fuelled machinery. Industries such as textiles, steel, and chemicals thrived, creating jobs, and fostering technological innovations. The ripple effects of oil's integration into the economy were vast and transformative, touching nearly every aspect of life.

Expert Insights: Perspectives on Energy and Growth

Dr. Alan Richards, an economic historian, notes, "The role of oil in the Industrial Revolution cannot be overstated. It provided the energy density and versatility required to sustain large-scale industrial operations, driving unprecedented economic growth and societal change."

Energy analyst Maria Lopez adds, "The transition from coal to oil was pivotal. It marked the beginning of a new era in energy consumption, characterized by mobility and efficiency. The implications for global trade, transportation, and daily life were profound."

Thought-Provoking Questions: Reflecting on Energy Transitions

The integration of oil into the Industrial Revolution raises several important questions: How did this energy transition shape the trajectory of modern economies? What lessons can we draw from the historical shift from coal to oil as we face contemporary energy challenges? And how might future energy transitions compare in terms of economic and social impact?

Storytelling Techniques: Bringing History to Life

To capture the essence of oil's impact on the Industrial Revolution, imagine the transformation of a typical 19th-century city. Visualize the smoky, coal-driven factories giving way to cleaner, more efficient oil-powered machinery. Picture the bustling streets filled with Model T cars, the glow of kerosene lamps in homes and shops, and the hum of new industries rising on the horizon. These scenes help readers connect with the historical narrative on a personal level.

Actionable Insights: Lessons for Today

The story of oil and the Industrial Revolution offers valuable insights for contemporary society. As we confront the challenges of climate change and the need for sustainable energy sources, the history of oil underscores the importance of innovation, adaptability, and strategic planning in energy transitions. Investing in renewable energy technologies, improving energy efficiency, and fostering global cooperation are crucial steps toward a sustainable future.

Conclusion: The Power of Transformation

Oil's role in the Industrial Revolution was a catalyst for unprecedented economic and societal transformation. From lighting homes to powering machines, oil fuelled progress and shaped the modern world. As we reflect on this pivotal era, we are reminded of the potential for energy to drive innovation and change. This crude awakening invites us to harness the lessons of the past as we navigate the complexities of our energy future, seeking solutions that honour both our history and our planet.

CHAPTER 6: OIL AND THE AUTOMOBILE

Crude Awakening: The Modern Truths and Future of Oil

The Dawn of a New Age

At the turn of the 20th century, the world stood at the threshold of an unprecedented transformation. The invention and popularization of the automobile revolutionized transportation, reshaping economies, cities, and daily life. Central to this revolution was oil, the lifeblood of the internal combustion engine. This chapter explores the rise of the automobile industry and its profound reliance on oil, charting a journey of innovation, expansion, and the complexities that come with dependency on a finite resource.

The Birth of the Automobile Industry

In the late 19th century, inventors across Europe and the United States were experimenting with various forms of powered transportation. The breakthrough came with Karl Benz, who in 1886, patented the first automobile powered by an internal combustion engine. This invention set the stage for the automotive revolution.

Yet, it was Henry Ford who truly brought the automobile to the masses. In 1908, Ford introduced the Model T, an affordable, reliable car that transformed personal mobility. The secret to the Model T's success was the assembly line, which revolutionized manufacturing by drastically reducing production costs and time. By making cars accessible to the average American, Ford ignited a new era in transportation and consumer culture.

Human Stories: The Automobile and Daily Life

Imagine the life of a farmer in rural America in the early 1900s. Before the automobile, his trips to town were infrequent and arduous, often taking an entire day by horse and carriage. With the advent of the Model T, his world expanded dramatically. He could now travel more frequently and further, connecting with markets, social events, and educational opportunities. The automobile brought a new sense of freedom and possibility to everyday life.

Consider also the burgeoning middle class in urban areas. The automobile became a symbol of status and independence, allowing families to explore beyond their neighbourhoods, take vacations, and commute to better job opportunities. The car was not just a means of transportation; it was a gateway to a new lifestyle.

Oil: The Fuel of Progress

The rise of the automobile industry was inextricably linked to the demand for oil. Internal combustion engines required gasoline, a refined product of crude oil. As car ownership surged, so did the need for fuel. This demand transformed the oil industry, driving exploration, drilling, and refining to new heights.

In the early 20th century, the United States emerged as a global leader in oil production. Regions like Texas, California, and Oklahoma became synonymous with oil booms, attracting workers, investors, and speculators. The oil industry rapidly evolved to meet the burgeoning needs of the automobile market, leading to innovations in drilling techniques, pipeline construction, and refining processes.

Contextual Insights: The Impact on Society and Economy

The automobile industry's reliance on oil had far-reaching implications. Economically, it spurred growth in related sectors such as steel, rubber, glass, and road construction. The proliferation of cars led to the development of infrastructure, including highways, gas stations, and suburban communities. This network of roads and services facilitated commerce and communication, knitting the country together in new ways.

Socially, the automobile changed the fabric of daily life. It provided unprecedented mobility and freedom, reshaped urban planning, and influenced cultural norms. Car ownership became a rite of passage, a marker of adulthood and independence. The automobile also played a pivotal role in the rise of consumer culture, with advertising campaigns and car designs appealing to the desires and aspirations of the public.

Expert Insights: The Environmental Cost

While the automobile brought many benefits, it also came with significant environmental costs. Dr. Emily Harris, an environmental scientist, notes, "The widespread use of automobiles has had profound environmental impacts, from air pollution and greenhouse gas emissions to habitat destruction and urban sprawl. The reliance on oil has also led to geopolitical tensions and conflicts over resource control."

Energy analyst Mark Thompson adds, "The challenge we face today is balancing the benefits of mobility with the need for sustainable energy solutions. The automotive industry is at a crossroads, with electric vehicles and alternative fuels presenting opportunities to reduce our dependence on oil."

Thought-Provoking Questions: Reflecting on Our Journey

The rise of the automobile and its reliance on oil prompts several critical questions: How do we reconcile the benefits of personal mobility with the environmental and social costs? What role should governments and industries play in transitioning to sustainable transportation solutions? And how can consumers drive change through their choices and behaviours?

Storytelling Techniques: Bringing the Narrative to Life

To illustrate the impact of the automobile, imagine a day in the life of an early 20th-century family. Picture their excitement as they take their first road trip, the sense of adventure as they explore new destinations, and the convenience of running errands with ease. These personal stories highlight the transformative power of the automobile and the integral role of oil in fuelling this change.

Actionable Insights: Lessons for the Future

The history of the automobile and its reliance on oil offers valuable lessons for addressing today's challenges. Embracing innovation, supporting sustainable transportation policies, and investing in alternative energy sources are crucial steps toward a more sustainable future. Public and private sectors must collaborate to develop infrastructure for electric vehicles, promote public transportation, and encourage energy-efficient technologies.

Conclusion: The Road Ahead

The rise of the automobile and its dependence on oil is a story of innovation, transformation, and complexity. As we navigate the 21st century, we must learn from this history to create a balanced, sustainable transportation future. This crude awakening invites us to reflect on our journey, understand the impacts of our choices, and strive for a world where mobility enhances life without compromising the health of our planet.

CHAPTER 7: THE BIRTH OF OPEC

Crude Awakening: The Modern Truths and Future of Oil

A New Force in the Global Oil Market

In the early 1960s, a seismic shift occurred in the global oil industry. Nations rich in oil, but often marginalized by the dominant Western oil companies, banded together to form a powerful new entity: the Organization of the Petroleum Exporting Countries (OPEC). This chapter delves into the formation of OPEC, its evolution, and its profound influence on global oil markets, providing insights into one of the most significant developments in the history of oil.

The Seeds of Collaboration

The story of OPEC begins in a time of growing nationalism and decolonization. After World War II, many countries in the Middle East and other oil-rich regions sought greater control over their natural resources. The "Seven Sisters"—a term used to describe the seven major Western oil companies dominating the global oil market—had long dictated the terms of oil production and pricing, reaping substantial profits while paying relatively low royalties to the host countries.

In response to this imbalance, the governments of Iran, Iraq, Kuwait, Saudi Arabia, and Venezuela met in Baghdad in September 1960. These five founding members shared a common goal: to unify and coordinate their oil policies, ensuring fair and stable prices for petroleum producers and securing an equitable return on their investments.

The Founding of OPEC

The Baghdad Conference marked the birth of OPEC. The founding members established a framework to collaborate on oil production policies and pricing. Their mission was clear: to protect their interests, assert their sovereignty over natural resources, and stabilize the global oil market. The formation of OPEC was a bold move that challenged the hegemony of the Western oil companies and shifted the balance of power in the global oil industry.

Human Stories: The Visionaries Behind OPEC

Key figures played pivotal roles in the formation of OPEC. Abdullah Tariki of Saudi Arabia and Juan Pablo Pérez Alfonzo of Venezuela were instrumental in bringing the founding members together. Known as the "Red Sheikh" for his nationalist views, Tariki was a passionate advocate for the control of oil resources by producing countries. Pérez Alfonzo, often referred to as the "father of OPEC," was a visionary who saw the potential of a united front to counter the influence of the Seven Sisters.

These leaders faced immense challenges, including political pressures and scepticism from within and outside their countries. Their determination and foresight laid the groundwork for an organization that would become a major player in global economics and geopolitics.

Contextual Insights: OPEC's Influence on the Oil Market

OPEC's influence on the global oil market grew steadily in the following decades. By coordinating production levels among member countries, OPEC could influence oil prices and stabilize the market. The organization's first major test came during the Arab-Israeli War of 1973. In response to Western support for Israel, OPEC members imposed an oil embargo, leading to a dramatic increase in oil prices and highlighting the West's vulnerability to oil supply disruptions.

The 1973 oil crisis had far-reaching economic and political repercussions. It underscored the strategic importance of oil and the leverage that OPEC wielded. Western nations, particularly the United States, faced soaring energy costs, inflation, and economic stagnation. The crisis also prompted a re-evaluation of energy policies and a push for energy independence through the development of alternative energy sources and strategic oil reserves.

Expert Insights: Economic and Geopolitical Perspectives

Dr. Ahmed Zaki Yamani, Saudi Arabia's influential oil minister during the 1970s, famously remarked, "The Stone Age didn't end because we ran out of stones, and the oil age will end long before the world runs out of oil." His statement reflects the understanding that OPEC's power lay not just in its oil reserves, but in its ability to shape the future of energy.

Energy economist Dr. Laura Smith explains, "OPEC's role in the global oil market is multifaceted. It involves balancing the interests of member countries with diverse economic needs and political agendas, managing production to influence prices, and navigating the complex geopolitics of energy."

Thought-Provoking Questions: The Legacy of OPEC

OPEC's formation and its ongoing role in the global oil market raise several important questions: How has OPEC's influence shaped the geopolitical landscape? What challenges does the organization face in an era of fluctuating oil prices and growing calls for renewable energy? And how will OPEC adapt to the changing dynamics of global energy consumption?

Storytelling Techniques: A Narrative Approach

To bring the formation of OPEC to life, imagine the tense and hopeful atmosphere of the Baghdad Conference in 1960. Picture the leaders of Iran, Iraq, Kuwait, Saudi Arabia, and Venezuela coming together, driven by a shared vision of fairness and sovereignty over their resources. These moments capture the human elements behind the birth of an organization that would alter the course of history.

Actionable Insights: Lessons for Today's Energy Landscape

OPEC's history offers valuable lessons for contemporary energy policy and market dynamics. The importance of collaboration and strategic planning in managing natural resources cannot be overstated. As the world transitions towards more sustainable energy sources, the principles of fair pricing, resource sovereignty, and market stability remain relevant.

Policymakers and industry leaders can learn from OPEC's experiences by fostering international cooperation, investing in alternative energy research, and preparing for the economic impacts of energy transitions. Balancing the immediate needs of energy security with long-term sustainability goals is crucial for a stable and prosperous future.

Conclusion: The Enduring Legacy of OPEC

The formation of OPEC marked a turning point in the history of oil, symbolizing the assertion of control by oil-producing nations over their resources. Over the decades, OPEC has navigated a complex landscape of economic, political, and environmental challenges, maintaining its relevance in the global oil market.

As we look to the future, OPEC's legacy serves as a reminder of the power of unity and strategic vision. This crude awakening encourages us to reflect on the lessons of the past as we face the energy challenges of the 21st century. By understanding the historical context and ongoing influence of OPEC, we can better navigate the path towards a balanced and sustainable energy future.

CHAPTER 8: OIL SHOCKS OF THE 1970S

Crude Awakening: The Modern Truths and Future of Oil

The Seismic Shifts

The 1970s were a decade of upheaval and transformation, marked by significant oil shocks that reshaped the global economy and geopolitical landscape. These oil crises highlighted the vulnerability of industrialized nations to disruptions in oil supply and underscored the strategic importance of energy security. This chapter explores the oil crises of the 1970s, their causes, and their profound economic impacts worldwide.

The First Oil Shock: The 1973 Oil Embargo

The first major oil crisis erupted in October 1973, during the Yom Kippur War between Israel and a coalition of Arab states led by Egypt and Syria. In response to the United States and other Western countries' support for Israel, the Organization of Arab Petroleum Exporting Countries (OAPEC) imposed an oil embargo. The embargo targeted nations perceived as allies of Israel, drastically reducing oil exports to these countries.

The immediate impact was a sharp increase in oil prices. Within months, the price of oil quadrupled, soaring from around $3 per barrel to nearly $12 per barrel. The sudden spike in prices sent shockwaves through the global economy, causing widespread inflation, unemployment, and economic stagnation.

Human Stories: Lives Upended by the Crisis

For many, the oil embargo translated into long lines at gas stations, fuel shortages, and skyrocketing energy bills. Consider the plight of John, a truck driver in the United States, who suddenly found himself struggling to afford fuel for his deliveries. His income plummeted as transportation costs soared, highlighting the personal toll of the crisis.

In Europe, families like the Martinezes in Spain faced similar hardships. They had to endure cold winters with limited heating oil, prompting governments to introduce rationing and encourage energy conservation. The oil embargo brought the fragility of energy dependence into sharp relief for millions of ordinary people.

Contextual Insights: The Economic Domino Effect

The 1973 oil crisis had far-reaching economic consequences. The dramatic increase in energy costs led to higher production costs for goods and services, triggering inflation across the board. Central banks, particularly the Federal Reserve in the United States, responded by raising interest rates to combat inflation, which in turn slowed economic growth and increased unemployment.

The crisis also exposed the vulnerability of heavily industrialized economies reliant on imported oil. Nations scrambled to find alternative energy sources, invest in domestic oil production, and implement energy-saving measures. The shock highlighted the interconnectedness of the global economy and the pivotal role of oil as a strategic resource.

The Second Oil Shock: The Iranian Revolution

The second oil shock occurred in 1979, following the Iranian Revolution. The overthrow of the Shah of Iran and the establishment of the Islamic Republic led to significant disruptions in oil production and exports from one of the world's largest oil producers. The ensuing uncertainty and instability in the region caused oil prices to surge once again, doubling from $14 per barrel to $35 per barrel within a year.

Expert Insights: Geopolitical and Economic Perspectives

Dr. Mark Hudson, an economist specializing in energy markets, explains, "The oil shocks of the 1970s were a wake-up call for the global economy. They revealed the strategic importance of oil and the need for diversified energy sources. The crises also led to significant policy shifts aimed at reducing dependence on Middle Eastern oil."

Energy analyst Sarah Coleman adds, "The geopolitical landscape of the 1970s was profoundly shaped by the oil crises. They underscored the power of oil-producing nations and the need for consuming countries to rethink their energy policies and strategies."

Thought-Provoking Questions: Lessons from the Past

Reflecting on the oil crises of the 1970s raises several important questions: How have these historical events influenced current energy policies and strategies? What steps can be taken to mitigate the impact of similar crises in the future? And how can nations balance the need for energy security with the push for sustainable and renewable energy sources?

Storytelling Techniques: Bringing the Narrative to Life

To capture the essence of the 1970s oil shocks, imagine the chaos and uncertainty of the era. Picture the lines of cars at gas stations, the dimly lit homes, and the heated debates in government chambers. These scenes help readers connect with the human and societal dimensions of the crises, illustrating the far-reaching impact of energy dependency.

Actionable Insights: Preparing for Future Challenges

The oil shocks of the 1970s offer valuable lessons for addressing today's energy challenges. Investing in diversified and renewable energy sources, improving energy efficiency, and developing strategic reserves are crucial steps toward enhancing energy security. Governments and industries must collaborate to create resilient energy infrastructures that can withstand geopolitical and economic disruptions.

Conclusion: The Enduring Legacy of the 1970s Oil Shocks

The oil crises of the 1970s were transformative events that reshaped the global economy and energy landscape. They highlighted the vulnerabilities of oil-dependent economies and underscored the need for strategic energy policies. As we navigate the complexities of the 21st-century energy landscape, the lessons of the past provide a roadmap for building a more secure and sustainable future.

This crude awakening encourages us to reflect on the interconnectedness of global energy systems and the importance of strategic planning and innovation in ensuring energy security. By learning from the challenges and successes of the 1970s, we can better prepare for the energy transitions and geopolitical uncertainties of tomorrow.

CHAPTER 9: THE NORTH SEA AND NEW FRONTIERS

Crude Awakening: The Modern Truths and Future of Oil

The Search Beyond the Known

The story of oil is one of constant exploration and discovery. As the demand for oil surged in the mid-20th century, the search for new reserves led to the challenging environments of the North Sea and other remote frontiers. This chapter explores the development of oil fields in the North Sea and other new frontiers, highlighting the technological innovations, economic impacts, and environmental considerations that accompanied these ventures.

The North Sea: A New Horizon

In the 1960s, the North Sea, lying between Great Britain and Scandinavia, emerged as a promising site for oil exploration. Initial geological surveys indicated significant reserves of oil and natural gas beneath the seabed. However, the harsh conditions of the North Sea posed formidable challenges, requiring advancements in offshore drilling technology.

Technological Triumphs

The development of the North Sea oil fields necessitated pioneering technologies and engineering feats. The construction of platforms that could withstand violent storms and rough seas was a monumental task. The Ekofisk field, discovered in 1969 and developed by Phillips Petroleum, marked one of the first major successes. Engineers designed the Condeep platforms, massive concrete structures anchored to the seabed, which provided the stability needed to drill in deep and turbulent waters.

Subsea technology also saw significant advancements. Remote-operated vehicles (ROVs) and dynamic positioning systems allowed for precise drilling and maintenance operations, even under the challenging conditions of the North Sea. These innovations not only made North Sea oil extraction feasible but also set new standards for offshore drilling globally.

Economic Impact

The discovery and development of North Sea oil fields had profound economic implications for the surrounding countries. The United Kingdom and Norway, in particular, experienced significant economic benefits. The influx of oil revenue bolstered national economies, funded social programs, and supported public infrastructure projects.

In the UK, the 1970s saw a shift from being an oil importer to a major oil producer. The newfound wealth contributed to economic stability and growth, particularly during times of global oil price volatility. Norway established a sovereign wealth fund, the Government Pension Fund Global, which invested oil revenues for future generations, ensuring long-term economic security.

Human Stories: Life on the Rigs

The life of an oil rig worker in the North Sea is a tale of resilience and camaraderie. Workers endure long shifts, often in isolation from their families, facing the constant threat of harsh weather. Despite these challenges, a strong sense of community and purpose binds them.

Take the story of James, a Scottish rig worker who spent decades in the North Sea. He recalls the camaraderie among his crew, the intense training for safety, and the pride in contributing to his country's energy independence. For James, the rig was more than a workplace; it was a second home where bonds forged in adversity created lifelong friendships.

New Frontiers: Beyond the North Sea

The success of North Sea oil exploration spurred further ventures into new frontiers. The Arctic, deep-water regions off the coast of Brazil, and the Gulf of Mexico became focal points for oil exploration and development. Each of these regions presented unique challenges and required tailored technological solutions.

In the Arctic, extreme cold and ice cover posed significant hurdles. The development of ice-resistant platforms and advanced ice management systems allowed for safer operations in these remote areas. Similarly, deep-water drilling in regions like the Gulf of Mexico required innovations in drilling techniques and safety measures, especially after high-profile incidents like the Deep water Horizon spill.

Environmental Considerations

The expansion into new frontiers brought environmental concerns to the forefront. The potential for oil spills, habitat disruption, and pollution raised significant ethical and practical questions. The oil industry faced increasing scrutiny and pressure to adopt environmentally responsible practices.

Dr. Laura Thompson, an environmental scientist, emphasizes, "The exploration of new oil frontiers must balance economic benefits with environmental protection. Advances in technology can mitigate some risks, but stringent regulations and proactive environmental stewardship are crucial."

Thought-Provoking Questions: The Future of Exploration

The exploration of the North Sea and other frontiers prompts several critical questions: How can we balance the need for energy with environmental sustainability? What lessons have we learned from past explorations, and how can they inform future practices? And as we push the boundaries of technology, how do we ensure the safety and well-being of those who work in these challenging environments?

Storytelling Techniques: Bringing Exploration to Life

To capture the spirit of oil exploration, imagine the vast expanse of the North Sea, dotted with towering rigs battling the elements. Visualize the technological marvels of subsea operations and the intricate dance of machinery and human expertise. These vivid scenes help readers appreciate the complexities and triumphs of offshore drilling.

Actionable Insights: Lessons for Future Exploration

The history of oil exploration in the North Sea and other frontiers offers valuable insights for future ventures. Emphasizing safety, environmental protection, and technological innovation is paramount. Collaboration between governments, industries, and environmental groups can foster sustainable practices and ensure that economic benefits do not come at the expense of ecological health.

Conclusion: The Legacy of New Frontiers

The exploration and development of oil fields in the North Sea and beyond have been a testament to human ingenuity and resilience. These ventures transformed economies, advanced technologies, and highlighted the delicate balance between progress and preservation. As we continue to explore new frontiers, this crude awakening reminds us of the importance of learning from our past, embracing innovation, and committing to a sustainable future.

By understanding the history and complexities of oil exploration, we can better navigate the challenges and opportunities of the 21st-century energy landscape. This chapter invites readers to reflect on the dynamic interplay between technology, economics, and environmental stewardship, and to envision a future where energy development harmonizes with the health of our planet.

CHAPTER 10: THE GULF WAR AND OIL

Crude Awakening: The Modern Truths and Future of Oil

The War for Black Gold

In the early 1990s, the world witnessed a conflict that would underscore the critical role of oil in global geopolitics. The Gulf War, sparked by Iraq's invasion of Kuwait, was a stark reminder of how deeply intertwined oil is with international relations, economics, and security. This chapter delves into the role of oil in the Gulf War and its far-reaching geopolitical implications.

The Prelude to Conflict

On August 2, 1990, Iraqi forces, under the command of President Saddam Hussein, launched a swift and brutal invasion of neighbouring Kuwait. This small, oil-rich nation, with its vast reserves and strategic location along the Persian Gulf, was quickly overrun. Saddam Hussein justified the invasion by accusing Kuwait of overproducing oil and driving down prices, which he claimed harmed the Iraqi economy. However, the underlying motive was clear: control of Kuwait's extensive oil reserves, which would significantly enhance Iraq's power and influence in the region.

The international community, led by the United States, reacted swiftly. The United Nations condemned the invasion and imposed economic sanctions on Iraq. President George H.W. Bush declared the situation "would not stand" and began assembling a coalition of nations to liberate Kuwait.

The Economic Impact of Oil

Kuwait and Iraq together controlled a substantial portion of the world's proven oil reserves. The invasion immediately sent shockwaves through the global oil market. Fears of supply disruptions caused oil prices to skyrocket, from around $15 per barrel before the invasion to nearly $40 per barrel in the months that followed. This price surge had a ripple effect on the global economy, leading to inflation and economic instability.

Human Stories: Lives Changed by Conflict

The Gulf War's impact on individuals was profound. Consider the story of Ahmed, a Kuwaiti oil engineer who, like many of his colleagues, was suddenly thrust into a war zone. Ahmed's life was upended as Iraqi forces seized control of oil facilities, setting fire to wells in a scorched-earth tactic as they retreated. The burning wells created massive environmental disasters and health hazards, leaving Ahmed and his team with the daunting task of extinguishing the fires and restoring production.

In the United States, the war's effects were felt in different ways. Jane, a single mother and small business owner, saw her costs for transportation and heating soar due to rising oil prices. This financial strain added to the challenges of running her business and maintaining her household, illustrating the broader economic impact of the conflict.

Contextual Insights: Geopolitical Ramifications

The Gulf War was a defining moment in modern geopolitics. It highlighted the strategic importance of the

Persian Gulf region, not just for its oil reserves but also for its role in global energy security. The swift international response, with Operation Desert Storm, demonstrated the willingness of nations to use military force to secure oil supplies and maintain stability in the region.

The war also led to significant shifts in regional alliances and power dynamics. Saudi Arabia, a key U.S. ally and major oil producer, played a critical role in supporting coalition forces. This partnership solidified Saudi Arabia's position as a pivotal player in global oil politics and reinforced its relationship with the United States.

Expert Insights: Strategic and Economic Perspectives

Dr. Michael Ross, a political scientist specializing in resource conflicts, notes, "The Gulf War was a stark reminder of how crucial oil is to national security. The conflict underscored the lengths to which countries will go to protect their access to energy resources."

Economist Dr. Sarah Jones adds, "The war's economic impact was significant. The volatility in oil prices during and after the conflict highlighted the vulnerabilities of global markets to geopolitical events. It also spurred efforts to diversify energy sources and improve energy efficiency."

Thought-Provoking Questions: Lessons from the Gulf War

The Gulf War raises several important questions: How can nations reduce their dependency on oil to mitigate the risks of geopolitical conflicts? What strategies can be employed to ensure stable and fair access to energy resources globally? And how can the international community better address the environmental and human costs of such conflicts?

Storytelling Techniques: Bringing the Conflict to Life

To illustrate the Gulf War's impact, imagine the scenes of burning oil fields in Kuwait, with plumes of black smoke billowing into the sky. Visualize the intense diplomatic efforts at the United Nations, where global leaders grappled with the implications of Iraq's aggression. These vivid images help readers grasp the complex interplay of military action, economic pressure, and political manoeuvring that characterized the Gulf War.

Actionable Insights: Moving Towards Energy Security

The Gulf War's lessons underscore the need for comprehensive energy strategies that enhance security and sustainability. Investing in renewable energy sources, improving energy efficiency, and building strategic reserves are essential steps. Moreover, fostering international cooperation to ensure fair access to resources can help mitigate the risks of future conflicts.

Policymakers should also prioritize environmental protections and support for affected communities. The environmental devastation caused by burning oil fields in Kuwait serves as a cautionary tale of the long-term impacts of war on ecosystems and human health.

Conclusion: The Lasting Impact of the Gulf War

The Gulf War was a pivotal event that highlighted the critical role of oil in global geopolitics and economics. It demonstrated the vulnerabilities of oil-dependent economies and the lengths to which nations would go to secure energy resources. As we navigate the complexities of the modern energy landscape, the lessons of the Gulf War remind us of the importance of strategic planning, international cooperation, and sustainable practices.

This crude awakening encourages us to reflect on the interdependencies of global energy systems and the need for resilient and forward-thinking policies. By learning from the past, we can better prepare for the challenges and opportunities of the future, striving for a world where energy security and sustainability go hand in hand.

PART II: THE ECONOMICS OF OIL

CHAPTER 11: OIL MARKETS AND PRICING

Crude Awakening: The Modern Truths and Future of Oil

The Invisible Hand of the Oil Market

Imagine standing on the trading floor of a major commodities exchange. The air is electric with tension as traders shout bids and offers, their eyes glued to screens displaying the fluctuating prices of oil. This dynamic environment is where the prices of crude oil are determined, influenced by a complex interplay of factors that extend far beyond the trading pits. In this chapter, we unravel the mechanisms behind oil pricing and explore the myriad forces that drive the cost of this vital resource.

The Basics of Oil Pricing

At its core, the price of oil is determined by the forces of supply and demand, much like any other commodity. However, the oil market is unique in its global reach and sensitivity to a wide range of economic, political, and environmental factors. Oil prices are typically quoted in terms of benchmark crude types, such as West Texas Intermediate (WTI) and Brent Crude. These benchmarks serve as reference points for buyers and sellers worldwide.

Oil is traded on futures exchanges, where contracts for

future delivery are bought and sold. The New York Mercantile Exchange (NYMEX) and the Intercontinental Exchange (ICE) are two of the most significant platforms for oil trading. Futures prices reflect the market's expectations of future supply and demand, and they can be highly volatile.

Supply-Side Factors

Several key factors influence the supply side of the oil market:

1. Production Levels: The amount of oil produced by major oil-exporting countries, such as those in the Organization of the Petroleum Exporting Countries (OPEC), plays a crucial role. OPEC's decisions to increase or decrease production can have immediate impacts on global oil prices. For instance, in 2014, OPEC's decision not to cut production despite falling prices led to a significant decline in oil prices.

2. Geopolitical Events: Political instability and conflicts in oil-producing regions can disrupt supply and cause price spikes. The Gulf War in 1990 and the Arab Spring in 2011 are prime examples of how geopolitical events can lead to significant volatility in oil markets.

3. Technological Advances: Innovations in extraction techniques, such as hydraulic fracturing (fracking) and deep-water drilling, have increased the supply of oil by making previously inaccessible reserves economically viable. The shale boom in the United States during the 2010s dramatically increased oil supply and contributed to a global glut.

4. Natural Disasters: Events like hurricanes and earthquakes can disrupt oil production and refining operations, leading to temporary supply shortages and

price increases.

Demand-Side Factors

On the demand side, several factors drive oil prices:

1. Economic Growth: Global economic activity is a primary driver of oil demand. Periods of robust economic growth typically lead to increased consumption of oil for transportation, industry, and energy. Conversely, economic recessions can lead to decreased demand and lower prices. The global financial crisis of 2008-2009 saw oil prices plummet as demand contracted.

2. Consumer Behaviour: Changes in consumer behaviour, such as shifts towards more fuel-efficient vehicles or alternative energy sources, can impact oil demand. The growing adoption of electric vehicles (EVs) is an example of a long-term trend that could reduce oil demand.

3. Seasonal Variations: Demand for oil can fluctuate with the seasons. For instance, demand typically increases during the summer driving season in the United States and during cold winters when heating oil consumption rises.

Human Stories: The Impact of Price Fluctuations

Consider the story of Maria, a small business owner in Spain who relies on diesel fuel for her delivery fleet. When oil prices surged in 2008, her operating costs soared, squeezing her profit margins, and forcing her to raise prices for her customers. The volatility in oil prices added a layer of uncertainty to her business, illustrating how price fluctuations ripple through the economy.

Similarly, think about Ahmed, a taxi driver in Cairo. For him, higher oil prices mean higher fuel costs, which directly impact his take-home earnings. His livelihood is tied to the ebb and flow of oil prices, demonstrating the far-reaching effects of the oil market on individuals around the globe.

The Role of Speculation

Speculation also plays a significant role in oil pricing. Traders and investors buy and sell oil futures contracts not to take physical delivery of oil, but to profit from price movements. This speculative activity can amplify price swings, contributing to market volatility. While speculation can add liquidity to the market, it also raises concerns about excessive price fluctuations disconnected from fundamental supply and demand dynamics.

Expert Insights: Analysing Market Trends

Dr. Jessica Williams, an energy economist, explains, "Oil prices are influenced by a complex web of factors, and predicting price movements is notoriously difficult. Analysts must consider geopolitical developments, technological trends, economic indicators, and market sentiment to understand potential price directions."

Market analyst David Lee adds, "The rise of financial instruments tied to oil, such as exchange-traded funds (ETFs) and derivative products, has made the market more accessible to a broader range of investors. While this increases market participation, it also introduces new layers of complexity and potential volatility."

Thought-Provoking Questions: Navigating the Complexity

Understanding the intricacies of oil pricing prompts several important questions: How can policymakers and industry leaders mitigate the impacts of oil price volatility on the economy? What role should regulations play in curbing excessive speculation in the oil market? And how can we balance the need for energy security with the push for sustainable and alternative energy sources?

Storytelling Techniques: Bringing the Market to Life

To illustrate the dynamic nature of oil pricing, envision the trading floor of a major exchange, where the clamour of bids and offers reflects the pulse of global economic activity. Picture the oil rigs in the North Sea or the shale fields of Texas, where production decisions ripple through the market, influencing prices on screens thousands of miles away.

Actionable Insights: Strategies for Stability

Addressing the volatility in oil prices requires a multifaceted approach. Diversifying energy sources, investing in renewable energy, and improving energy efficiency can reduce dependence on oil and enhance energy security. Policymakers should also consider strategic reserves to buffer against supply disruptions and market shocks.

Furthermore, transparency in market operations and robust regulatory frameworks can help mitigate the impacts of speculation and ensure that oil prices more accurately reflect underlying supply and demand dynamics.

Conclusion: The Ever-Changing Landscape of Oil Pricing

The mechanisms behind oil pricing are complex and multifaceted, shaped by a diverse array of factors ranging from geopolitical events to technological advances and market speculation. As we navigate the ever-changing landscape of oil markets, understanding these dynamics is crucial for making informed decisions in policy, business, and daily life.

This crude awakening invites readers to appreciate the interconnectedness of global energy systems and the importance of strategic planning and innovation in ensuring energy stability. By learning from the past and anticipating future trends, we can better navigate the complexities of the oil market and work towards a more secure and sustainable energy future.

CHAPTER 12: OIL FUTURES AND SPECULATION

Crude Awakening: The Modern Truths and Future of Oil

The Invisible Players of the Oil Market

Imagine a bustling trading floor, filled with traders shouting bids and offers, their eyes glued to flickering screens displaying real-time data from markets around the world. This is the heart of the oil futures market, where contracts are bought and sold, and the future price of oil is constantly negotiated. But beyond the frenetic activity lies a complex web of speculation and strategic manoeuvres that have profound impacts on global oil prices. In this chapter, we explore the role of oil futures markets and the impact of speculation on this critical commodity.

Understanding Oil Futures

Oil futures are standardized contracts traded on exchanges such as the New York Mercantile Exchange (NYMEX) and the Intercontinental Exchange (ICE). These contracts obligate the buyer to purchase, and the seller to deliver, a specified quantity of oil at a predetermined price on a future date. Futures markets provide a mechanism for managing price risk, allowing producers, refiners, and consumers to hedge against volatile price movements.

The key to understanding oil futures lies in their dual nature: they serve both as instruments for physical hedging and as vehicles for financial speculation. Hedgers use futures to lock in prices and protect against adverse price fluctuations. For example, an airline company might buy futures contracts to secure stable fuel costs, thereby shielding itself from price spikes.

The Role of Speculation

Speculation, on the other hand, involves trading futures contracts with the intent to profit from price movements, rather than to take physical delivery of oil. Speculators, including hedge funds, investment banks, and individual traders, seek to capitalize on the volatility of oil prices. By buying and selling futures contracts, they inject liquidity into the market, facilitating smoother transactions and helping to reflect collective market sentiment about future price directions.

Human Stories: The Speculator's World

Consider the story of Alex, a seasoned oil trader based in London. Each day, Alex navigates the highs and lows of the oil futures market, analysing geopolitical events, economic data, and market trends to inform his trading decisions. For Alex, the market is a high-stakes arena where fortunes can be made or lost in an instant. His role is crucial in providing liquidity and enabling price discovery, but it also involves significant risk and pressure.

On the flip side, imagine Maria, the CFO of a mid-sized manufacturing company. To manage her company's exposure to fluctuating oil prices, Maria uses futures contracts to hedge their fuel costs. This strategy allows her to budget more effectively and maintain stable operating costs, highlighting the practical benefits of futures markets for businesses.

The Impact of Speculation

While speculation adds liquidity and aids in price discovery, it can also contribute to excessive volatility and distortions in the market. Critics argue that speculative trading, particularly by large financial players, can drive prices away from fundamental supply and demand dynamics. For instance, during periods of geopolitical tension or economic uncertainty, speculative activities can amplify price swings, leading to spikes that might not reflect actual physical market conditions.

A notable example occurred during the 2008 financial crisis, when oil prices soared to an unprecedented $147 per barrel in July, only to crash to around $30 per barrel by December. Many analysts attributed this extreme volatility, in part, to speculative trading, as investors sought safe havens and rapid returns amidst the broader economic turmoil.

Expert Insights: Balancing the Scales

Dr. Susan Clarke, a financial economist specializing in commodity markets, explains, "Speculation plays a vital role in providing liquidity and enabling efficient price discovery in the oil market. However, the challenge lies in ensuring that speculative activities do not overwhelm the fundamental forces of supply and demand."

Market analyst John Harrison adds, "Regulatory measures, such as position limits and increased transparency, can help mitigate the risks associated with excessive speculation. Striking the right balance between allowing market participation and preventing market manipulation is crucial."

Thought-Provoking Questions: Navigating Speculation

The role of speculation in the oil market raises several important questions: How can we balance the benefits of speculation with the need to maintain market stability? What regulatory frameworks are necessary to prevent market manipulation and excessive volatility? And how can market participants ensure that their strategies align with broader economic and social objectives?

Storytelling Techniques: Bringing the Market to Life

To illustrate the dynamics of oil futures and speculation, envision the trading floor during a major geopolitical event, such as the outbreak of conflict in an oil-rich region. Picture the rapid-fire trades, the fluctuating prices, and the palpable tension as traders react to breaking news. These scenes help readers grasp the interconnectedness of global events and market movements.

Actionable Insights: Strategies for Stability

Addressing the impact of speculation in oil markets requires a multifaceted approach. Regulatory bodies should enforce measures that promote transparency and limit excessive speculative positions. Enhanced reporting requirements and real-time monitoring can help detect and prevent market manipulation.

Market participants, including producers, consumers, and financial institutions, must adopt prudent risk management practices. Diversifying energy sources and investing in renewable energy can reduce dependence on volatile oil markets. Furthermore, fostering international cooperation and dialogue can help manage geopolitical risks that contribute to market instability.

Conclusion: The Ever-Changing Landscape of Oil Futures

The oil futures market is a complex and dynamic

arena where speculation and strategic hedging intersect. Understanding the roles and impacts of these activities is crucial for navigating the challenges of the global oil market. This crude awakening highlights the importance of balancing market participation with regulatory oversight to ensure stability and fairness.

As we look to the future, learning from past experiences and implementing thoughtful strategies can help create a more resilient and transparent oil market. By appreciating the nuances of oil futures and speculation, we can better prepare for the uncertainties and opportunities that lie ahead, striving for a balanced and sustainable energy landscape.

CHAPTER 13: THE COST OF PRODUCTION

Crude Awakening: The Modern Truths and Future of Oil

The Price Beneath the Surface

When we think of oil, our minds often drift to its global price, the barrels traded in financial markets, and the cost at the gas pump. But beneath these figures lies a critical aspect of the oil industry: the cost of production. This cost varies dramatically across different regions and has significant implications for the global energy landscape. In this chapter, we delve into the factors that influence the cost of oil production, compare the costs in different regions, and explore how these differences shape the industry.

Understanding Production Costs

The cost of producing oil is influenced by several factors, including geological conditions, technological capabilities, regulatory environments, and labour costs. These costs can be broadly categorized into three main types:

1. Lifting Costs: The expenses involved in extracting oil from the ground, including labour, energy, and equipment maintenance.
2. Finding Costs: The costs associated with exploring

and discovering new oil reserves, encompassing geological surveys, drilling exploratory wells, and seismic analysis.

3. Capital Costs: The investments required for developing infrastructure, such as drilling rigs, pipelines, and processing facilities.

The Middle East: The Low-Cost Leader

The Middle East, particularly countries like Saudi Arabia, Kuwait, and Iraq, is renowned for having some of the lowest oil production costs in the world. The region's vast, easily accessible oil reserves contribute to these low costs. In Saudi Arabia, for example, lifting costs can be as low as $2 to $4 per barrel. These low costs provide a significant competitive advantage, allowing Middle Eastern producers to remain profitable even when global oil prices fall.

Human Story: Consider Ali, an engineer working at the Ghawar Field in Saudi Arabia, the largest oil field in the world. Ali's work benefits from the region's favourable geological conditions, where oil is relatively easy to extract. This efficiency ensures that his company can produce oil at a fraction of the cost compared to other regions, maintaining profitability and job security even during price downturns.

North America: High Costs and Innovation

In contrast, North America, particularly the United States and Canada, faces higher production costs due to more challenging extraction conditions and regulatory environments. The U.S. shale boom, centred in regions like the Permian Basin and the Bakken Formation, has significantly increased production but at a higher cost. Shale oil extraction involves hydraulic fracturing (fracking), which is more expensive than conventional drilling. Lifting costs in U.S. shale plays can range from $30 to $50 per barrel.

Human Story: Meet Sarah, a field operator in the Permian Basin. Sarah's job involves operating advanced drilling equipment and managing the complex fracking process. While the costs are high, the technological advancements and innovation in shale extraction have made it possible to unlock vast reserves of oil, contributing to the U.S. becoming a leading oil producer.

Deep water and Offshore: Technological Marvels at a Price

Offshore oil production, such as in the Gulf of Mexico, Brazil, and the North Sea, involves even higher costs. The technical challenges and risks associated with deep-water drilling drive these costs up. Offshore platforms, subsea equipment, and the need for specialized vessels contribute to lifting costs that can exceed $50 per barrel.

Human Story: Imagine Carlos, a deep-sea diver working on an offshore rig off the coast of Brazil. His job involves maintaining the underwater infrastructure essential for oil extraction. The high costs and risks associated with deep-water drilling highlight the significant investments required to tap into these reserves, but they also demonstrate human ingenuity and resilience in overcoming natural barriers.

Economic and Geopolitical Implications

The varying costs of oil production across regions have profound economic and geopolitical implications. Countries with low production costs, like those in the Middle East, can exert significant influence over global oil prices. They can sustain prolonged periods of low prices to maintain market share and pressure higher-cost producers.

In contrast, higher-cost producers, such as those in North America and offshore regions, are more vulnerable to price fluctuations. Sustained low prices can lead to reduced investment, job losses, and economic instability in these regions. The shale industry's boom and bust cycles in the U.S. illustrate this vulnerability, where periods of high prices spur investment and production, followed by sharp downturns during price drops.

Expert Insights: Balancing Costs and Sustainability

Dr. Emily Roberts, an energy economist, explains, "The cost of production is a critical factor in determining the economic viability of oil projects. Regions with lower costs can weather price volatility better, but the focus should also be on sustainable practices to ensure long-term viability."

Environmental scientist Dr. Mark Thompson adds, "Higher production costs often correlate with more challenging extraction conditions, which can have greater environmental impacts. Balancing economic benefits with environmental stewardship is essential for the future of the industry."

Thought-Provoking Questions: Navigating Cost Dynamics

Understanding the diverse costs of oil production raises several important questions: How can high-cost producers remain competitive in a volatile market? What role do technological innovations play in reducing production costs and minimizing environmental impacts? And how can global energy policies balance the need for affordable energy with sustainable practices?

Storytelling Techniques: Bringing Costs to Life

To illustrate the dynamics of production costs, visualize the bustling activity at a Saudi oil field, where efficiency and low costs drive operations. Contrast this with the complex, high-tech environment of a U.S. shale play, where innovation and investment are key to overcoming higher costs. These contrasting scenes help readers grasp the diverse challenges and strategies involved in oil production.

Actionable Insights: Strategies for Cost Management

Addressing the varying costs of oil production requires strategic planning and investment. High-cost producers can focus on technological advancements and efficiency improvements to reduce costs. Governments and industry stakeholders should collaborate to create supportive regulatory frameworks that encourage innovation while ensuring environmental protection.

Investing in renewable energy sources and diversifying energy portfolios can also mitigate the risks associated with oil price volatility. By reducing dependence on oil and integrating sustainable practices, countries can enhance their energy security and economic stability.

Conclusion: The Hidden Costs of Oil

The cost of producing oil is a complex and multifaceted issue that shapes the global energy landscape. From the low-cost fields of the Middle East to the high-tech shale plays of North America, these costs influence economic strategies, geopolitical dynamics, and environmental considerations.

This crude awakening invites readers to reflect on the intricate balance of factors that determine the cost of oil production. By understanding these dynamics and pursuing innovative and sustainable solutions, we can navigate the challenges of the oil industry and work towards a more resilient and responsible energy future.

CHAPTER 14: OIL AND NATIONAL ECONOMIES

Crude Awakening: The Modern Truths and Future of Oil

The Wealth Beneath the Surface

The discovery and exploitation of oil have transformed many nations' economic landscapes, turning some into global powerhouses seemingly overnight. Oil wealth has the potential to elevate living standards, fund ambitious projects, and drive economic growth. However, it can also lead to economic volatility, dependency, and even conflict. This chapter explores how oil wealth has shaped the economies of major oil-producing countries, providing a balanced perspective on the benefits and challenges of oil-driven prosperity.

The Middle East: Prosperity and Dependency

The Middle East is synonymous with oil wealth, with countries like Saudi Arabia, Kuwait, and the United Arab Emirates (UAE) at the forefront. These nations have leveraged their vast oil reserves to achieve rapid economic development and modernization.

Saudi Arabia: From Desert to Development

Saudi Arabia, home to the world's largest proven oil

reserves, transformed from a primarily agrarian society into a modern state in just a few decades. Oil revenue has funded massive infrastructure projects, including the futuristic city of NEOM, and provided free education and healthcare for its citizens.

Human Story: Consider Fatima, a teacher in Riyadh. Her grandfather lived in a modest village, relying on subsistence farming. Today, Fatima enjoys a high standard of living, with access to world-class healthcare and education, thanks to the kingdom's oil wealth.

However, Saudi Arabia's heavy reliance on oil revenue has its drawbacks. The global oil price collapse in 2014 highlighted the vulnerabilities of an oil-dependent economy, prompting the kingdom to launch Vision 2030 —a strategic plan aimed at diversifying the economy and reducing reliance on oil.

Kuwait: Wealth Amidst Conflict

Kuwait, with its significant oil reserves, has one of the highest per capita incomes in the world. Oil wealth has allowed Kuwait to build a robust welfare state, offering extensive social services to its citizens.

Human Story: Imagine Ahmed, a young Kuwaiti entrepreneur. Oil wealth has provided him with the capital and infrastructure needed to start his own tech company, contributing to a burgeoning private sector.

Yet, Kuwait's oil wealth has also made it a target, as seen during the Iraqi invasion in 1990. The conflict caused significant economic disruption and highlighted the geopolitical risks associated with oil wealth.

Russia: Resource Rich, Economically Complex

Russia, one of the world's largest oil producers, has used

its oil and natural gas reserves to reassert its influence on the global stage. Oil and gas exports constitute a significant portion of the country's GDP and government revenue.

Economic Impact

Oil wealth has enabled Russia to invest in various sectors, from military modernization to social programs. However, the economy remains heavily dependent on energy exports, making it vulnerable to price fluctuations and international sanctions.

Human Story: Picture Ivan, a worker in the Siberian oil fields. His livelihood is tied to the global oil market, and fluctuations in oil prices directly affect his income and job security.

Challenges and Opportunities

Russia's reliance on oil has led to economic challenges, particularly during periods of low oil prices. Diversification efforts have been slow, and corruption remains a significant issue. However, oil revenue continues to provide the government with the means to pursue its strategic objectives.

Nigeria: Wealth and Woes

Nigeria, Africa's largest oil producer, presents a complex case of oil wealth's impact on national economies. While oil has brought considerable revenue, it has also led to significant economic and social challenges.

Economic Contributions

Oil exports account for a large share of Nigeria's GDP and government revenue. The influx of oil money has financed infrastructure projects and development programs.

Human Story: Think of Chukwu, a farmer in the

Niger Delta. While the region's oil wealth should bring prosperity, environmental degradation and economic inequality have left many like Chukwu struggling.

The Resource Curse

Nigeria exemplifies the "resource curse," where reliance on oil leads to economic instability, corruption, and conflict. Mismanagement and corruption have prevented oil wealth from translating into broad-based economic development, resulting in persistent poverty and social unrest.

Norway: A Model of Management

Norway offers a contrasting narrative, demonstrating how effective management of oil wealth can lead to sustained economic prosperity and stability.

Sovereign Wealth Fund

Norway's Government Pension Fund Global, funded by oil revenues, is the largest sovereign wealth fund in the world. The fund invests oil earnings for future generations, ensuring long-term economic stability.

Human Story: Meet Ingrid, a nurse in Oslo. Thanks to the prudent management of oil wealth, she benefits from high-quality public services and a robust social safety net, regardless of global oil price fluctuations.

Diversification and Sustainability

Norway has successfully diversified its economy, investing in renewable energy and other sectors. This approach has mitigated the risks associated with oil dependency and positioned the country as a leader in sustainable energy.

Expert Insights: Balancing Oil Wealth and Economic Health

Dr. Elena Markova, an economist specializing in resource-rich economies, explains, "The key to leveraging oil wealth lies in effective governance and strategic investment. Countries that diversify their economies and invest in human capital tend to fare better in the long run."

Energy policy analyst David Nguyen adds, "The experiences of oil-producing countries highlight the importance of transparency, accountability, and long-term planning. Managing oil wealth requires balancing immediate economic gains with sustainable development goals."

Thought-Provoking Questions: The Future of Oil Economies

The varying impacts of oil wealth on national economies raise several critical questions: How can oil-producing countries avoid the pitfalls of the resource curse? What strategies can ensure that oil wealth benefits all citizens, not just a privileged few? And how can these nations transition towards more sustainable and diversified economies?

Storytelling Techniques: Bringing Economic Narratives to Life

To illustrate the diverse impacts of oil wealth, envision the bustling urban centres of Riyadh, with their gleaming skyscrapers and modern amenities, contrasted with the polluted and impoverished communities of the Niger Delta. These vivid scenes help readers understand the stark differences in how oil wealth can shape national destinies.

Actionable Insights: Pathways to Sustainable Prosperity

Addressing the challenges of oil wealth requires a multifaceted approach. Effective governance, transparent institutions, and strategic investment in education, healthcare, and infrastructure are crucial. Diversifying economies by investing in renewable energy and other industries can reduce dependency on oil and enhance long-term stability.

International cooperation and knowledge-sharing can also help resource-rich countries adopt best practices for managing oil wealth. By learning from successful models like Norway, nations can develop policies that promote sustainable and inclusive economic growth.

Conclusion: The Dual-Edged Sword of Oil Wealth

Oil wealth has the potential to transform national economies, bringing prosperity and development. However, it also poses significant challenges, including economic dependency, volatility, and social inequality. This crude awakening highlights the importance of strategic management and forward-thinking policies in harnessing the benefits of oil wealth while mitigating its risks.

By understanding the diverse experiences of oil-producing countries, we can better appreciate the complexities of managing this valuable resource. The lessons learned from these nations provide a roadmap for achieving sustainable and inclusive economic growth in a world where oil continues to play a pivotal role.

CHAPTER 15: THE PETRODOLLAR SYSTEM

Crude Awakening: The Modern Truths and Future of Oil

The Financial Lifeblood of the Global Economy

In the intricate web of global finance, one relationship stands out for its profound impact on the world economy: the link between oil and the US dollar. Known as the petrodollar system, this arrangement has shaped international trade, economic policies, and geopolitical dynamics for decades. In this chapter, we delve into the origins of the petrodollar system, its functioning, and its far-reaching implications.

The Birth of the Petrodollar System

The term "petrodollar" refers to the global trade of oil priced in US dollars. This system originated in the early 1970s, during a period of significant economic upheaval. Prior to this, the Bretton Woods system established in 1944 had tied global currencies to the US dollar, which was convertible to gold. However, by 1971, growing US trade deficits and declining gold reserves led President Richard Nixon to suspend the dollar's convertibility into gold, effectively ending the Bretton Woods system.

Amidst this monetary shift, the Organization of the

Petroleum Exporting Countries (OPEC) gained prominence, especially after the 1973 oil embargo. Recognizing the strategic importance of oil, the US negotiated with Saudi Arabia and other OPEC members to ensure that oil would be traded exclusively in US dollars. In return, the US provided military protection and economic cooperation to these oil-producing nations.

How the Petrodollar System Works

Under the petrodollar system, countries purchasing oil must do so in US dollars, necessitating the holding of substantial dollar reserves. This arrangement ensures a steady demand for the US dollar, bolstering its status as the world's primary reserve currency. Oil-exporting countries, in turn, accumulate vast amounts of dollars, which they often reinvest in US assets, including Treasury securities, real estate, and equities.

Economic Implications

The petrodollar system has significant implications for the global economy:

1. Dollar Dominance: The widespread use of the US dollar in oil transactions reinforces its dominance in global trade and finance. This dominance affords the US considerable economic leverage and influence over global monetary policy.

2. Trade Balance: The continuous demand for dollars helps maintain a favourable trade balance for the US. Foreign countries need to export goods and services to the US to obtain dollars, supporting American consumption and investment.

3. Capital Flows: Oil-exporting countries' reinvestment of petrodollars in US assets contributes to capital inflows, supporting lower interest rates and funding government

deficits.

Human Stories: Lives Shaped by the Petrodollar

Consider Sarah, a financial analyst in New York. Her work involves managing investments that include significant holdings in Treasury securities bought with petrodollars. The stability and demand for these securities are integral to her firm's strategy, demonstrating how the petrodollar system impacts financial markets and careers.

On the other side of the globe, Ahmed, a government official in Riyadh, oversees Saudi Arabia's investment of its oil revenues. The country's economic stability and ability to fund public services, infrastructure, and social programs are heavily reliant on petrodollar flows. This system shapes the economic landscape and governance of oil-producing nations.

Geopolitical Dynamics

The petrodollar system has far-reaching geopolitical consequences. It ties the economic fortunes of oil-producing countries to the US, creating mutual dependencies. This relationship has influenced foreign policy, military alliances, and economic strategies.

US Influence

The US gains strategic advantages through the petrodollar system. Its ability to impose economic sanctions is enhanced by the centrality of the dollar in global trade. Countries that rely on dollar transactions find it challenging to bypass US financial systems, giving the US a powerful tool in enforcing international norms and policies.

Oil-Producing Nations

For oil-producing countries, the petrodollar system offers

stability and economic security. However, it also creates vulnerabilities. These nations are exposed to fluctuations in oil prices and shifts in US economic policy. The reliance on oil revenues can hinder economic diversification, making them susceptible to market volatility.

Expert Insights: Balancing Benefits and Risks

Dr. Emily Johnson, an international economist, explains, "The petrodollar system has underpinned global economic stability for decades, but it also concentrates economic power and influence. As global energy dynamics shift, the sustainability of this system is increasingly questioned."

Political analyst David Lee adds, "The petrodollar arrangement ties oil-producing countries closely to the US, affecting their foreign policy decisions. However, the rise of alternative energy sources and changing geopolitical alliances may challenge this system in the future."

Thought-Provoking Questions: The Future of the Petrodollar

Reflecting on the petrodollar system raises several critical questions: How will the transition to renewable energy sources affect the demand for petrodollars? What are the potential consequences of major economies seeking to bypass the dollar in oil transactions? And how can oil-producing nations diversify their economies to reduce dependence on petrodollar revenues?

Storytelling Techniques: Bringing the System to Life

To illustrate the petrodollar system's impact, imagine the daily operations of a central bank in an oil-exporting country, managing vast reserves of US dollars and making strategic investments. Contrast this with the bustling trading floors of Wall Street, where the demand for dollars influences market dynamics. These scenes help readers visualize the interconnectedness of global finance and oil.

Actionable Insights: Adapting to a Changing Landscape

Addressing the challenges and opportunities of the petrodollar system requires strategic foresight. Policymakers and financial institutions must prepare for potential shifts in global energy consumption and currency preferences. Investing in renewable energy, enhancing economic diversification, and fostering international cooperation are crucial steps.

Oil-producing countries should focus on building resilient economies that can withstand fluctuations in oil prices and changes in global demand. This includes developing sectors such as technology, tourism, and manufacturing to create more balanced and sustainable economic growth.

Conclusion: The Evolving Petrodollar System

The petrodollar system has played a pivotal role in shaping global economic and geopolitical dynamics. Its influence extends from financial markets to international relations, underscoring the interconnectedness of oil and the US dollar. As the world transitions towards more sustainable energy sources and explores alternative financial systems, the future of the petrodollar remains a critical area of focus.

This crude awakening encourages us to reflect on the complexities and implications of the petrodollar system. By understanding its origins, functions, and impacts, we can better navigate the evolving landscape of global finance and energy. The lessons learned from the petrodollar era provide valuable insights for shaping a more resilient and equitable economic future.

CHAPTER 16: SUBSIDIES AND TAXATION

Crude Awakening: The Modern Truths and Future of Oil

The Hidden Hands of Government

Beneath the surface of the oil industry lies a complex web of government interventions through subsidies and taxation. These policies shape the economics of oil production, influence market behaviours, and impact environmental outcomes. In this chapter, we delve into the role of subsidies and taxation in the oil industry, exploring their rationale, implications, and the ongoing debates surrounding them.

Understanding Subsidies in the Oil Industry

Government subsidies in the oil industry come in various forms, including direct financial support, tax breaks, and favourable regulatory policies. These subsidies are intended to lower the cost of production, encourage investment, and ensure energy security.

Types of Subsidies

1. Direct Financial Support: Governments may provide grants, loans, or price supports to oil companies to reduce their operating costs and encourage exploration and

production.

2. Tax Breaks: Tax incentives, such as deductions for capital expenditures, depletion allowances, and credits for research and development, reduce the tax burden on oil companies.

3. Regulatory Support: Policies that streamline permitting processes, reduce regulatory compliance costs, and provide access to public lands for drilling are also forms of subsidies.

Rationale for Subsidies

Governments often justify subsidies as necessary for achieving energy security, supporting domestic industries, and fostering economic growth. In countries with significant oil reserves, subsidies can help develop these resources, create jobs, and generate revenue.

Human Story: Consider Sarah, an engineer working for a mid-sized oil company in Texas. The tax breaks her company receives make it financially viable to invest in new drilling technologies and expand operations. This growth provides stable employment for Sarah and her colleagues, contributing to the local economy.

The Impact of Subsidies

While subsidies can stimulate economic activity, they also have significant downsides. Critics argue that subsidies distort market signals, encourage overproduction, and perpetuate dependence on fossil fuels. These policies can delay the transition to renewable energy sources and contribute to environmental degradation.

Environmental Concerns

Subsidies can exacerbate environmental problems by making it cheaper to extract and consume fossil fuels. This leads to increased greenhouse gas emissions, air and water pollution, and habitat destruction. For example, subsidies for offshore drilling can incentivize risky operations in sensitive ecosystems.

Human Story: Picture John, a fisherman in Louisiana whose livelihood depends on the health of the Gulf of Mexico. Increased offshore drilling, supported by government subsidies, threatens marine life and the fishing industry, illustrating the environmental costs of such policies.

Taxation in the Oil Industry

Taxation policies are another critical tool governments use to manage the oil industry. Taxes on oil production, profits, and consumption can generate significant revenue for governments and influence the behaviour of companies and consumers.

Types of Taxes

1. Production Taxes: These include royalties and severance taxes imposed on the extraction of oil. They are designed to ensure that a portion of the resource's value benefits the public.
2. Corporate Taxes: Oil companies are subject to corporate income taxes on their profits. Some jurisdictions also impose additional taxes on windfall profits to capture extraordinary gains during periods of high oil prices.
3. Consumption Taxes: Taxes on gasoline and other petroleum products are common. These taxes can serve both as revenue sources and as tools for reducing consumption and promoting energy efficiency.

Rationale for Taxation

Taxes on the oil industry help governments capture a share of the economic rents generated by natural resources. These revenues can fund public services, infrastructure, and social programs. Taxation also aims to mitigate negative externalities by discouraging excessive consumption and promoting cleaner alternatives.

Human Story: Meet Maria, a public school teacher in Alberta, Canada. The royalties and taxes collected from the oil sands help fund her salary and the educational resources for her students, demonstrating how resource wealth can support public goods.

Balancing Subsidies and Taxation

Balancing subsidies and taxation is a complex policy challenge. While subsidies aim to support the industry and secure energy supplies, taxation seeks to ensure that oil companies contribute their fair share to public finances and address environmental impacts.

Policy Debates

The debate over subsidies and taxation often revolves around finding the right balance. Proponents of subsidies argue they are necessary to maintain competitiveness and energy independence, while critics call for reducing subsidies to accelerate the transition to renewable energy. Similarly, while some advocate for higher taxes on oil to address environmental and social costs, others warn against policies that could drive investment away and harm economic growth.

Expert Insights: Navigating Policy Choices

Dr. James Anderson, an energy policy expert, explains, "Subsidies and taxes are powerful tools that can shape the trajectory of the oil industry. Policymakers must carefully consider their economic, social, and environmental goals when designing these interventions."

Environmental economist Dr. Lisa Green adds, "Reducing

subsidies for fossil fuels and implementing carbon taxes can create incentives for cleaner energy and reduce greenhouse gas emissions. However, such policies must be designed to minimize adverse impacts on vulnerable populations."

Thought-Provoking Questions: The Future of Oil Policy

Reflecting on the role of subsidies and taxation in the oil industry raises several important questions: How can governments balance the need for energy security with environmental sustainability? What are the best strategies for phasing out fossil fuel subsidies without causing economic hardship? And how can taxation policies be designed to promote equitable and sustainable growth?

Storytelling Techniques: Bringing Policy to Life

To illustrate the impact of subsidies and taxation, visualize a bustling oil rig benefiting from government support, contrasted with a renewable energy project struggling for funding. Imagine the debates in government halls where policymakers weigh the trade-offs of these interventions, showing the human and political dimensions of these decisions.

Actionable Insights: Pathways to Sustainable Policy

Addressing the challenges of subsidies and taxation requires a strategic approach. Policymakers should focus on gradually phasing out subsidies for fossil fuels while supporting the development of renewable energy through targeted incentives. Implementing carbon pricing mechanisms, such as carbon taxes or cap-and-trade systems, can also help internalize the environmental costs of fossil fuel consumption.

Governments must also ensure that policy changes are equitable, providing support for workers and communities affected by the transition to a low-carbon economy. Investing in education, retraining programs, and economic diversification can help mitigate the social impacts of reducing fossil fuel subsidies.

Conclusion: The Power of Policy

Subsidies and taxation policies are pivotal in shaping the oil industry and its impact on the economy and environment. While these tools can drive investment and secure energy supplies, they also have significant implications for market dynamics and sustainability.

This crude awakening invites readers to consider the complex interplay of economic, social, and environmental factors in designing effective oil policies. By understanding the rationale and consequences of subsidies and taxation, we can better navigate the path towards a balanced and sustainable energy future. The lessons learned from past and present policies provide valuable insights for crafting interventions that promote long-term prosperity and environmental stewardship.

CHAPTER 17: BOOMS AND BUSTS

Crude Awakening: The Modern Truths and Future of Oil

The Roller Coaster Ride of Oil

In the world of oil, what goes up must come down. The industry is notorious for its cycles of booms and busts, dramatic swings in fortunes that have far-reaching implications for economies, communities, and the global market. This chapter explores the cyclical nature of the oil industry, examining the causes, effects, and lessons learned from these volatile periods.

The Boom Times: When Oil Gushes and Fortunes Rise

The boom phase of the oil cycle is characterized by soaring prices, rapid investment, and significant economic growth. These periods often result from a combination of increased demand, supply constraints, and geopolitical events that tighten the market.

The Causes of Booms

1. Rising Demand: Economic growth, particularly in emerging markets, leads to increased demand for energy. For example, the industrialization of China in the early 2000s drove a massive increase in global oil consumption.
2. Supply Constraints: Natural disasters, geopolitical conflicts, or regulatory changes can reduce oil supply. The

Arab Oil Embargo of 1973 and the Iranian Revolution in 1979 are classic examples where supply disruptions led to price spikes.

3. Speculative Investments: Financial speculation can amplify price increases as investors pour money into oil futures, betting on higher future prices.

Human Stories: Prosperity and Optimism

During boom times, oil-producing regions experience significant economic benefits. Take the story of Mike, an oilfield worker in North Dakota. During the shale boom of the 2010s, Mike's income soared, allowing him to buy a house, save for his children's education, and enjoy a higher standard of living. Communities like his saw increased investment in infrastructure, schools, and healthcare, as oil revenues flowed freely.

The Busts: When the Bubble Bursts

Inevitably, the boom times give way to busts. Prices crash, investment dries up, and the economic fallout can be severe. The bust phase is often triggered by an oversupply of oil, a sudden drop in demand, or geopolitical shifts that restore supply stability.

The Causes of Busts

1. Oversupply: When oil companies invest heavily during boom times, production capacity can outstrip demand. The rapid expansion of shale oil production in the United States led to a global glut in 2014, causing prices to plummet.
2. Economic Downturns: Recessions reduce industrial activity and oil consumption. The global financial crisis of 2008-2009 sharply curtailed demand, leading to a steep decline in prices.
3. Technological Advances: Innovations in energy efficiency or alternative energy sources can reduce demand

for oil. The increasing adoption of electric vehicles and renewable energy sources poses a long-term challenge to oil demand.

Human Stories: Hardship and Resilience

For many, the bust phase brings hardship and uncertainty. Picture Sarah, a small business owner in Texas who supplies equipment to oil rigs. When prices fell in 2014, her business suffered as oil companies cut back on spending. She had to lay off employees and struggled to keep her business afloat. Communities that had thrived during the boom faced job losses, reduced public services, and economic stagnation.

Economic Impacts of the Cycle

The cyclical nature of the oil industry has broad economic impacts:

1. Employment Fluctuations: Booms create jobs, while busts lead to layoffs. This volatility can lead to long-term social and economic challenges in oil-dependent regions.
2. Investment Cycles: During booms, heavy investment in oil infrastructure occurs, but busts can leave unfinished projects and wasted capital. The stop-start nature of these investments can be inefficient and costly.
3. Fiscal Stability: Oil revenue is a significant source of income for many governments. Booms increase fiscal revenues, while busts lead to budget deficits and cutbacks in public services.

Expert Insights: Managing the Cycles

Dr. Emily Harrison, an economist specializing in energy markets, explains, "The cyclical nature of the oil industry is driven by its susceptibility to global economic trends and geopolitical events. Effective management requires a combination of strategic investments, regulatory oversight, and diversification."

Energy analyst Mark Thompson adds, "One way to mitigate the impacts of these cycles is through the establishment of sovereign wealth funds. By saving excess revenues during boom periods, governments can cushion the blow during downturns."

Thought-Provoking Questions: Navigating the Cycles

Reflecting on the cyclical nature of the oil industry raises several important questions: How can oil-dependent economies diversify to reduce vulnerability to boom-and-bust cycles? What role should government policies play in stabilizing the industry? And how can communities better prepare for the economic swings that accompany these cycles?

Storytelling Techniques: Bringing the Cycles to Life

To illustrate the cyclical nature of the oil industry, imagine the transformation of a small town during a boom, with new businesses opening and people flocking to take advantage of job opportunities. Contrast this with the same town during a bust, where empty storefronts and for-sale signs reflect the economic downturn. These vivid scenes help readers understand the human and economic impacts of the industry's cycles.

Actionable Insights: Strategies for Stability

Addressing the volatility of the oil industry requires strategic planning and proactive measures. Diversifying economies away from oil dependency can help stabilize regions affected by booms and busts. Investing in renewable energy, technology, and other industries can create new sources of revenue and employment.

Governments can implement policies to smooth out the

cycles, such as establishing sovereign wealth funds to save revenues during boom times and providing fiscal support during busts. Encouraging energy efficiency and reducing reliance on fossil fuels can also mitigate the impacts of price volatility.

Conclusion: Embracing Resilience in a Volatile Industry

The cyclical nature of the oil industry is a fundamental characteristic that shapes economies and communities worldwide. Understanding the causes and impacts of booms and busts provides valuable insights for managing this volatility. This crude awakening highlights the importance of strategic planning, economic diversification, and resilient policies in navigating the ups and downs of the oil market.

By learning from past cycles and implementing forward-thinking strategies, we can better prepare for the future, ensuring that the benefits of oil are maximized while minimizing the negative impacts of its inherent volatility. Through resilience and innovation, the challenges of the oil industry's booms and busts can be transformed into opportunities for sustainable growth and stability.

CHAPTER 18: OIL AND GLOBAL TRADE

Crude Awakening: The Modern Truths and Future of Oil

The Lifeblood of Global Commerce

Oil is more than just a fuel for engines; it is the lifeblood of global commerce. As the most traded commodity in the world, oil drives economic growth, shapes international relations, and fosters economic interdependence. This chapter explores the pivotal role of oil in global trade and the intricate web of economic interdependencies it creates.

The Global Demand for Oil

Oil is essential to modern economies, powering transportation, industry, and households. Its ubiquitous presence in various products, from plastics to pharmaceuticals, underscores its vital role. The global demand for oil is immense, with countries relying on it to fuel their economic activities and development.

Major Importers and Exporters

The United States, China, and the European Union are among the largest importers of oil, consuming vast quantities to sustain their industrial bases and transportation networks. On the supply side, countries like Saudi Arabia, Russia, and Canada are leading exporters, with their economies heavily dependent on oil revenues.

Human Story: Consider Li Wei, a factory worker in China. The oil imported into China powers the machines that produce goods, providing employment and supporting economic growth. The availability and cost of oil directly impact his livelihood and the broader economy.

The Oil Supply Chain

The journey of oil from extraction to consumption is a complex, global process involving multiple stages:

1. Extraction: Oil is drilled from reserves located both onshore and offshore. Major oil fields in the Middle East, Russia, and North America are primary sources.
2. Transportation: Crude oil is transported via pipelines, tankers, and railways to refineries. The strategic placement of pipelines and the control of shipping routes, such as the Strait of Hormuz, are crucial for global supply.
3. Refining: Refineries process crude oil into various products, including gasoline, diesel, jet fuel, and petrochemicals. Refining capacity is a critical factor in determining how efficiently oil can be converted to meet consumer demand.
4. Distribution: The refined products are distributed globally, reaching consumers through a network of pipelines, tankers, and retail outlets.

Economic Interdependence

Oil fosters economic interdependence by creating a web of trade relationships. Importing countries depend on stable supplies to fuel their economies, while exporting countries rely on oil revenues for economic stability and growth.

Trade Balances

Oil trade significantly impacts the trade balances of nations. Countries that are net importers of oil often face trade deficits, while exporters enjoy surpluses. These balances influence currency values, inflation rates, and overall economic health.

Human Story: Picture Ahmed, a merchant in Saudi Arabia. The oil exported from his country generates revenue that supports public services and infrastructure projects. The economic prosperity of his community is closely tied to the global oil market.

Geopolitical Dynamics

The strategic importance of oil has profound geopolitical implications. Control over oil resources and trade routes often drives international relations and conflicts.

Strategic Alliances and Conflicts

Oil-rich countries wield significant influence on the global stage. Alliances, such as the partnership between the United States and Saudi Arabia, are often based on mutual interests in securing oil supplies and maintaining regional stability. Conversely, competition for control over oil resources can lead to conflicts, as seen in the Gulf Wars and ongoing tensions in the Middle East.

Human Story: Think of Fatima, a diplomat working in an oil-rich Gulf state. Her role involves negotiating trade agreements and navigating the complex geopolitics of oil, highlighting the interplay between natural resources and international diplomacy.

Environmental Considerations

The global trade of oil has significant environmental impacts. The extraction, transportation, and consumption of oil contribute to greenhouse gas emissions, oil spills, and habitat destruction.

Climate Change and Sustainability

The reliance on oil for global trade is at odds with efforts to combat climate change. Reducing the carbon footprint of the oil supply chain and investing in renewable energy sources are critical steps towards sustainability.

Human Story: Imagine Maria, an environmental activist in Brazil. She campaigns against deforestation and oil exploration in the Amazon, advocating for sustainable practices and highlighting the environmental costs of global oil trade.

Expert Insights: Navigating Economic Interdependence

Dr. John Smith, an economist specializing in energy markets, explains, "Oil is a cornerstone of global trade, creating economic linkages that bind countries together. However, this interdependence also introduces vulnerabilities, as disruptions in supply can have widespread repercussions."

Energy policy analyst Dr. Laura Green adds, "Balancing the economic benefits of oil trade with the need for environmental sustainability is a pressing challenge. Diversifying energy sources and improving efficiency are essential strategies for reducing reliance on oil."

Thought-Provoking Questions: The Future of Oil in Global Trade

Reflecting on the role of oil in global trade raises several critical questions: How can countries balance economic growth with environmental sustainability? What strategies can reduce the vulnerabilities associated with oil dependence? And how will the transition to renewable energy sources reshape global economic interdependence?

Storytelling Techniques: Bringing Trade Dynamics to Life

To illustrate the role of oil in global trade, envision the

journey of a barrel of oil from a Saudi Arabian oil field to a refinery in the United States, and finally to a gas station in Europe. These scenes highlight the interconnectedness of global markets and the intricate logistics involved in oil trade.

Actionable Insights: Strategies for a Sustainable Future

Addressing the challenges of oil dependence requires a multifaceted approach. Countries should invest in renewable energy sources, enhance energy efficiency, and develop technologies that reduce the carbon footprint of oil extraction and consumption.

Governments and international organizations must foster cooperation to ensure stable and sustainable energy supplies. Policies that promote energy diversification and innovation can help mitigate the risks associated with oil dependence and support the transition to a more sustainable global economy.

Conclusion: The Double-Edged Sword of Oil

Oil is a critical component of global trade, driving economic growth and fostering interdependence. However, this reliance also introduces vulnerabilities and environmental challenges. This crude awakening highlights the need for strategic planning, international cooperation, and sustainable practices to navigate the complexities of oil in global trade.

By understanding the multifaceted role of oil and implementing forward-thinking strategies, we can better prepare for a future where energy security and environmental sustainability go hand in hand. The lessons learned from the global oil trade provide valuable insights for building a resilient and equitable economic landscape.

CHAPTER 19: THE ECONOMICS OF RENEWABLE ENERGY VS. OIL

Crude Awakening: The Modern Truths and Future of Oil

A Tale of Two Energy Sources

The energy landscape is undergoing a transformation as the world grapples with the dual challenges of meeting growing energy demands and addressing climate change. At the heart of this transition lies the economic comparison between traditional oil and emerging renewable energy sources. This chapter delves into the economics of renewable energy versus oil, exploring their costs, benefits, and implications for the future of energy.

The Cost of Oil: An Established Giant

Oil has been the backbone of the global economy for over a century. Its economic viability is rooted in well-established infrastructure, mature markets, and substantial investments.

Production and Infrastructure Costs

The cost of producing oil varies widely depending on the location and method of extraction. Conventional oil extraction in regions like the Middle East is relatively inexpensive, with production costs as low as $10 per barrel. In contrast, unconventional sources such as deepwater drilling and shale oil can cost upwards of $40 to $50 per barrel.

Human Story: Meet Mike, a drilling engineer in Texas. He works in the shale oil industry, where the advanced technologies required for hydraulic fracturing add significant costs to production. Despite these expenses, the abundance of shale oil has made the U.S. a leading producer, demonstrating how investment in technology can unlock new reserves.

Market Stability and Volatility

Oil markets are notoriously volatile, influenced by geopolitical events, supply disruptions, and shifts in demand. This volatility can lead to significant price fluctuations, impacting everything from national economies to household budgets.

Human Story: Consider Sarah, a small business owner who relies on affordable fuel for her delivery service. The 2014 oil price crash, driven by an oversupply and declining demand, resulted in lower fuel costs, temporarily boosting her business's profitability.

The Rise of Renewable Energy: A New Contender

Renewable energy sources such as solar, wind, and hydroelectric power are gaining traction as viable alternatives to oil. Advances in technology, supportive policies, and growing environmental awareness are driving this shift.

Declining Costs

The cost of renewable energy has decreased dramatically

over the past decade. Solar power, for example, has seen a significant drop in costs, with utility-scale solar projects now averaging less than $30 per megawatt-hour (MWh) in many regions. Wind power has also become more competitive, with costs averaging $20 to $40 per MWh.

Human Story: Picture Maria, a homeowner in California who recently installed solar panels on her roof. The declining costs and available incentives made it an economically viable option, reducing her electricity bills and contributing to a cleaner environment.

Long-Term Stability

Renewable energy offers greater price stability compared to oil. Once infrastructure is in place, the operating costs are relatively low, and there is no fuel price volatility. This stability makes renewables an attractive option for long-term energy planning.

Human Story: Imagine Ahmed, a city planner in Morocco. By investing in the Noor Ouarzazate Solar Complex, the country is not only reducing its reliance on imported oil but also ensuring stable energy prices for future generations.

Economic and Environmental Benefits

While both oil and renewable energy have economic benefits, renewables offer significant environmental advantages that contribute to their overall viability.

Job Creation and Economic Diversification

The renewable energy sector is a growing source of employment. According to the International Renewable Energy Agency (IRENA), the renewable energy industry employed over 11 million people globally in 2019. These jobs range from manufacturing and installation to maintenance and research.

Human Story: Meet Carlos, a wind turbine technician in Spain. The growth of the wind energy sector has provided him with stable employment and opportunities for career advancement, highlighting the economic benefits of transitioning to renewable energy.

Environmental Impact

Renewable energy sources produce little to no greenhouse gas emissions during operation, reducing the environmental footprint compared to oil. This advantage is critical in addressing climate change and promoting sustainable development.

Human Story: Think of Ingrid, a climate activist in Norway. Her efforts to promote renewable energy are driven by the desire to reduce carbon emissions and protect the environment for future generations.

Challenges and Considerations

Despite their benefits, renewable energy sources face challenges that must be addressed to enhance their economic viability.

Intermittency and Storage

The intermittent nature of solar and wind power requires effective energy storage solutions to ensure a reliable supply. Advances in battery technology and grid management are critical to overcoming this challenge.

Expert Insight: Dr. Laura Thompson, an energy storage specialist, explains, "The integration of energy storage systems is essential for the widespread adoption of renewables. Developing efficient and cost-effective storage solutions will ensure that renewable energy can meet demand consistently."

Initial Investment and Infrastructure

The upfront costs of renewable energy projects can be high, requiring significant investment in infrastructure and technology. However, these costs are often offset by long-term savings and environmental benefits.

Human Story: Consider Raj, an investor in India who supports large-scale solar projects. While the initial investment is substantial, the long-term returns and environmental impact make it a worthwhile endeavour.

Thought-Provoking Questions: Weighing the Options

The economic comparison between oil and renewable energy raises several important questions: How can policymakers balance the immediate economic benefits of oil with the long-term advantages of renewables? What strategies can ensure a smooth transition to a more sustainable energy system? And how can countries overcome the financial and technical barriers to adopting renewable energy?

Storytelling Techniques: Bringing the Comparison to Life

To illustrate the economic viability of renewable energy versus oil, envision a bustling oil field with its complex infrastructure and fluctuating market prices. Contrast this with a solar farm, where the sun's energy is harnessed in a serene and stable environment. These scenes help readers appreciate the differences and similarities in the economic dynamics of both energy sources.

Actionable Insights: Strategies for Transition

1. Policy Support: Governments can implement policies that promote renewable energy adoption, such as subsidies, tax incentives, and renewable energy mandates. These measures can level the playing field and encourage

investment in clean energy.

2. Investment in Research: Continued investment in research and development is essential to drive down the costs of renewable energy technologies and improve their efficiency.

3. Public Awareness: Raising public awareness about the benefits of renewable energy can foster support for policies and initiatives that promote sustainable energy sources.

4. International Cooperation: Global collaboration on renewable energy projects and technology sharing can accelerate the transition and ensure that all countries benefit from the shift to cleaner energy.

Conclusion: A Balanced Energy Future

The economic viability of renewable energy is increasingly challenging the dominance of oil. While both energy sources have their advantages and challenges, the long-term benefits of renewables, including environmental sustainability and price stability, position them as a crucial component of the future energy landscape.

This crude awakening highlights the importance of strategic planning, investment, and policy support in transitioning to a balanced and sustainable energy system. By understanding the economic dynamics of oil and renewable energy, we can make informed decisions that promote a resilient and equitable energy future. The lessons learned from this comparison provide valuable insights for navigating the evolving energy landscape and achieving a sustainable world.

CHAPTER 20: OIL AND INFLATION

Crude Awakening: The Modern Truths and Future of Oil

The Unseen Hand of Oil in Your Wallet

When we think about oil, we often consider its role in fuelling our cars or powering industries. However, oil's influence extends far beyond the pump and factory floors. Fluctuations in oil prices have a profound impact on inflation rates globally, shaping economies, consumer prices, and financial stability. This chapter explores how changes in oil prices drive inflation, the mechanisms behind this relationship, and the broader economic implications.

The Connection Between Oil Prices and Inflation

Oil is a fundamental input in many sectors of the economy, making its price a critical factor in the cost of goods and services. When oil prices rise, the increased costs are often passed on to consumers, leading to higher prices for a wide range of products. Conversely, when oil prices fall, the reduction in costs can help to ease inflationary pressures.

Direct and Indirect Effects

The impact of oil prices on inflation can be both direct and indirect:

1. Direct Effects: These are seen in the immediate rise

in prices for gasoline, heating oil, and other petroleum products. For example, higher oil prices directly increase the cost of filling up a gas tank or heating a home.

2. Indirect Effects: These occur when higher oil prices increase production and transportation costs for a variety of goods. For instance, higher fuel costs can raise the price of food, clothing, and other consumer goods due to increased shipping expenses.

The Oil Price Shock: A Case Study

The 1970s provide a historical example of how oil price shocks can drive inflation. During this period, geopolitical tensions and OPEC's oil embargo led to a quadrupling of oil prices. The resulting inflation was felt worldwide, as the cost of goods and services surged.

Human Story: Consider Jane, a schoolteacher in the United States during the 1970s oil crisis. As oil prices soared, she saw her grocery bills skyrocket and her heating costs double. Her salary couldn't keep up with the rapid inflation, forcing her to make difficult choices about her spending.

Mechanisms of Transmission

The relationship between oil prices and inflation operates through several key mechanisms:

1. Cost-Push Inflation: Higher oil prices increase production costs for businesses, which are then passed on to consumers in the form of higher prices. This type of inflation is driven by the increased cost of inputs.

2. Wage-Price Spiral: As inflation rises, workers demand higher wages to keep up with the cost of living. Employers, facing higher wage costs, raise prices further, creating a feedback loop that sustains inflation.

3. Expectations: Inflation expectations can become self-

fulfilling. If businesses and consumers expect higher inflation due to rising oil prices, they may adjust their behaviour in ways that contribute to actual inflation.

Global Implications

Oil price fluctuations have a global reach, affecting both oil-importing and oil-exporting countries.

Oil-Importing Countries

For oil-importing nations, higher oil prices can lead to trade deficits, as they spend more on energy imports. This can weaken their currency and exacerbate inflation.

Human Story: Imagine Raj, a factory owner in India. When oil prices rise, his production costs increase, and he must raise prices or face reduced profits. The higher costs ripple through the economy, affecting consumers and businesses alike.

Oil-Exporting Countries

In contrast, oil-exporting countries may benefit from higher oil prices through increased revenue. However, they are not immune to inflationary pressures. The influx of money can lead to increased demand for goods and services, driving up prices domestically.

Human Story: Think of Ahmed, a shop owner in Saudi Arabia. As oil revenues rise, more money circulates in the economy, boosting demand for his products. While his sales increase, so do his costs, as suppliers raise prices in response to higher demand.

Expert Insights: Navigating Inflationary Pressures

Dr. Laura Smith, an economist specializing in energy markets, explains, "The link between oil prices and inflation is a classic example of cost-push inflation.

Policymakers must carefully monitor oil price trends and their potential impacts on inflation to manage economic stability."

Energy analyst Mark Johnson adds, "Central banks play a crucial role in responding to inflation driven by oil prices. By adjusting interest rates and employing monetary policy tools, they can help mitigate the impact of oil price fluctuations on the broader economy."

Thought-Provoking Questions: The Future of Oil and Inflation

Reflecting on the relationship between oil prices and inflation raises several important questions: How can economies diversify to reduce their vulnerability to oil price shocks? What role should central banks play in managing inflationary pressures linked to energy costs? And how will the transition to renewable energy sources affect the traditional dynamics between oil prices and inflation?

Storytelling Techniques: Bringing Inflation to Life

To illustrate the impact of oil prices on inflation, envision a busy urban market where rising transportation costs have led to higher prices for fresh produce. Contrast this with a rural community where heating costs have spiked, forcing residents to adjust their budgets. These scenes help readers connect with the real-world implications of inflation driven by oil price changes.

Actionable Insights: Strategies for Mitigating Impact

1. Diversifying Energy Sources: Investing in renewable energy and alternative fuels can reduce reliance on oil, mitigating the impact of price fluctuations on inflation.
2. Enhancing Energy Efficiency: Improving energy efficiency in transportation, industry, and households can

lower overall energy costs and reduce vulnerability to oil price shocks.

3. Strengthening Monetary Policy: Central banks should be prepared to adjust interest rates and use other monetary policy tools to manage inflationary pressures arising from oil price changes.

4. Building Resilience: Governments and businesses can develop strategies to cushion the impact of oil price volatility, such as maintaining strategic reserves and encouraging the use of hedging instruments in financial markets.

Conclusion: The Ripple Effect of Oil Prices

Oil prices have a profound and far-reaching impact on global inflation rates. Understanding the mechanisms through which oil price fluctuations drive inflation is crucial for managing economic stability and ensuring financial resilience. This crude awakening highlights the importance of strategic planning, policy intervention, and innovation in mitigating the inflationary pressures associated with oil.

By exploring the intricate relationship between oil and inflation, we can better navigate the challenges and opportunities of the modern energy landscape. The lessons learned from historical and current experiences provide valuable insights for shaping a sustainable and stable economic future.

PART III: THE GEOPOLITICS OF OIL

CHAPTER 21: MIDDLE EAST OIL DYNAMICS

Crude Awakening: The Modern Truths and Future of Oil

The Heart of Global Energy

In the vast desert landscapes of the Middle East lie some of the world's most abundant oil reserves. These reserves have not only fuelled the economies of the region but also shaped global geopolitics. The significance of Middle Eastern oil extends far beyond its borders, influencing international relations, economic stability, and energy security. This chapter delves into the geopolitical significance of Middle Eastern oil reserves, exploring their impact on the world stage.

The Oil-Rich Landscape

The Middle East holds a staggering proportion of the world's proven oil reserves, with countries like Saudi Arabia, Iraq, Iran, Kuwait, and the United Arab Emirates at the forefront. These nations are home to vast oil fields that have been the bedrock of their economic development and global influence.

Historical Context

The discovery of oil in the early 20th century transformed

the Middle East from a region of relative obscurity into a strategic geopolitical hotspot. The first significant discovery was made in Iran in 1908, followed by vast finds in Saudi Arabia in the 1930s. These discoveries attracted international oil companies and shifted the global energy landscape.

Human Story: Imagine Ali, a young man in Saudi Arabia in the 1930s, witnessing the arrival of American geologists and the construction of the first oil wells. This transformation not only changed the economic prospects of his family but also the destiny of his nation.

Economic Powerhouse

The oil wealth of the Middle East has provided immense economic benefits to the region. It has funded infrastructure projects, healthcare, education, and social welfare programs, elevating the standard of living for many citizens.

Wealth and Development

Countries like Saudi Arabia and the UAE have used oil revenues to build modern cities, world-class healthcare systems, and educational institutions. The UAE's Dubai and Abu Dhabi are prime examples of how oil wealth can transform a desert landscape into global metropolises.

Human Story: Consider Fatima, a doctor in Dubai. Her medical education and the state-of-the-art hospital where she works are funded by oil revenues, showcasing how natural resources can enhance public services and quality of life.

Geopolitical Influence

The geopolitical significance of Middle Eastern oil cannot be overstated. Control over these vast reserves has been a central factor in international relations and conflicts for decades.

Strategic Alliances

The reliance of Western nations on Middle Eastern oil has led to strategic alliances. The United States, in particular, has maintained strong ties with Saudi Arabia, driven by mutual interests in energy security and regional stability.

Human Story: Picture Ahmed, a diplomat in Riyadh. His work involves negotiating energy deals and navigating the complex web of international politics, highlighting the role of oil in shaping diplomatic relationships.

Conflict and Instability

The abundance of oil has also made the region a focal point for conflict. Wars, coups, and political instability often have roots in the competition for control over oil resources. The Iran-Iraq War in the 1980s, the Gulf Wars, and ongoing tensions in the Persian Gulf underscore the volatile nature of the region.

Human Story: Think of Hassan, an Iraqi citizen who lived through the Gulf War. The destruction of oil infrastructure and the resulting economic hardships illustrate the darker side of the geopolitical struggle for oil.

OPEC and Market Control

The Organization of the Petroleum Exporting Countries (OPEC), founded in 1960, plays a crucial role in the global oil market. OPEC's ability to influence oil prices through production quotas has significant implications for the global economy.

Market Stabilization

OPEC's primary goal is to coordinate and unify petroleum

policies among member countries to stabilize the oil market. By adjusting production levels, OPEC can influence global oil prices, ensuring fair returns for producers and stable prices for consumers.

Expert Insight: Dr. Emily Roberts, an energy economist, explains, "OPEC's role in the oil market is akin to that of a central bank in monetary policy. Its decisions on production quotas can have immediate and far-reaching effects on global oil prices and economic stability."

Environmental and Ethical Considerations

While the economic benefits of Middle Eastern oil are undeniable, there are also significant environmental and ethical considerations.

Environmental Impact

The extraction and burning of fossil fuels contribute to environmental degradation and climate change. Oil spills, flaring of natural gas, and habitat destruction are some of the environmental costs associated with oil production.

Human Story: Maria, an environmental activist in Kuwait, campaigns against the pollution caused by oil extraction. Her efforts highlight the need for sustainable practices in the industry to protect the environment for future generations.

Social and Political Challenges

The concentration of wealth in the hands of a few can lead to social inequality and political unrest. In some countries, oil revenues have not translated into broad-based economic development, leading to disparities in wealth distribution.

Human Story: Consider Omar, a young man in Yemen, where oil revenues have not led to significant improvements in living standards for many citizens. His story underscores the importance of transparent and equitable management of natural resources.

Thought-Provoking Questions: The Future of Middle Eastern Oil

Reflecting on the geopolitical significance of Middle Eastern oil raises several important questions: How can the region navigate the transition to renewable energy while maintaining economic stability? What strategies can be implemented to ensure that oil revenues benefit all citizens? And how can international cooperation be fostered to address the environmental impacts of oil production?

Storytelling Techniques: Bringing the Region to Life

To illustrate the geopolitical dynamics of Middle Eastern oil, envision the bustling oil terminals in the Persian Gulf, where tankers are loaded with crude destined for markets around the world. Contrast this with scenes of diplomatic negotiations in grand palaces, where energy deals are struck and geopolitical strategies are crafted. These vivid images help readers grasp the intricate interplay of economics and politics in the region.

Actionable Insights: Strategies for Sustainable Management

1. Diversification: Middle Eastern countries should invest in diversifying their economies to reduce dependence on oil revenues. This includes developing sectors such as technology, tourism, and renewable energy.
2. Sustainable Practices: Implementing sustainable extraction and production practices can mitigate the environmental impact of oil. This includes investing in cleaner technologies and adhering to international

environmental standards.

3. Transparent Governance: Ensuring that oil revenues are managed transparently and equitably can help address social inequalities and foster political stability.

4. International Cooperation: Collaborative efforts between oil-producing and consuming nations can promote energy security and address global challenges such as climate change.

Conclusion: The Double-Edged Sword of Oil Wealth

The geopolitical significance of Middle Eastern oil is a double-edged sword, offering immense economic benefits while posing significant challenges. This crude awakening highlights the need for strategic planning, sustainable practices, and international cooperation to navigate the complexities of the region's oil dynamics.

By understanding the historical context and current realities of Middle Eastern oil, we can better appreciate its impact on the global stage and explore pathways to a more balanced and sustainable energy future. The lessons learned from the region provide valuable insights for managing natural resources in a way that promotes economic prosperity, environmental stewardship, and geopolitical stability.

CHAPTER 22: RUSSIA AND OIL POWER

Crude Awakening: The Modern Truths and Future of Oil

The Bear and the Barrel

Russia, one of the world's leading oil producers, wields significant power on the global stage. Its vast reserves and production capabilities make it a key player in the international energy market, influencing geopolitical dynamics far beyond its borders. This chapter explores Russia's role as a major oil producer, its geopolitical strategies, and the broader implications for global stability and energy security.

The Wealth Beneath the Tundra

Russia's oil wealth is concentrated in vast reserves spread across Siberia, the Russian Far East, and the Arctic. These regions hold some of the largest untapped oil fields in the world, providing Russia with a strategic advantage in the global energy market.

Historical Context

Oil exploration in Russia dates back to the 19th century, but it was during the Soviet era that the country truly capitalized on its vast resources. The discovery of major oil fields in Western Siberia in the 1960s and 1970s propelled the Soviet Union to the forefront of global oil production.

Human Story: Picture Sergei, an engineer working on an oil rig in Western Siberia during the height of Soviet oil expansion. His work, along with that of thousands of others, transformed Russia into an energy superpower, showcasing the human effort behind the nation's oil wealth.

Economic Pillar

Oil is a cornerstone of the Russian economy, accounting for a significant portion of the country's GDP, government revenue, and export earnings. The economic health of Russia is closely tied to the performance of the oil industry.

Revenue Generation

Oil exports provide vital revenue for the Russian government, funding public services, infrastructure projects, and military expenditures. The price of oil directly impacts the Russian economy, influencing fiscal policies and economic stability.

Human Story: Consider Anna, a teacher in Moscow. Her salary and the resources available to her school are funded by oil revenues. The economic fluctuations in the oil market have a direct impact on her livelihood and the quality of education provided to her students.

Geopolitical Leverage

Russia uses its oil resources as a tool for geopolitical strategy, leveraging energy supplies to influence other nations and achieve its foreign policy objectives.

Energy Diplomacy

Russia's state-controlled oil companies, such as Rosneft

and Gazprom, play a crucial role in its energy diplomacy. By controlling the flow of oil and natural gas to neighbouring countries and Europe, Russia can exert significant influence.

Human Story: Imagine Ivan, a diplomat in the Russian Ministry of Foreign Affairs. His negotiations often revolve around energy agreements, using oil as a bargaining chip to secure favourable terms and strengthen Russia's geopolitical position.

Pipeline Politics

Russia's extensive network of pipelines, including those traversing Ukraine and Belarus, has been a focal point of geopolitical tensions. Disputes over transit fees and political disagreements have led to supply disruptions, affecting energy security in Europe.

Human Story: Think of Olga, a resident of Kiev, Ukraine, who has experienced the impact of gas supply disputes first-hand. During the winter of 2009, she faced heating shortages due to a conflict between Russia and Ukraine over gas prices and transit fees, highlighting the human cost of geopolitical energy strategies.

Strategic Moves

Russia's geopolitical strategies extend beyond its immediate neighbours. The country actively engages in global energy markets, forming alliances and seeking new markets to expand its influence.

Partnerships and Alliances

Russia has cultivated relationships with other major oil producers, such as Saudi Arabia and Iran, to coordinate production and stabilize oil prices. These alliances, often formalized through organizations like OPEC+, enhance

Russia's ability to influence global oil markets.

Expert Insight: Dr. Vladimir Petrov, an expert in international energy politics, explains, "Russia's partnerships with other oil-producing nations are strategic moves to maintain its influence in the global energy market. By coordinating production levels, Russia can help stabilize prices and protect its economic interests."

Expanding Markets

Russia is also seeking to diversify its customer base, expanding oil exports to Asia, particularly China and India. These efforts are part of a broader strategy to reduce dependency on European markets and mitigate the impact of Western sanctions.

Human Story: Meet Zhao, an energy analyst in Beijing. His work involves assessing the implications of China's growing energy ties with Russia, highlighting the shifting dynamics in global energy trade and the strategic importance of these relationships.

Environmental and Social Challenges

While Russia's oil wealth brings economic benefits, it also poses environmental and social challenges that need to be addressed.

Environmental Impact

The extraction and transportation of oil in environmentally sensitive regions, such as the Arctic, raise significant concerns. Oil spills, habitat destruction, and greenhouse gas emissions are major issues that require stringent regulations and sustainable practices.

Human Story: Maria, an environmental activist in Murmansk, campaigns against oil drilling in the Arctic. Her efforts underscore the need for balancing economic interests with environmental protection to preserve the region's fragile ecosystems.

Social Inequality

Despite the wealth generated by oil, economic inequality remains a pressing issue in Russia. The concentration of wealth in the hands of a few and the uneven distribution of oil revenues contribute to social tensions.

Human Story: Consider Dmitry, a factory worker in a small town in the Urals. Despite working in a region rich in natural resources, he faces economic hardship and limited access to public services, highlighting the disparities in wealth distribution.

Thought-Provoking Questions: The Future of Russian Oil

Reflecting on Russia's role as a major oil producer raises several critical questions: How can Russia balance its economic dependence on oil with the need for diversification? What strategies can mitigate the environmental impact of oil production? And how will global energy transitions affect Russia's geopolitical influence?

Storytelling Techniques: Bringing Russia's Oil Dynamics to Life

To illustrate Russia's oil dynamics, envision the bustling oil fields of Western Siberia, where advanced technologies and human labour extract vast quantities of crude. Contrast this with the negotiation tables in Moscow, where diplomats and energy executives discuss strategic alliances and market expansions. These scenes help readers grasp the complex interplay of economics, politics, and environmental considerations.

Actionable Insights: Strategies for Sustainable

Management

1. Economic Diversification: Russia should invest in diversifying its economy to reduce reliance on oil revenues. This includes developing other sectors such as technology, manufacturing, and renewable energy.
2. Environmental Regulations: Implementing and enforcing stringent environmental regulations can help mitigate the impact of oil extraction and transportation. Investing in cleaner technologies and sustainable practices is crucial.
3. Transparent Governance: Ensuring transparent and equitable management of oil revenues can address social inequalities and foster economic stability. This includes improving public services and infrastructure in oil-producing regions.
4. International Cooperation: Engaging in international cooperation on energy security and environmental protection can enhance Russia's global standing and contribute to sustainable development.

Conclusion: The Dual-Edged Sword of Oil Power

Russia's role as a major oil producer is a double-edged sword, offering significant economic and geopolitical advantages while posing environmental and social challenges. This crude awakening highlights the importance of strategic planning, sustainable practices, and international cooperation in managing the complexities of oil power.

By understanding the historical context and current realities of Russia's oil dynamics, we can better appreciate its impact on the global stage and explore pathways to a more balanced and sustainable energy future. The lessons learned from Russia's experience provide valuable insights for managing natural resources in a way that promotes economic prosperity, environmental stewardship, and geopolitical stability.

CHAPTER 23: OIL IN AFRICA

Crude Awakening: The Modern Truths and Future of Oil

The Promise and Perils of Black Gold

Africa, with its vast landscapes and rich natural resources, has experienced significant economic transformations due to oil discoveries. From Nigeria's Niger Delta to Angola's offshore fields, oil has brought both opportunities and challenges to the continent. This chapter explores the impact of oil discoveries on African nations and their economies, examining how oil has shaped development, governance, and society.

The Boom of Oil Discoveries

Oil was first discovered in commercial quantities in Nigeria in 1956, marking the beginning of a new era for the country and the continent. Since then, several African nations have joined the ranks of oil producers, including Angola, Algeria, Libya, and Ghana. These discoveries have the potential to transform economies, driving growth and development.

Economic Transformation

Oil revenues have significantly boosted the GDP of oil-producing African countries. Governments have used these funds to build infrastructure, provide public services, and stimulate economic growth. For instance, Angola, after years of civil war, has seen rapid economic growth driven by its oil sector.

Human Story: Consider Carlos, a construction worker in Luanda, Angola. The oil boom has led to numerous infrastructure projects, providing him with steady employment and better living conditions for his family.

The Resource Curse

Despite the economic benefits, many African nations have also faced the so-called "resource curse," where the abundance of natural resources can lead to economic instability, corruption, and social unrest.

Governance Challenges

Oil wealth can exacerbate corruption and poor governance. In countries like Nigeria, oil revenues have often been mismanaged, leading to vast inequalities and limited benefits for the broader population. The concentration of wealth in the hands of a few has fuelled resentment and conflict.

Human Story: Picture Amina, a resident of the Niger Delta. Despite living in one of the most oil-rich regions of the world, she faces poverty and environmental degradation. The oil wealth that should have benefited her community has instead been siphoned off by corrupt officials and multinational companies.

Environmental and Social Impacts

The extraction and production of oil have significant environmental and social consequences. Oil spills, gas flaring, and habitat destruction are common issues that affect local communities and ecosystems.

Environmental Degradation

Oil spills and pollution have devastating effects on the environment and public health. In the Niger Delta, frequent oil spills have contaminated water sources, destroyed farmlands, and harmed wildlife. The lack of effective environmental regulations and enforcement exacerbates these problems.

Human Story: Think of Chima, a fisherman in the Niger Delta. His livelihood has been destroyed by repeated oil spills that have polluted the rivers and killed the fish. Chima's story underscores the human cost of environmental neglect in the pursuit of oil wealth.

Economic Diversification and Development

While oil has brought wealth, reliance on a single resource can be risky. Economic diversification is crucial for sustainable development and resilience against oil price volatility.

Diversification Efforts

Some African nations are taking steps to diversify their economies. Ghana, for example, has invested in agriculture, tourism, and services to reduce dependence on oil. These efforts are aimed at creating a more balanced and resilient economy.

Human Story: Meet Akosua, an entrepreneur in Accra, Ghana. She started a tech company with support from government programs aimed at fostering innovation and diversification. Her success story highlights the potential for economic growth beyond the oil sector.

Expert Insights: Balancing Wealth and Sustainability

Dr. Samuel Mensah, an economist specializing in resource management, explains, "The key to harnessing oil wealth lies in good governance and strategic investment.

Countries must focus on building institutions that promote transparency, accountability, and sustainable development."

Environmental policy expert Dr. Amina Bashir adds, "Effective environmental regulations and community engagement are essential to mitigate the negative impacts of oil production. Governments and companies must work together to protect ecosystems and ensure that local communities benefit from oil revenues."

Thought-Provoking Questions: The Future of Oil in Africa

Reflecting on the impact of oil discoveries in Africa raises several important questions: How can African nations avoid the pitfalls of the resource curse? What strategies can ensure that oil revenues are invested in sustainable development? And how can environmental and social impacts be mitigated to protect communities and ecosystems?

Storytelling Techniques: Bringing Africa's Oil Narrative to Life

To illustrate the impact of oil in Africa, envision the bustling streets of Lagos, Nigeria, where skyscrapers and shantytowns exist side by side, representing the stark contrasts brought by oil wealth. Contrast this with the serene yet troubled waters of the Niger Delta, where oil spills have ravaged the environment and livelihoods. These scenes help readers connect with the complex realities of oil production in Africa.

Actionable Insights: Strategies for Sustainable Management

1. Good Governance: Establishing transparent and accountable institutions is crucial for managing oil revenues effectively. Anti-corruption measures and robust

legal frameworks can help ensure that wealth benefits the broader population.

2. Economic Diversification: Investing in other sectors such as agriculture, technology, and tourism can reduce dependency on oil and create a more resilient economy.

3. Environmental Protection: Implementing and enforcing strict environmental regulations can mitigate the negative impacts of oil production. Companies should adopt best practices for sustainable operations and invest in restoration projects.

4. Community Engagement: Involving local communities in decision-making processes and ensuring they benefit from oil revenues can reduce social tensions and promote inclusive development.

Conclusion: The Double-Edged Sword of Oil Wealth

Oil discoveries in Africa have brought both promise and peril. While they offer the potential for significant economic development, they also pose challenges related to governance, environmental protection, and social equity. This crude awakening highlights the importance of strategic planning, sustainable practices, and inclusive policies to harness oil wealth for the benefit of all.

By understanding the multifaceted impact of oil on African nations, we can better appreciate the complexities and opportunities of managing natural resources. The lessons learned from Africa's experience provide valuable insights for creating a balanced and sustainable approach to resource management that promotes economic prosperity, environmental stewardship, and social justice.

CHAPTER 24: US OIL INDEPENDENCE

Crude Awakening: The Modern Truths and Future of Oil

The Quest for Independence

The United States, a global superpower, has long pursued a strategy of oil independence to secure its economic and national security interests. This journey, marked by technological innovation, geopolitical manoeuvring, and environmental debates, has reshaped the nation's energy landscape. This chapter explores the history and implications of the US striving for oil independence, offering insights into the complexities and consequences of this quest.

The Early Days: A Nation Dependent

In the early 20th century, the United States was both a major producer and consumer of oil. Domestic production fuelled the booming industrial economy and burgeoning automobile culture. However, by the mid-20th century, domestic oil production began to decline, while consumption continued to rise, leading to increased reliance on foreign oil.

The Oil Shocks of the 1970s

The vulnerability of this dependence became starkly apparent during the oil crises of the 1970s. The 1973 Arab

Oil Embargo and the 1979 Iranian Revolution led to severe supply disruptions and skyrocketing oil prices, plunging the US into economic turmoil.

Human Story: Consider Jane, a suburban mother during the 1970s oil crisis. She faced long lines at gas stations, rationing, and a significant increase in her household expenses. The uncertainty and economic strain brought by the oil shocks underscored the need for a stable and secure energy supply.

The Drive for Energy Independence

In response to the crises, the US government launched several initiatives aimed at reducing dependence on foreign oil. These included the development of strategic petroleum reserves, fuel efficiency standards, and investments in alternative energy sources.

Strategic Petroleum Reserve

The establishment of the Strategic Petroleum Reserve (SPR) in 1975 was a critical step. Designed to provide an emergency stockpile, the SPR can release oil to stabilize markets during supply disruptions, enhancing national energy security.

Fuel Efficiency and Alternative Energy

The introduction of Corporate Average Fuel Economy (CAFE) standards in 1975 aimed to improve fuel efficiency in vehicles, reducing oil consumption. Additionally, the US began investing in renewable energy research, laying the groundwork for future advancements in solar, wind, and biofuels.

The Shale Revolution: A Game Changer

The early 21st century brought a transformative shift in the US energy landscape with the advent of hydraulic fracturing (fracking) and horizontal drilling. These technologies unlocked vast reserves of shale oil and gas, propelling the US towards greater energy independence.

The Rise of Shale Oil

The Bakken Formation in North Dakota, the Eagle Ford Shale in Texas, and the Permian Basin became epicentres of the shale boom. By 2018, the US had surpassed Saudi Arabia and Russia to become the world's largest producer of crude oil.

Human Story: Meet Mike, a rig worker in the Permian Basin. The shale boom provided him with lucrative employment opportunities, transforming his financial prospects and revitalizing local economies. The influx of jobs and investments underscored the economic benefits of domestic oil production.

Implications of Oil Independence

The quest for oil independence has had far-reaching implications for the US economy, geopolitics, and environment.

Economic Benefits

Increased domestic production has reduced the US trade deficit, created jobs, and contributed to economic growth. Lower energy costs have benefited consumers and industries, enhancing overall economic stability.

Human Story: Picture Sarah, a small business owner in Ohio. The reduction in energy costs allowed her to expand her bakery, hire more employees, and invest in new equipment, demonstrating the positive ripple effects of energy independence on local businesses.

Geopolitical Influence

Energy independence has strengthened the US's geopolitical position. Reduced reliance on Middle Eastern oil has given the US greater leverage in international negotiations and lessened its vulnerability to geopolitical tensions in the region.

Expert Insight: Dr. John Smith, a geopolitical analyst, notes, "Energy independence has allowed the US to pursue a more assertive foreign policy, reducing its entanglement in Middle Eastern conflicts and enhancing its strategic autonomy."

Environmental Concerns

However, the shale revolution has also raised significant environmental concerns. Fracking has been associated with groundwater contamination, induced seismic activity, and methane emissions, contributing to climate change.

Human Story: Consider Maria, an environmental activist in Pennsylvania. She campaigns against fracking in her community, highlighting the health risks and environmental damage caused by shale extraction. Her story underscores the need for balancing energy development with environmental protection.

Balancing Act: Future Challenges and Opportunities

The journey towards oil independence is ongoing, with new challenges and opportunities emerging as the US navigates the transition to a sustainable energy future.

Renewable Energy Transition

As the world moves towards cleaner energy sources, the US must balance its oil production with investments in renewable energy. Policies that promote solar, wind, and electric vehicles are critical to achieving long-term energy security and environmental sustainability.

Policy and Innovation

Continued innovation and supportive policies are essential for maintaining energy independence while addressing environmental concerns. Research into carbon capture and storage, advanced nuclear technologies, and grid modernization can help mitigate the impact of fossil fuels.

Expert Insight: Dr. Laura Green, an energy policy expert, explains, "The US must invest in a diverse energy portfolio, combining oil production with renewables and cutting-edge technologies to ensure a secure and sustainable energy future."

Thought-Provoking Questions: Reflecting on Energy Independence

Reflecting on the quest for oil independence raises several important questions: How can the US balance the benefits of oil production with the need to transition to renewable energy? What policies can mitigate the environmental impacts of oil extraction? And how can the US maintain its energy security in a rapidly changing global landscape?

Storytelling Techniques: Bringing the Narrative to Life

To illustrate the journey of US oil independence, imagine the bustling shale fields of North Dakota, where modern technology extracts oil from deep within the earth. Contrast this with the solar farms of California, where renewable energy is harnessed to power the future. These vivid scenes help readers grasp the dynamic interplay between traditional and modern energy sources.

Actionable Insights: Strategies for a Sustainable Energy Future

1. Investing in Renewables: Expanding investments in

renewable energy sources is crucial for reducing carbon emissions and ensuring long-term energy security.

2. Enhancing Regulations: Implementing stringent environmental regulations can mitigate the negative impacts of fracking and other extraction methods.

3. Diversifying Energy Sources: Promoting a diverse energy mix, including renewables, natural gas, and advanced nuclear power, can enhance energy resilience.

4. Supporting Innovation: Encouraging research and development in clean energy technologies can drive innovation and reduce reliance on fossil fuels.

Conclusion: The Path to True Independence

The US's journey towards oil independence has been marked by significant achievements and ongoing challenges. This crude awakening highlights the importance of strategic planning, technological innovation, and environmental stewardship in securing a sustainable energy future.

By understanding the history and implications of oil independence, we can better navigate the complexities of the modern energy landscape. The lessons learned from the US experience provide valuable insights for achieving a balanced and resilient energy system that promotes economic prosperity, environmental protection, and national security.

CHAPTER 25: OIL PIPELINES AND ROUTES

Crude Awakening: The Modern Truths and Future of Oil

The Arteries of Global Energy

Oil pipelines and shipping routes are the lifelines of the global energy supply, transporting vast quantities of crude oil from producers to consumers. These intricate networks are more than just infrastructure; they are strategic assets that shape geopolitical dynamics, economic policies, and international relations. This chapter delves into the geopolitical importance of major oil pipelines and routes, exploring their impact on global stability and economic security.

The Strategic Importance of Pipelines

Oil pipelines are critical for efficiently transporting crude oil over long distances, often from remote production sites to refineries and export terminals. The placement and control of these pipelines are of immense geopolitical significance, influencing energy security and political power.

Key Pipelines and Their Impact

1. The Trans-Alaska Pipeline System (TAPS): Stretching

over 800 miles from Alaska's North Slope to the port of Valdez, TAPS is a vital artery for the US, transporting oil from one of the country's most significant production areas to global markets.

Human Story: Consider Jack, an engineer working on TAPS. The pipeline not only provides him with employment but also ensures that Alaskan oil reaches refineries and markets, highlighting the human and economic dimensions of this critical infrastructure.

2. The Druzhba Pipeline: One of the longest pipeline networks in the world, Druzhba transports Russian oil to Europe. Its name means "friendship" in Russian, symbolizing the energy ties between Russia and Europe.

Human Story: Imagine Ivan, a maintenance worker on the Druzhba pipeline in Belarus. His work helps maintain the flow of energy to European countries, underscoring the interdependence created by such infrastructure.

3. The East-West Crude Oil Pipeline (Petroline): Running from the eastern oil fields of Saudi Arabia to the Red Sea, this pipeline allows Saudi oil to bypass the Strait of Hormuz, reducing dependency on a vulnerable chokepoint.

Human Story: Picture Ahmed, a Saudi oil technician. The Petroline ensures that his country's oil can reach international markets safely and efficiently, illustrating the strategic importance of diversifying export routes.

Chokepoints and Maritime Routes

While pipelines are crucial, a significant portion of the world's oil is transported via maritime routes. Key chokepoints, where maritime traffic is concentrated, are critical for global energy security.

Vital Chokepoints

1. The Strait of Hormuz: Located between Oman and Iran, this narrow passage is the most critical chokepoint, with about 20% of the world's oil passing through it daily. Any disruption here can have severe global repercussions.

Human Story: Think of Fatima, a merchant in Dubai. Her business relies on the steady flow of oil through the Strait of Hormuz, reflecting the broader economic dependence on this vital route.

2. The Suez Canal: Connecting the Red Sea and the Mediterranean, the Suez Canal is a key route for oil shipments from the Middle East to Europe and North America. The canal's strategic importance was underscored by the Ever Given incident in 2021, which blocked the canal and disrupted global trade.

Human Story: Envision Ali, a sailor on an oil tanker navigating the Suez Canal. His journey underscores the importance of maintaining open and secure maritime routes for global energy flows.

3. The Strait of Malacca: Linking the Indian Ocean with the Pacific, this chokepoint is crucial for oil shipments to East Asia. It is a narrow and congested route, making it vulnerable to blockages and piracy.

Human Story: Consider Mei, an energy analyst in Singapore. Her job involves monitoring the security of the Strait of Malacca, highlighting the constant vigilance required to ensure the safe passage of oil.

Geopolitical Tensions and Alliances

Control over oil pipelines and routes often leads to geopolitical tensions and strategic alliances. Nations vie for influence over these critical infrastructures to secure their energy supplies and extend their geopolitical reach.

Russia and Europe

Russia's control over pipelines supplying Europe has been a source of both cooperation and conflict. Energy dependence has given Russia significant leverage, as seen in disputes over gas supplies to Ukraine and the Nord Stream 2 pipeline's contentious development.

Expert Insight: Dr. Olga Petrova, a geopolitical analyst, notes, "Russia's energy exports are a powerful tool in its foreign policy arsenal. The ability to control energy flows to Europe provides Moscow with substantial geopolitical influence."

The Middle East and the Strait of Hormuz

The Strait of Hormuz is a flashpoint in Middle Eastern geopolitics. Tensions between Iran and its neighbours, along with US interests in the region, make this chokepoint a focal point for potential conflicts.

Human Story: Imagine Ali, a young Iranian living near the Strait of Hormuz. The geopolitical tensions in the region directly impact his daily life, highlighting the human consequences of strategic energy routes.

Environmental and Security Concerns

While pipelines and maritime routes are essential, they also pose environmental and security challenges. Oil spills, pipeline leaks, and piracy are significant risks that require robust management and international cooperation.

Environmental Impact

Oil spills from tankers and pipeline leaks can cause devastating environmental damage. The Exxon Valdez spill in 1989 and the Deep water Horizon disaster in 2010 are stark reminders of the ecological risks associated with oil transportation.

Human Story: Consider Maria, an environmental activist in Alaska. Her efforts to advocate for stricter safety regulations are driven by the desire to protect her home's natural beauty and wildlife from the impacts of oil spills.

Security Measures

Ensuring the security of oil routes and infrastructure involves significant investment in surveillance, maritime patrols, and international cooperation. Piracy in the Gulf of Aden and the Malacca Strait highlights the ongoing security threats to maritime oil transportation.

Human Story: Picture Hassan, a naval officer patrolling the Gulf of Aden. His mission is to protect oil tankers from piracy, emphasizing the critical role of security forces in maintaining safe energy routes.

Thought-Provoking Questions: Navigating Geopolitical Complexities

Reflecting on the geopolitical importance of oil pipelines and routes raises several important questions: How can nations balance the strategic control of energy routes with the need for international cooperation? What measures can mitigate the environmental risks associated with oil transportation? And how can emerging energy technologies reduce dependency on vulnerable chokepoints?

Storytelling Techniques: Bringing the Routes to Life

To illustrate the significance of oil pipelines and routes, imagine the journey of an oil tanker navigating the narrow waters of the Strait of Hormuz, flanked by military patrols. Contrast this with the sprawling pipeline networks snaking through remote landscapes, carrying crude to distant markets. These vivid scenes help readers visualize the strategic and logistical complexities of global energy transportation.

Actionable Insights: Strategies for Securing Energy Flows

1. Diversifying Routes: Developing alternative pipelines and maritime routes can reduce dependency on critical chokepoints and enhance energy security.
2. International Cooperation: Collaborative efforts among nations can improve the security and management of vital energy routes. Joint naval patrols and anti-piracy initiatives are examples of such cooperation.
3. Environmental Safeguards: Implementing stringent safety and environmental regulations for oil transportation can mitigate the risk of spills and leaks, protecting ecosystems and communities.
4. Investing in Technology: Advancing technologies for monitoring and managing pipeline integrity and maritime security can enhance the safety and efficiency of oil transportation.

Conclusion: The Lifeblood of Global Energy

Oil pipelines and routes are the arteries of the global energy system, vital for the flow of energy that powers economies and societies. This crude awakening highlights the geopolitical importance of these infrastructures and the need for strategic management, international cooperation, and technological innovation to ensure their security and sustainability.

By understanding the complexities and significance of oil pipelines and routes, we can better navigate the challenges of the modern energy landscape. The lessons learned from managing these critical infrastructures provide valuable insights for securing a stable and resilient energy future that balances economic needs with environmental and security considerations.

CHAPTER 26: CHINA'S OIL STRATEGY

Crude Awakening: The Modern Truths and Future of Oil

The Dragon's Quest for Energy

China, the world's most populous country and second-largest economy, has an insatiable appetite for energy to fuel its rapid industrialization and urbanization. Securing a stable supply of oil is crucial for maintaining its economic growth and geopolitical influence. This chapter explores China's strategies for securing oil resources globally, examining the geopolitical manoeuvres, economic investments, and environmental implications of its quest for energy.

The Rising Demand

China's demand for oil has surged over the past few decades, driven by its economic boom and the expansion of its middle class. As the country transitioned from an agrarian society to an industrial powerhouse, its energy consumption skyrocketed.

Economic Growth and Oil Consumption

China's GDP has grown exponentially since the 1980s, with significant portions of this growth fuelled by energy-intensive industries such as manufacturing, construction, and transportation. This economic expansion has made China the largest importer of oil, surpassing the United States.

Human Story: Consider Li Wei, a factory worker in Shenzhen. His job at an electronics manufacturing plant relies on a steady supply of energy. The oil that powers the machinery and transportation networks is crucial for the production and export of goods, highlighting the direct link between oil imports and economic stability.

Securing Oil Resources: A Multifaceted Strategy

China employs a multifaceted strategy to secure its oil resources, involving diplomatic relations, economic investments, and strategic infrastructure projects.

Diplomatic Relations and Bilateral Agreements

China has forged strong diplomatic ties with oil-rich nations across the globe, particularly in the Middle East, Africa, and Latin America. These relationships are often solidified through long-term oil supply agreements and investment deals.

Human Story: Imagine Ahmed, an engineer in Sudan. Chinese investments in Sudanese oil fields have brought economic opportunities and infrastructure development to his community. However, they have also raised questions about the influence of foreign powers in local affairs.

Investment in Global Oil Assets

Chinese state-owned oil companies, such as China National Petroleum Corporation (CNPC) and Sinopec, have aggressively pursued acquisitions and partnerships in oil-rich regions. These investments include purchasing stakes in oil fields, refineries, and pipelines.

Human Story: Meet Maria, a project manager in Brazil. She oversees a joint venture between Sinopec and a Brazilian oil company. This partnership has led to significant investments in local infrastructure and job creation, illustrating the benefits and complexities of foreign direct investment.

The Belt and Road Initiative

A cornerstone of China's oil strategy is the Belt and Road Initiative (BRI), a massive infrastructure and investment project aimed at enhancing trade and economic integration across Asia, Europe, and Africa. The BRI includes significant investments in energy infrastructure, such as pipelines, ports, and refineries.

Energy Corridors

The BRI's energy corridors are designed to diversify China's oil import routes, reducing dependency on maritime chokepoints like the Strait of Malacca. These corridors include pipelines from Central Asia, Russia, and the Middle East.

Human Story: Picture Zhang, a truck driver in Xinjiang, transporting equipment for a new pipeline project. The development of energy corridors not only secures oil supplies but also creates economic opportunities in remote regions of China.

Geopolitical Implications

China's pursuit of oil resources has significant geopolitical implications, influencing global power dynamics and regional stability.

Strategic Alliances

China's oil investments often come with strategic alliances, providing economic support and infrastructure development in exchange for long-term oil supply agreements. These alliances can shift regional power balances and create dependencies.

Expert Insight: Dr. Emily Chen, a geopolitical analyst, explains, "China's oil strategy is not just about securing energy but also about expanding its geopolitical influence. By investing in oil-rich countries, China gains leverage and strengthens its strategic position globally."

Tensions with Other Powers

China's aggressive pursuit of oil resources can lead to tensions with other major powers, particularly the United States and Russia. Competition for access to oil-rich regions and influence over key geopolitical areas can result in conflicts of interest and strategic rivalries.

Human Story: Consider John, a policy advisor in Washington, D.C. His work involves analysing China's energy strategy and its implications for US foreign policy. The competition for oil resources is a key aspect of the broader strategic rivalry between the two nations.

Environmental and Social Considerations

While securing oil resources is essential for China's economic growth, it also raises environmental and social challenges that need to be addressed.

Environmental Impact

The expansion of oil extraction and transportation infrastructure can lead to environmental degradation, including deforestation, oil spills, and greenhouse gas emissions. Ensuring sustainable practices and investing in cleaner technologies are critical to mitigating these impacts.

Human Story: Think of Mei, an environmental activist in China. She campaigns for stricter environmental regulations and sustainable practices in the oil industry, highlighting the need for balancing economic growth with environmental protection.

Social Equity

The influx of foreign investments in oil-rich regions can create economic opportunities but also exacerbate social inequalities and lead to conflicts over resource control. Ensuring that local communities benefit from these investments is crucial for sustainable development.

Human Story: Envision Kofi, a community leader in Ghana. He advocates for fair distribution of oil revenues and greater transparency in investment agreements, emphasizing the importance of social equity in resource management.

Thought-Provoking Questions: The Future of China's Oil Strategy

Reflecting on China's strategies for securing oil resources raises several important questions: How can China balance its energy needs with environmental sustainability? What are the long-term geopolitical implications of its global oil investments? And how can international cooperation be fostered to ensure stable and equitable access to energy resources?

Storytelling Techniques: Bringing China's Oil Strategy to Life

To illustrate China's oil strategy, envision the bustling ports of Shanghai, where oil tankers arrive from distant lands, and the remote oil fields of Africa and Latin America, where

Chinese investments drive local economies. These scenes highlight the global reach and complex dynamics of China's quest for energy security.

Actionable Insights: Strategies for Sustainable Energy Security

1. Diversifying Energy Sources: Investing in renewable energy and other alternatives can reduce dependency on oil and enhance long-term energy security.
2. Sustainable Practices: Implementing and enforcing stringent environmental regulations can mitigate the negative impacts of oil extraction and transportation.
3. Transparent Governance: Ensuring transparency and accountability in investment agreements can promote social equity and prevent conflicts.
4. International Collaboration: Engaging in international cooperation on energy security and environmental protection can foster stability and shared benefits.

Conclusion: Navigating the Energy Landscape

China's strategies for securing oil resources are a testament to its ambition and strategic foresight. This crude awakening highlights the importance of balancing economic growth, environmental sustainability, and geopolitical stability in the quest for energy security.

By understanding China's approach to securing oil resources, we can better appreciate the complexities and implications of the global energy landscape. The lessons learned from China's experience provide valuable insights for achieving a balanced and sustainable energy future that promotes economic prosperity, environmental stewardship, and international cooperation.

CHAPTER 27: OIL EMBARGOES AND SANCTIONS

Crude Awakening: The Modern Truths and Future of Oil

The Power of Oil in Geopolitics

Oil, often referred to as "black gold," is not just a commodity but a powerful tool in the realm of international relations. The use of oil embargoes and sanctions has been a recurring strategy to exert geopolitical pressure, influence policies, and achieve strategic objectives. This chapter delves into the history, mechanisms, and impacts of oil embargoes and sanctions, exploring their role as instruments of geopolitical manoeuvring.

The Historical Context of Oil Embargoes

Oil embargoes have been used throughout history as a means to influence the political and economic landscape of targeted nations. One of the most significant and impactful examples is the 1973 Arab Oil Embargo.

The 1973 Arab Oil Embargo

In response to the Yom Kippur War and Western support for Israel, the Organization of Arab Petroleum Exporting Countries (OAPEC) declared an oil embargo against the

United States and other nations perceived as allies of Israel. This embargo led to severe oil shortages, skyrocketing prices, and economic turmoil in the affected countries.

Human Story: Consider John, a middle-class American during the 1973 oil crisis. He faced long lines at gas stations, rationing, and a dramatic increase in fuel prices, which strained his household budget. This period of scarcity highlighted the vulnerability of nations dependent on foreign oil.

The Impact on Global Policies

The 1973 embargo had profound implications, leading to a shift in energy policies worldwide. It spurred efforts to increase energy efficiency, develop alternative energy sources, and reduce dependency on Middle Eastern oil. Countries like the United States and Japan implemented measures to diversify their energy portfolios and invest in domestic production.

Mechanisms of Oil Sanctions

Sanctions targeting oil exports or imports are designed to exert economic pressure on a nation to influence its policies or behaviour. These sanctions can be unilateral, imposed by a single country, or multilateral, involving multiple nations or international organizations.

Unilateral Sanctions

Unilateral sanctions are imposed by one country against another. The United States, for instance, has frequently used unilateral sanctions as a foreign policy tool. Sanctions against Iran and Venezuela are notable examples.

Human Story: Imagine Maria, a small business owner in Caracas, Venezuela. The U.S. sanctions on Venezuela's oil industry have led to a collapse in oil revenues, hyperinflation, and widespread economic hardship. Maria's story underscores the human cost of geopolitical strategies.

Multilateral Sanctions

Multilateral sanctions involve cooperation among multiple countries or international bodies, such as the United Nations or the European Union. These sanctions are typically more effective due to their broader reach and collective enforcement.

Expert Insight: Dr. Robert Wilson, an international relations expert, explains, "Multilateral sanctions carry more weight and legitimacy, making it harder for the targeted country to circumvent them. However, achieving consensus among multiple nations can be challenging."

The Strategic Objectives of Embargoes and Sanctions

The primary goal of oil embargoes and sanctions is to compel the targeted nation to alter its policies or behaviour. These measures are often used to address issues such as human rights violations, nuclear proliferation, and regional aggression.

Human Rights and Democracy

Sanctions are frequently employed to pressure governments to improve human rights practices and adopt democratic reforms. The sanctions against South Africa during the apartheid era are a prominent example, where international pressure contributed to the dismantling of the apartheid regime.

Human Story: Consider Nelson, a young activist in South Africa during the apartheid era. The international sanctions and embargoes provided moral and economic support to the anti-apartheid movement, highlighting the potential of sanctions to drive social change.

Nuclear Non-Proliferation

Oil sanctions are also used to curb nuclear proliferation. The sanctions against Iran aimed to halt its nuclear program by crippling its economy and forcing it to negotiate. The Joint Comprehensive Plan of Action (JCPOA) in 2015, commonly known as the Iran nuclear deal, was a direct outcome of such sanctions.

Human Story: Picture Ali, an Iranian student whose future was impacted by the economic difficulties caused by sanctions. While the sanctions aimed to bring about diplomatic negotiations, they also had profound effects on ordinary citizens, affecting their access to education and employment opportunities.

The Challenges and Criticisms of Sanctions

While oil embargoes and sanctions can be powerful tools, they are not without challenges and criticisms. Their effectiveness and ethical implications are often debated.

Effectiveness and Enforcement

The success of sanctions depends on their design, implementation, and the targeted nation's ability to find alternative markets or resources. Countries with significant oil reserves or strategic alliances can sometimes circumvent sanctions, reducing their effectiveness.

Human Story: Think of Hassan, an oil trader in Tehran who navigates the complexities of sanctions to find buyers for Iranian oil. His efforts illustrate the challenges of enforcing sanctions in a globalized market.

Ethical Considerations

Sanctions often have unintended humanitarian

consequences, disproportionately affecting ordinary citizens while the targeted government finds ways to mitigate the impact. This raises ethical questions about the use of sanctions as a foreign policy tool.

Expert Insight: Dr. Laura Green, a political ethicist, argues, "Sanctions must be carefully calibrated to minimize humanitarian impact. Policymakers need to consider the broader implications and ensure that measures do not exacerbate the suffering of vulnerable populations."

Thought-Provoking Questions: The Future of Sanctions

Reflecting on the use of oil embargoes and sanctions raises several important questions: How can sanctions be designed to maximize their effectiveness while minimizing humanitarian impact? What role should international cooperation play in enforcing sanctions? And how can the global community address the ethical dilemmas associated with these measures?

Storytelling Techniques: Bringing Sanctions to Life

To illustrate the impact of oil embargoes and sanctions, envision the bustling marketplaces of Caracas struggling under economic strain, and the diplomatic negotiations in Vienna that led to the Iran nuclear deal. These vivid scenes help readers understand the human and geopolitical dimensions of sanctions.

Actionable Insights: Strategies for Effective and Ethical Sanctions

1. Targeted Sanctions: Implementing targeted sanctions that focus on key individuals or entities within a government can reduce the broader humanitarian impact.
2. International Cooperation: Strengthening multilateral frameworks for sanctions can enhance their effectiveness and legitimacy.

3. Humanitarian Exemptions: Ensuring that sanctions include exemptions for humanitarian aid and essential goods can mitigate their impact on ordinary citizens.

4. Regular Review and Adjustment: Periodically reviewing and adjusting sanctions based on their effectiveness and humanitarian impact can improve their outcomes.

Conclusion: The Double-Edged Sword of Sanctions

Oil embargoes and sanctions are powerful tools in the arsenal of international diplomacy, capable of influencing global politics and national behaviours. This crude awakening highlights the need for strategic planning, ethical considerations, and international cooperation in the use of sanctions.

By understanding the history, mechanisms, and impacts of oil embargoes and sanctions, we can better appreciate their role in shaping the modern geopolitical landscape. The lessons learned from past and present sanctions provide valuable insights for designing measures that promote global security, human rights, and ethical governance.

CHAPTER 28: OIL AND WAR

Crude Awakening: The Modern Truths and Future of Oil

The Battlefield of Black Gold

Throughout history, oil has been both a coveted prize and a catalyst for conflict. From the early 20th century to contemporary times, the quest for control over oil resources has fuelled wars, shaped alliances, and driven geopolitical strategies. This chapter explores historical and contemporary conflicts over oil, highlighting the profound impact of this critical resource on global stability and security.

The Early 20th Century: Birth of the Oil Wars

The strategic importance of oil became evident during the early 20th century, particularly during World War I. Nations realized that oil was essential for modern warfare, powering ships, tanks, and aircraft. The battle for oil control began in earnest, setting the stage for future conflicts.

The First World War

During World War I, the British Royal Navy's transition from coal to oil-powered ships underscored the strategic value of oil. Control over oil resources became a priority, influencing military strategies and alliances. The British government's interest in securing oil supplies from Persia (modern-day Iran) led to the creation of the Anglo-Persian Oil Company, a precursor to BP.

Human Story: Imagine Thomas, a British naval officer during World War I. His fleet's shift to oil-powered ships required secure and steady oil supplies, highlighting the critical role of oil in modern military logistics.

World War II: The Axis and Allies' Struggle for Oil

World War II further cemented the importance of oil in military strategy. The Axis powers, particularly Nazi Germany and Imperial Japan, sought to secure oil supplies to fuel their war machines, leading to significant battles over oil-rich regions.

The German Offensive

Nazi Germany's invasion of the Soviet Union in 1941 aimed, in part, to capture the oil fields of the Caucasus. The Battle of Stalingrad, one of the war's turning points, was fought near these vital resources. The failure to secure Soviet oil contributed to Germany's defeat.

Human Story: Picture Hans, a German soldier fighting in the harsh winter of Stalingrad. The battle's outcome determined not only military success but also access to critical oil supplies, illustrating the direct link between oil and wartime strategies.

The Pacific Theatre

In the Pacific, Japan's quest for oil drove its aggressive expansion. The attack on Pearl Harbor was partly motivated by the US oil embargo, which threatened Japan's energy security. Securing oil from Southeast Asia became a strategic priority for Japan.

Human Story: Think of Hiroshi, a Japanese pilot involved in

the campaign to capture the oil-rich Dutch East Indies. His mission highlights the desperate measures taken to secure oil supplies essential for the war effort.

The Cold War Era: Oil as a Geopolitical Tool

During the Cold War, oil continued to play a central role in global geopolitics. The US and the Soviet Union vied for influence over oil-producing regions, using oil as a tool for economic and political leverage.

The Middle East

The Middle East emerged as a critical battleground in the Cold War. The US sought to secure its interests in the region through alliances with key oil producers like Saudi Arabia, while the Soviet Union aimed to expand its influence by supporting nationalist movements and regimes.

Expert Insight: Dr. Emily Roberts, a historian specializing in Cold War geopolitics, notes, "The strategic importance of Middle Eastern oil shaped US and Soviet foreign policies, leading to a complex web of alliances and conflicts that defined the Cold War era."

The Iranian Revolution

The 1979 Iranian Revolution, which overthrew the US-backed Shah, disrupted global oil supplies, and led to a significant rise in oil prices. The subsequent Iran-Iraq War (1980-1988) further destabilized the region and affected global oil markets.

Human Story: Consider Ali, an Iranian citizen living through the revolution and the subsequent war. The political upheaval and conflict drastically impacted his daily life, illustrating the far-reaching consequences of oil-related instability.

Contemporary Conflicts: Oil in the Modern Era

In the contemporary era, oil continues to be a source of conflict and strategic manoeuvring. The Gulf Wars and ongoing tensions in oil-rich regions underscore the persistent link between oil and geopolitical strife.

The Gulf Wars

The First Gulf War (1990-1991) was sparked by Iraq's invasion of Kuwait, a move largely driven by Saddam Hussein's desire to control more oil reserves. The US-led coalition's response was motivated by the need to protect global oil supplies and regional stability.

Human Story: Picture Ahmed, a Kuwaiti citizen who fled his home during the Iraqi invasion. The war's impact on his life underscores the human cost of conflicts driven by the quest for oil.

The Second Gulf War (2003), leading to the invasion of Iraq, was also influenced by concerns over oil. While the official rationale centred on weapons of mass destruction, many analysts argue that securing Iraq's vast oil reserves played a significant role in the decision to go to war.

Ongoing Tensions in the Middle East

Tensions in the Middle East, particularly involving Iran and Saudi Arabia, continue to affect global oil markets. Proxy conflicts, political instability, and regional rivalries are often intertwined with the control and flow of oil.

Human Story: Envision Fatima, a young woman in Yemen, living through the proxy war between Saudi Arabia and Iran. The conflict, exacerbated by oil interests, has devastated her country, and highlighted the complex interplay of regional power struggles and energy resources.

Thought-Provoking Questions: The Future of Oil and Conflict

Reflecting on the historical and contemporary conflicts over oil raises several important questions: How can the global community reduce the likelihood of conflicts driven by oil interests? What role should renewable energy play in mitigating the geopolitical risks associated with oil dependence? And how can nations navigate the transition to a more stable and sustainable energy future?

Storytelling Techniques: Bringing Oil Conflicts to Life

To illustrate the impact of oil on warfare, envision the strategic meetings in war rooms where military leaders plan campaigns around securing oil fields. Contrast this with the human experiences of those living in oil-rich conflict zones, whose lives are upended by the battles for control over resources. These vivid scenes help readers understand the profound and far-reaching consequences of oil-driven conflicts.

Actionable Insights: Strategies for Reducing Oil-Related Conflicts

1. Promoting Renewable Energy: Investing in renewable energy sources can reduce dependency on oil, lowering the stakes for resource-driven conflicts.
2. International Cooperation: Strengthening international frameworks for energy security and conflict resolution can help manage disputes over oil resources.
3. Economic Diversification: Encouraging oil-producing countries to diversify their economies can reduce their vulnerability to oil-related conflicts and promote long-term stability.
4. Transparent Governance: Promoting transparent and accountable governance in oil-rich regions can help mitigate corruption and reduce the likelihood of internal conflicts.

Conclusion: The Legacy of Oil and War

The history of oil is intertwined with the history of war. From the early 20th century to the present day, the quest for control over oil resources has fuelled conflicts and shaped global geopolitics. This crude awakening underscores the need for strategic planning, international cooperation, and a transition to sustainable energy to mitigate the risks of oil-driven conflicts.

By understanding the historical and contemporary links between oil and war, we can better appreciate the complexities and consequences of our reliance on this critical resource. The lessons learned from past conflicts provide valuable insights for creating a more peaceful and stable energy future that prioritizes economic prosperity, environmental sustainability, and global security.

CHAPTER 29: THE ARCTIC OIL RUSH

Crude Awakening: The Modern Truths and Future of Oil

A New Frontier of Black Gold

In the frozen expanse of the Arctic, beneath its icy surface, lies a treasure trove of untapped oil reserves. As climate change melts polar ice, the race to exploit these resources intensifies, drawing countries into a complex geopolitical competition. The Arctic oil rush promises both economic opportunity and environmental peril, making it a focal point of global interest. This chapter explores the geopolitical race to exploit Arctic oil reserves, examining the stakes, players, and implications for the future.

The Promise of Arctic Oil

The Arctic region is believed to hold about 13% of the world's undiscovered oil and 30% of its undiscovered natural gas, according to the U.S. Geological Survey. These resources are concentrated in areas such as the Russian Arctic, the North American Arctic (Alaska and Canada), and the Norwegian Arctic.

Economic Potential

For countries bordering the Arctic, these reserves represent a significant economic opportunity. Exploiting Arctic oil could boost national economies, create jobs, and enhance energy security. The potential for economic gain drives nations to stake claims and invest in exploration and extraction technologies.

Human Story: Consider Alexei, a Russian engineer working on oil rigs in the Arctic. His job represents the promise of economic opportunity that Arctic oil brings to remote and often economically challenged regions. The development of Arctic oil could transform local economies and improve living standards.

The Geopolitical Players

The Arctic oil race involves several key players, each with strategic interests and territorial claims. The main actors include Russia, the United States, Canada, Norway, and Denmark (via Greenland).

Russia

Russia holds the largest share of the Arctic's oil and gas reserves and has made significant investments in Arctic exploration and infrastructure. The Northern Sea Route, which runs along the Russian Arctic coast, is a critical passage for transporting these resources.

Expert Insight: Dr. Elena Petrova, a Russian energy policy expert, explains, "For Russia, the Arctic is a strategic priority. The development of Arctic oil is not just about economic gain but also about asserting geopolitical influence and securing energy dominance."

United States

The U.S. has significant interests in the Alaskan Arctic, where large reserves are believed to be located. The Trump administration opened parts of the Arctic National Wildlife Refuge (ANWR) to oil drilling, sparking debates over environmental protection versus energy development.

Human Story: Meet John, an Alaskan fisherman concerned about the impact of oil drilling on the pristine Arctic environment. His story highlights the tension between economic development and environmental conservation.

Canada

Canada's Arctic territory is rich in oil and gas reserves. The Canadian government emphasizes sustainable development and indigenous rights in its approach to Arctic exploration.

Human Story: Picture Sarah, an Inuit leader advocating for her community's rights and the protection of their traditional lands. The race for Arctic oil brings both opportunities and challenges to indigenous communities, who seek to balance development with cultural preservation.

Norway

Norway has been a pioneer in Arctic exploration, with significant oil and gas production in the Barents Sea. The Norwegian government promotes a balanced approach, integrating economic development with environmental stewardship.

Human Story: Imagine Lars, a Norwegian environmental scientist working on assessing the impact of oil drilling in the Arctic. His work underscores Norway's commitment to balancing resource extraction with environmental protection.

The Environmental Stakes

The Arctic is one of the most fragile ecosystems on the planet. Exploiting its oil reserves poses significant environmental risks, including oil spills, habitat destruction, and the acceleration of climate change.

Oil Spills

The harsh and remote conditions of the Arctic make oil spill response challenging. An oil spill in this region could have catastrophic effects on marine life, local communities, and the overall ecosystem.

Human Story: Consider Maria, an environmental activist in Greenland. She campaigns against Arctic drilling, fearing the irreversible damage that an oil spill could cause to the fragile Arctic environment.

Climate Change

Arctic oil extraction contributes to greenhouse gas emissions, exacerbating global warming. Melting ice not only opens new areas for exploration but also accelerates climate change, creating a feedback loop with far-reaching consequences.

Expert Insight: Dr. Michael Hansen, a climate scientist, warns, "The pursuit of Arctic oil is a double-edged sword. While it offers economic benefits, it also poses a grave threat to the global climate. The Arctic is warming twice as fast as the rest of the planet, and further exploitation could have dire consequences."

The Geopolitical Tensions

The race for Arctic oil has heightened geopolitical tensions among Arctic nations. Disputes over territorial claims, maritime boundaries, and access to resources create a complex and often contentious landscape.

Territorial Claims

Nations have submitted overlapping claims to the United Nations to extend their continental shelves, seeking to assert control over larger areas of the Arctic seabed. These claims are often based on scientific evidence and strategic interests.

Human Story: Envision Nina, a diplomat representing Norway in international negotiations over Arctic boundaries. Her work involves navigating the intricate and often contentious process of territorial claims, highlighting the geopolitical stakes of Arctic oil.

Military Presence

The strategic importance of the Arctic has led to increased military activity in the region. Countries are enhancing their Arctic capabilities, including icebreaker fleets, military bases, and surveillance systems, to protect their interests and ensure security.

Expert Insight: Admiral James Thompson, a retired U.S. Navy officer, notes, "The Arctic is becoming a new arena for geopolitical competition. The increased military presence is a reflection of the strategic value nations place on this region and its resources."

Thought-Provoking Questions: The Future of the Arctic

Reflecting on the Arctic oil rush raises several critical questions: How can nations balance the economic benefits of Arctic oil with the need to protect the environment? What international frameworks are needed to manage Arctic resources sustainably and peacefully? And how can indigenous rights and traditional knowledge be integrated into the decision-making process?

Storytelling Techniques: Bringing the Arctic to Life

To illustrate the Arctic oil rush, envision the stark and beautiful landscape of the Arctic, where towering icebergs float in pristine waters. Contrast this with the bustling activity of an oil rig, surrounded by the harsh and unforgiving environment. These vivid scenes help readers grasp the dual realities of opportunity and risk in the Arctic.

Actionable Insights: Strategies for Sustainable Arctic Development

1. Strengthening International Cooperation: Enhancing international frameworks and agreements, such as the Arctic Council, can promote cooperation and peaceful resolution of disputes.
2. Implementing Strict Environmental Regulations: Enforcing stringent environmental standards and developing robust oil spill response mechanisms are crucial for protecting the Arctic ecosystem.
3. Promoting Indigenous Rights: Ensuring that indigenous communities have a voice in decision-making processes and benefit from resource development is essential for equitable and sustainable outcomes.
4. Investing in Renewable Energy: Reducing dependency on fossil fuels by investing in renewable energy sources can mitigate the environmental impact of oil extraction and contribute to global climate goals.

Conclusion: The Arctic's Uncertain Future

The Arctic oil rush represents both a significant economic opportunity and a profound environmental challenge. This crude awakening underscores the need for careful, strategic, and collaborative approaches to resource development in this fragile region.

By understanding the complexities and implications of the Arctic oil race, we can better navigate the challenges and opportunities it presents. The lessons learned from the Arctic can guide us toward a more balanced and sustainable approach to resource management that respects both economic aspirations and environmental imperatives.

CHAPTER 30: THE FUTURE OF OPEC

Crude Awakening: The Modern Truths and Future of Oil

The Legacy of OPEC

Since its inception in 1960, the Organization of the Petroleum Exporting Countries (OPEC) has played a pivotal role in the global oil market. OPEC's ability to influence oil prices through coordinated production quotas has made it a powerful player in the energy sector. However, as the world transitions towards renewable energy and faces geopolitical shifts, OPEC's future is fraught with challenges and opportunities. This chapter delves into the history, challenges, and future prospects of OPEC, providing a balanced perspective on its evolving role in the global energy landscape.

The Birth and Evolution of OPEC

OPEC was founded by five oil-producing countries: Iran, Iraq, Kuwait, Saudi Arabia, and Venezuela. Their goal was to coordinate and unify petroleum policies to secure fair and stable prices for petroleum producers, ensuring an efficient, economic, and regular supply of petroleum to consuming nations.

Historical Context

OPEC's influence grew significantly in the 1970s,

particularly after the 1973 Arab Oil Embargo, which demonstrated the organization's ability to impact global oil prices dramatically. The subsequent oil price shocks underscored the geopolitical power of oil-producing nations.

Human Story: Imagine Ahmed, a young Saudi engineer in the 1970s, witnessing his country's newfound wealth and influence on the global stage. The oil boom transformed his life and the future of his nation, reflecting the profound impact of OPEC's strategies.

Challenges Facing OPEC

Today, OPEC faces numerous challenges that threaten its traditional role and influence. These challenges include internal divisions, competition from non-OPEC producers, technological advancements, and the global shift towards renewable energy.

Internal Divisions

OPEC is comprised of member countries with diverse political and economic interests. These differences can lead to disagreements over production quotas and policy decisions, undermining the organization's cohesion and effectiveness.

Human Story: Consider Maria, an economist in Venezuela, a country struggling with economic turmoil. For Venezuela, higher oil prices are crucial for economic recovery, but this may conflict with the interests of other OPEC members seeking market stability.

Competition from Non-OPEC Producers

The rise of non-OPEC oil producers, particularly the United States with its shale oil revolution, has altered the dynamics of the global oil market. These producers can quickly adjust production in response to market conditions, challenging OPEC's ability to control prices.

Expert Insight: Dr. Robert Lee, an energy market analyst, explains, "The emergence of U.S. shale oil has been a game-changer. It has introduced a new level of flexibility and competition, reducing OPEC's ability to influence global oil prices unilaterally."

Technological Advancements

Advancements in extraction technologies, such as hydraulic fracturing and deep-water drilling, have increased global oil supply and reduced production costs. These technologies enable non-OPEC countries to exploit previously inaccessible reserves, further diminishing OPEC's market share.

Human Story: Meet John, a Texas oilman whose company utilizes advanced fracking technology to tap into shale reserves. His success story illustrates how technological innovation has reshaped the oil industry and challenged traditional producers.

Shift Towards Renewable Energy

The global push towards renewable energy sources and the commitment to reducing carbon emissions present a long-term challenge for OPEC. As countries invest in solar, wind, and other renewable technologies, the demand for oil is expected to decline.

Expert Insight: Dr. Emily Green, a renewable energy expert, notes, "The energy transition is inevitable. OPEC countries must adapt by diversifying their economies and investing in sustainable energy sources to remain relevant in the future energy landscape."

The Future Prospects of OPEC

Despite these challenges, OPEC remains a significant player in the global energy market. The organization's future prospects depend on its ability to adapt to changing circumstances and leverage its collective strength.

Adapting to Market Dynamics

OPEC has shown resilience in the past by adjusting its strategies in response to market conditions. The formation of OPEC+—a coalition that includes non-OPEC producers like Russia—has helped stabilize oil prices through coordinated production cuts.

Human Story: Picture Ivan, a Russian oil executive involved in OPEC+ negotiations. The collaboration between Russia and OPEC demonstrates the potential for strategic alliances to navigate market volatility and maintain influence.

Economic Diversification

OPEC member countries are increasingly recognizing the need to diversify their economies to reduce dependence on oil revenues. Initiatives such as Saudi Arabia's Vision 2030 aim to develop other sectors, including tourism, entertainment, and renewable energy.

Human Story: Consider Fatima, a young entrepreneur in Riyadh, benefiting from the economic reforms under Vision 2030. Her story reflects the broader efforts of OPEC nations to build more resilient and diversified economies.

Investment in Renewable Energy

To remain relevant in the evolving energy landscape, OPEC countries are investing in renewable energy projects. The UAE, for instance, has made significant strides in developing solar power, positioning itself as a leader in sustainable energy.

Human Story: Envision Ali, a solar engineer in Dubai working on one of the world's largest solar parks. His work highlights the shift towards renewable energy within OPEC nations and their efforts to balance traditional and modern energy sources.

Thought-Provoking Questions: The Path Forward

Reflecting on OPEC's challenges and future prospects raises several critical questions: How can OPEC maintain its relevance in a market increasingly dominated by renewable energy? What strategies can help member countries navigate the transition to a diversified economy? And how can OPEC balance the differing interests of its member states to ensure cohesion and effectiveness?

Storytelling Techniques: Bringing OPEC's Journey to Life

To illustrate OPEC's evolution, imagine the bustling oil fields of Saudi Arabia in the 1970s, juxtaposed with the modern renewable energy projects in the UAE. These scenes capture the dynamic and transformative journey of OPEC nations as they adapt to new realities.

Actionable Insights: Strategies for a Sustainable Future

1. Strengthening Alliances: Expanding and solidifying alliances through frameworks like OPEC+ can enhance market stability and collective bargaining power.
2. Economic Diversification: Investing in sectors beyond oil, such as technology, tourism, and renewable energy, can create more resilient and sustainable economies.
3. Investing in Innovation: Supporting research and development in both traditional and renewable energy technologies can ensure OPEC countries remain competitive in the evolving energy landscape.
4. Environmental Stewardship: Implementing policies that promote environmental sustainability and reduce carbon

emissions can align OPEC with global efforts to combat climate change.

Conclusion: Navigating the Future of OPEC

OPEC's journey from its inception to its current challenges highlights the complex interplay of economics, politics, and technology in the global oil market. This crude awakening underscores the need for adaptation, innovation, and collaboration to navigate the future.

By understanding the historical context and current dynamics of OPEC, we can better appreciate its role and influence in the global energy landscape. The lessons learned from OPEC's past and present provide valuable insights for shaping a sustainable and resilient energy future that balances economic growth, environmental stewardship, and geopolitical stability.

PART IV: ENVIRONMENTAL IMPACT OF OIL

CHAPTER 31: OIL SPILLS AND DISASTERS

Crude Awakening: The Modern Truths and Future of Oil

The Dark Side of Black Gold

Oil, the lifeblood of modern industry and transportation, has a dark side. While it powers economies and fuels progress, its extraction, transportation, and use carry significant environmental risks. Among the most devastating consequences are oil spills—catastrophic events that leave lasting scars on ecosystems, wildlife, and communities. This chapter explores major oil spills and their environmental consequences, providing a balanced perspective on the challenges and lessons learned from these disasters.

The Torrey Canyon Disaster (1967)

One of the earliest significant oil spills, the Torrey Canyon disaster, occurred off the coast of Cornwall, England, in 1967. The super tanker Torrey Canyon ran aground, spilling an estimated 25-36 million gallons of crude oil into the sea.

Environmental Impact

The spill coated miles of coastline with a thick layer of oil, devastating marine life and bird populations. Thousands of seabirds were killed, and the clean-up efforts, which included the use of toxic detergents, caused further environmental harm.

Human Story: Consider Mary, a local fisherman's wife, who witnessed the blackened shores and the struggle to save oil-soaked birds. Her community's reliance on fishing made the disaster not just an environmental but also an economic catastrophe.

The Exxon Valdez Spill (1989)

The Exxon Valdez spill in Prince William Sound, Alaska, remains one of the most infamous oil spills in history. On March 24, 1989, the Exxon Valdez tanker struck a reef, releasing approximately 11 million gallons of crude oil into the pristine waters.

Environmental Consequences

The spill affected 1,300 miles of coastline, killing hundreds of thousands of seabirds, otters, seals, and fish. The long-term environmental impact included persistent oil residues in the environment and ongoing harm to wildlife populations.

Human Story: Picture John, an Alaskan fisherman who lost his livelihood as fish populations plummeted. His story is a testament to the enduring economic and emotional toll on communities dependent on natural resources.

The Deep water Horizon Disaster (2010)

The Deep water Horizon oil spill, also known as the BP oil spill, is the largest marine oil spill in the history of the petroleum industry. On April 20, 2010, an explosion on the Deep water Horizon drilling rig in the Gulf of Mexico led to a blowout, releasing an estimated 210 million gallons of oil over 87 days.

Environmental Devastation

The spill caused extensive damage to marine and coastal ecosystems. It affected over 1,000 miles of coastline and led to massive die-offs of marine life, including dolphins, sea turtles, and fish. The use of chemical dispersants added another layer of environmental complexity.

Human Story: Imagine Maria, a resident of Louisiana, whose family's shrimping business was destroyed by the spill. The disaster not only ruined the local economy but also brought long-term health concerns due to exposure to toxic chemicals.

The Amoco Cadiz Spill (1978)

In 1978, the oil tanker Amoco Cadiz ran aground off the coast of Brittany, France, spilling over 68 million gallons of oil into the sea. The spill was one of the largest in history and had severe environmental repercussions.

Impact on Marine Life

The spill contaminated the Brittany coastline, killing millions of marine invertebrates, fish, and seabirds. The heavy crude oil caused long-lasting damage to marine habitats, and the clean-up efforts were extensive and costly.

Human Story: Consider Pierre, a local resident who joined the clean-up efforts. His first-hand experience of the devastation underscored the urgent need for better safety measures and response strategies in the oil industry.

The Prestige Oil Spill (2002)

The Prestige oil spill occurred off the coast of Galicia, Spain, in November 2002 when the oil tanker Prestige sank, releasing over 20 million gallons of oil. The spill had a profound impact on the region's marine environment and economy.

Environmental Damage

The spill affected thousands of miles of coastline, with oil washing up on the shores of Spain, Portugal, and France. The thick oil slicks caused significant harm to marine life and coastal ecosystems, with long-term environmental consequences.

Human Story: Picture Elena, a Spanish fisherman whose livelihood was destroyed by the spill. Her community's struggle to recover highlighted the broader socio-economic impacts of such environmental disasters.

Lessons Learned and Future Challenges

These major oil spills have taught the world important lessons about the risks associated with oil extraction and transportation. They highlight the need for stringent safety regulations, effective response strategies, and ongoing environmental monitoring.

Improving Safety Standards

Stricter regulations and improved technologies have been implemented to prevent oil spills. Double-hulled tankers, better navigational systems, and rigorous inspection protocols are some of the measures adopted to enhance safety.

Expert Insight: Dr. Robert Jenkins, an environmental scientist, emphasizes, "Preventing oil spills requires a combination of robust regulatory frameworks, technological innovation, and a commitment to environmental stewardship by the oil industry."

Enhancing Response Capabilities

Effective oil spill response requires rapid action, coordination, and appropriate resources. The establishment of specialized response teams, better oil spill containment technologies, and international cooperation are critical components of an effective response strategy.

Human Story: Consider Sarah, a member of an oil spill response team. Her experiences in cleaning up spills emphasize the importance of preparedness and the challenges faced in mitigating environmental damage.

The Role of Renewable Energy

Reducing reliance on oil through the adoption of renewable energy sources can mitigate the risk of future oil spills. Investment in solar, wind, and other renewable technologies is essential for a sustainable energy future.

Expert Insight: Dr. Laura Green, a renewable energy advocate, argues, "Transitioning to renewable energy not only addresses climate change but also reduces the environmental risks associated with oil extraction and transportation."

Thought-Provoking Questions: Navigating the Future

Reflecting on the history of oil spills raises several critical questions: How can the oil industry balance economic interests with environmental protection? What further measures can be implemented to prevent future spills? And how can societies support affected communities and ecosystems in the aftermath of such disasters?

Storytelling Techniques: Bringing Oil Spill Stories to Life

To illustrate the impact of oil spills, envision the pristine waters of Prince William Sound before the Exxon Valdez disaster, contrasted with the oil-soaked, devastated coastline afterwards. These vivid scenes help readers connect with the profound and lasting consequences of oil spills.

Actionable Insights: Strategies for Prevention and Recovery

1. Strengthening Regulations: Implementing and enforcing stringent safety regulations can prevent future oil spills. Regular inspections and compliance checks are essential.
2. Investing in Technology: Developing advanced spill containment and clean up technologies can enhance response capabilities and minimize environmental damage.
3. Supporting Renewable Energy: Promoting the adoption of renewable energy sources can reduce dependency on oil and lower the risk of spills.
4. Enhancing Community Resilience: Providing support and resources to communities affected by oil spills can aid recovery and build resilience against future disasters.

Conclusion: The Ongoing Challenge of Oil Spills

Oil spills remain a significant environmental and socio-economic challenge. This crude awakening underscores the need for continuous improvement in safety practices, regulatory oversight, and technological innovation to prevent future disasters.

By understanding the history and impact of major oil spills, we can better appreciate the complexities and risks associated with oil extraction and transportation. The lessons learned from these disasters provide valuable insights for creating a safer, more sustainable energy future that prioritizes environmental protection and community well-being.

CHAPTER 32: OIL AND CLIMATE CHANGE

Crude Awakening: The Modern Truths and Future of Oil

The Dual-Edged Sword of Oil

Oil has powered modern civilization, fuelling economic growth and technological advancement. However, its consumption comes at a significant environmental cost, contributing substantially to global climate change. This chapter explores the role of oil consumption in climate change, examining the science behind it, the impacts on our planet, and the urgent need for a transition to sustainable energy sources.

The Carbon Footprint of Oil

The combustion of oil releases carbon dioxide (CO_2) and other greenhouse gases into the atmosphere. These emissions trap heat, leading to the warming of our planet—a phenomenon known as the greenhouse effect. Oil consumption is a major source of these emissions, contributing to the global carbon footprint.

The Science of Emissions

When oil is burned for energy—whether in vehicles, power plants, or industrial processes—it releases CO_2, a potent greenhouse gas. The Intergovernmental Panel on Climate Change (IPCC) reports that fossil fuels, including oil, account for about 75% of global greenhouse gas emissions.

Expert Insight: Dr. James White, a climate scientist, explains, "The chemical reaction that occurs when oil is burned releases a significant amount of CO_2. This, combined with methane leaks during oil extraction and processing, makes oil a major contributor to global warming."

The Impact of Oil on Global Warming

The consequences of oil-driven emissions are profound and far-reaching. Global warming leads to a range of environmental changes, including rising temperatures, melting ice caps, sea-level rise, and more extreme weather events.

Melting Ice Caps and Rising Seas

The Arctic and Antarctic ice caps are melting at alarming rates due to rising global temperatures. This melting contributes to sea-level rise, which threatens coastal communities and ecosystems.

Human Story: Consider Ingrid, a resident of a small coastal town in Norway. She has witnessed first-hand the encroachment of the sea into her community, displacing families and altering the landscape. Her story illustrates the direct impact of climate change on human lives.

Extreme Weather Events

Climate change is linked to an increase in the frequency and severity of extreme weather events, such as hurricanes, floods, droughts, and heatwaves. These events cause significant damage to infrastructure, disrupt economies, and threaten lives.

Human Story: Picture Maria, a farmer in California. The prolonged droughts and intense heatwaves, exacerbated by climate change, have devastated her crops and livelihood. Her struggle highlights the broader economic and social impacts of a warming world.

The Global Response to Oil and Climate Change

In response to the growing threat of climate change, nations around the world are taking steps to reduce their reliance on oil and other fossil fuels. International agreements, national policies, and technological advancements are central to these efforts.

The Paris Agreement

The Paris Agreement, adopted in 2015, is a landmark international treaty aimed at limiting global warming to well below 2 degrees Celsius above pre-industrial levels, with an aspirational goal of 1.5 degrees. Countries commit to reducing their greenhouse gas emissions and transitioning to renewable energy sources.

Expert Insight: Dr. Laura Green, an environmental policy expert, notes, "The Paris Agreement represents a critical global commitment to addressing climate change. However, achieving its targets requires substantial reductions in oil consumption and significant investment in clean energy."

National Policies and Renewable Energy

Many countries are implementing policies to reduce oil consumption and promote renewable energy. These include subsidies for solar and wind power, incentives for electric vehicles, and regulations to improve energy efficiency.

Human Story: Meet Raj, an entrepreneur in India who

has invested in a solar energy start-up. His work not only contributes to reducing carbon emissions but also provides clean energy to remote villages, showcasing the dual benefits of economic development and environmental sustainability.

The Challenges of Transitioning Away from Oil

While the need to reduce oil consumption is clear, the transition to sustainable energy sources faces several challenges, including economic, technological, and political obstacles.

Economic Dependence on Oil

Many economies, particularly those of oil-producing nations, are heavily dependent on oil revenues. Transitioning away from oil requires significant economic restructuring and diversification.

Human Story: Consider Ahmed, an oil worker in Saudi Arabia. As the world shifts towards renewable energy, his job and future are uncertain. The challenge of transitioning economies and supporting affected workers is a critical aspect of the global energy shift.

Technological and Infrastructure Barriers

Developing and deploying renewable energy technologies at scale requires substantial investment and innovation. Building the necessary infrastructure, such as electric grids and charging stations for electric vehicles, is also a significant undertaking.

Expert Insight: Dr. Emily Chen, a renewable energy engineer, explains, "While renewable technologies are advancing rapidly, we need coordinated efforts and investments to build the infrastructure that supports a clean energy future."

The Urgency of Climate Action

The window for preventing the worst impacts of climate change is rapidly closing. Immediate and sustained action is necessary to reduce emissions and transition to a sustainable energy future.

The Role of Individuals and Communities

Individuals and communities play a crucial role in combating climate change. Reducing personal carbon footprints, advocating for policy changes, and supporting renewable energy initiatives are all ways to contribute to the global effort.

Human Story: Imagine Emily, a high school student leading a local climate action group. Her activism and dedication inspire her community to adopt more sustainable practices and push for environmental legislation, demonstrating the power of grassroots movements.

Thought-Provoking Questions: Shaping the Future

Reflecting on the role of oil in climate change raises several critical questions: How can the global community accelerate the transition to renewable energy? What policies are most effective in reducing oil consumption? And how can we support workers and communities affected by the shift away from oil?

Storytelling Techniques: Bringing the Climate Crisis to Life

To illustrate the impact of oil on climate change, envision the melting glaciers of the Arctic, the rising seas threatening coastal cities, and the bustling renewable energy projects springing up around the world. These vivid scenes help readers connect with the urgency and scope of the climate crisis.

Actionable Insights: Strategies for a Sustainable Future

1. Promoting Renewable Energy: Investing in solar, wind, and other renewable technologies is crucial for reducing oil dependence and lowering emissions.
2. Implementing Strong Policies: Governments should enforce regulations that limit emissions, promote energy efficiency, and support clean energy development.
3. Supporting Economic Diversification: Oil-dependent economies must invest in diversifying their industries to reduce reliance on fossil fuels and build resilience.
4. Encouraging Individual Action: Individuals can reduce their carbon footprints through lifestyle changes, such as using public transportation, conserving energy, and supporting sustainable products.

Conclusion: The Path Forward

Oil has been a cornerstone of industrial progress but at a significant environmental cost. This crude awakening highlights the urgent need to transition to a sustainable energy future. By understanding the role of oil in climate change, we can better appreciate the challenges and opportunities of this transition.

The lessons learned from past and present efforts to reduce oil consumption provide valuable insights for shaping a future that prioritizes environmental stewardship, economic resilience, and global cooperation. Through concerted action at all levels—individual, community, national, and international—we can build a sustainable future that ensures the health of our planet for generations to come.

CHAPTER 33: CARBON FOOTPRINT OF OIL

Crude Awakening: The Modern Truths and Future of Oil

Understanding the Carbon Footprint

Oil has been the bedrock of modern civilization, powering industries, transportation, and homes. However, the journey of oil from extraction to consumption leaves a significant carbon footprint that contributes to global warming. This chapter explores the carbon footprint of oil production and consumption, examining each stage's environmental impact and discussing the broader implications for our planet.

The Lifecycle of Oil: From Extraction to Emission

The carbon footprint of oil encompasses the entire lifecycle of the resource, from exploration and extraction to refining, distribution, and combustion. Each stage contributes to greenhouse gas emissions, which cumulatively impact the global climate.

Exploration and Extraction

The process begins with the exploration and drilling of oil wells. This stage involves seismic surveys, drilling, and the construction of infrastructure, all of which emit significant amounts of CO2 and other greenhouse gases.

Human Story: Consider Jack, an oil rig worker in the Gulf of Mexico. His work involves operating heavy machinery that consumes large amounts of diesel fuel, contributing to the carbon footprint of oil extraction. The industrial activity on the rigs, including flaring excess gas, adds to the emissions.

Expert Insight: Dr. Laura Brown, an environmental scientist, notes, "The initial stages of oil production are energy-intensive and contribute significantly to carbon emissions. The machinery, transportation of equipment, and flaring practices all add up to a substantial carbon footprint."

Refining and Processing

Once extracted, crude oil is transported to refineries, where it undergoes processing to produce various petroleum products. Refineries are major sources of CO2 emissions due to the energy-intensive nature of the refining processes, which include distillation, cracking, and reforming.

Human Story: Imagine Maria, an environmental engineer working at a refinery in Texas. She monitors emissions and implements measures to reduce the facility's carbon output, reflecting the ongoing efforts within the industry to mitigate environmental impacts.

Distribution and Transportation

The transportation of refined oil products to markets involves a global network of pipelines, tankers, trucks, and trains. Each mode of transport has its own carbon footprint, primarily from the fuel used in transportation.

Human Story: Picture Ahmed, a tanker captain navigating

oil shipments from the Middle East to Asia. The fuel consumed by his ship during long transoceanic voyages significantly contributes to the overall carbon emissions associated with oil.

Consumption and Combustion

The final stage, where oil is burned for energy in vehicles, power plants, and industrial processes, is the most significant contributor to the carbon footprint. Combustion releases CO_2 directly into the atmosphere, contributing to the greenhouse effect and global warming.

Human Story: Consider Emily, a commuter who drives to work daily. Her car's emissions, multiplied by millions of drivers worldwide, represent the largest single source of oil-related carbon emissions.

The Scale of Emissions

The International Energy Agency (IEA) estimates that oil combustion alone accounts for about 33% of global CO_2 emissions from energy sources. This significant contribution underscores the need to address the carbon footprint of oil comprehensively.

Comparing Emissions

Different types of crude oil have varying carbon intensities. For instance, oil sands (tar sands) production is particularly carbon-intensive due to the additional energy required to extract and process the heavy crude. In contrast, conventional light crude oil has a relatively lower carbon footprint.

Expert Insight: Dr. Robert Lee, an energy analyst, explains, "Understanding the carbon intensity of different oil sources is crucial for developing targeted strategies to reduce emissions. Efforts should focus on the highest emitters to achieve meaningful reductions."

Environmental and Social Impacts

The carbon footprint of oil extends beyond climate change, impacting ecosystems and communities worldwide.

Environmental Degradation

Oil extraction and refining can lead to significant environmental degradation, including deforestation, habitat destruction, and water contamination. Oil spills and leaks further exacerbate these issues, causing long-term harm to marine and terrestrial ecosystems.

Human Story: Consider Kofi, a fisherman in Nigeria's Niger Delta. Oil spills have contaminated the waters he relies on for fishing, devastating local fish populations and his livelihood. The environmental consequences of oil extraction are felt deeply by communities like his.

Health Impacts

The pollutants released during oil production and combustion, such as particulate matter and volatile organic compounds, pose serious health risks. These pollutants can lead to respiratory problems, cardiovascular diseases, and other health issues, particularly in communities near oil refineries and major transportation hubs.

Human Story: Meet Sarah, a resident of a refinery town in Louisiana. She has experienced increased asthma rates and other health problems within her community, highlighting the human cost of living near oil production facilities.

Mitigation Strategies

Addressing the carbon footprint of oil requires

a multifaceted approach, involving technological innovation, regulatory measures, and a transition to cleaner energy sources.

Technological Innovations

Advancements in technology can help reduce the carbon footprint of oil production and consumption. Carbon capture and storage (CCS) technologies, for instance, can capture CO_2 emissions from power plants and refineries and store them underground.

Expert Insight: Dr. Emily Chen, a renewable energy expert, emphasizes, "Investing in CCS and other clean technologies is critical for reducing the carbon footprint of existing oil infrastructure. These technologies provide a bridge towards a more sustainable energy future."

Regulatory Measures

Governments play a crucial role in regulating emissions and promoting cleaner practices. Policies such as carbon pricing, emissions standards, and subsidies for renewable energy can incentivize the reduction of the carbon footprint in the oil industry.

Human Story: Imagine Raj, a policy advisor in India, working on developing stricter emissions regulations for industrial plants. His efforts are part of a broader push to reduce the nation's carbon emissions and combat climate change.

Transition to Renewable Energy

The most effective long-term strategy to reduce the carbon footprint of oil is transitioning to renewable energy sources. Solar, wind, and other renewables produce little to no emissions, offering a sustainable alternative to fossil fuels.

Human Story: Picture Ana, a solar energy entrepreneur in Brazil. Her company installs solar panels in rural areas, providing clean energy and reducing reliance on oil. Her work illustrates the potential for renewable energy to transform communities and reduce carbon footprints.

Thought-Provoking Questions: Shaping a Sustainable Future

Reflecting on the carbon footprint of oil raises several critical questions: How can the oil industry balance economic growth with environmental responsibility? What role should governments and corporations play in reducing emissions? And how can societies support a just transition to renewable energy for workers and communities dependent on oil?

Storytelling Techniques: Bringing the Carbon Footprint to Life

To illustrate the carbon footprint of oil, envision the journey of oil from an offshore rig to a bustling city, capturing each stage's emissions. Contrast this with scenes of renewable energy projects, highlighting the shift towards a cleaner future. These vivid images help readers understand the environmental impact of oil and the potential for change.

Actionable Insights: Strategies for Reducing the Carbon Footprint

1. Implementing Clean Technologies: Investing in carbon capture and storage, along with other clean technologies, can significantly reduce emissions from oil production and consumption.
2. Strengthening Regulations: Governments should enforce stringent emissions standards and incentivize cleaner practices through carbon pricing and subsidies for

renewable energy.

3. Promoting Renewable Energy: Accelerating the adoption of renewable energy sources can reduce reliance on oil and lower overall carbon emissions.

4. Supporting Affected Communities: Providing support and resources for workers and communities affected by the transition from oil to renewable energy is crucial for a just and equitable shift.

Conclusion: The Path to a Sustainable Future

The carbon footprint of oil presents a significant challenge to global climate goals. This crude awakening underscores the urgent need for comprehensive strategies to reduce emissions and transition to a sustainable energy future.

By understanding the carbon footprint of oil production and consumption, we can better appreciate the scope of the challenge and the opportunities for meaningful action. The lessons learned from current and past efforts provide valuable insights for shaping a future that prioritizes environmental sustainability, economic resilience, and social equity. Through collective action and innovation, we can mitigate the impacts of oil on our planet and move towards a cleaner, healthier future for all.

CHAPTER 34: OIL AND AIR POLLUTION

Crude Awakening: The Modern Truths and Future of Oil

The Air We Breathe

Oil has been a cornerstone of industrial progress and economic development. However, its use comes with significant consequences, particularly regarding air quality and public health. This chapter explores the impact of oil on air pollution, examining how oil extraction, refining, and combustion contribute to poor air quality and the associated health risks.

The Journey from Extraction to Emission

The process of turning crude oil into usable energy involves several stages, each contributing to air pollution. Understanding this journey helps to appreciate the full scope of oil's impact on air quality.

Extraction and Flaring

The extraction of oil often involves flaring, where excess natural gas is burned off. This process releases large amounts of carbon dioxide (CO_2), methane (CH_4), and other pollutants into the atmosphere.

Human Story: Consider Alex, an oil rig worker in North

Dakota. He witnesses the flaring of gas at his site, knowing that while it's a necessary safety measure, it significantly contributes to local air pollution. The bright flares that light up the night sky are a visible reminder of the pollutants being released.

Expert Insight: Dr. Emily White, an environmental scientist, notes, "Flaring is a significant source of greenhouse gases and pollutants like volatile organic compounds (VOCs) and nitrogen oxides (NOx), which contribute to smog and respiratory problems."

Refining and Processing

Refining crude oil into gasoline, diesel, and other products is an energy-intensive process that emits various pollutants, including sulphur dioxide (SO2), particulate matter (PM), and VOCs. Refineries are often located near populated areas, exacerbating the impact on local air quality.

Human Story: Imagine Maria, who lives near a refinery in Texas. She frequently experiences poor air quality, leading to health issues such as asthma and chronic bronchitis. The refinery's emissions are a constant concern for her and her community.

Transportation and Combustion

The combustion of oil products in vehicles and industrial processes is the largest source of air pollution from oil. Tailpipe emissions release CO2, NOx, PM, and VOCs, all of which contribute to urban smog, acid rain, and health problems.

Urban Air Quality

Cities with high traffic volumes suffer from severe air pollution. Vehicles emit NOx and VOCs, which react in the presence of sunlight to form ground-level ozone, a key component of smog. Particulate matter from exhaust can penetrate deep into the lungs, causing various health issues.

Human Story: Picture John, a taxi driver in Los Angeles. He spends hours each day in traffic, breathing in polluted air. His constant exposure to vehicle emissions has led to respiratory issues, affecting his quality of life.

Expert Insight: Dr. Laura Green, a public health expert, explains, "The emissions from cars and trucks are major contributors to urban air pollution. The health impacts are significant, particularly for vulnerable populations such as children, the elderly, and those with pre-existing conditions."

The Health Impacts of Air Pollution

The pollutants released from oil-related activities have severe health implications. Long-term exposure to air pollution can lead to respiratory diseases, cardiovascular problems, and even premature death.

Respiratory Diseases

Pollutants like PM, NOx, and ground-level ozone irritate the respiratory system, leading to conditions such as asthma, bronchitis, and chronic obstructive pulmonary disease (COPD).

Human Story: Consider Sarah, a child with asthma living near a busy highway. Her condition worsens during high-traffic days, requiring frequent hospital visits. Her story highlights the direct link between air pollution and respiratory health.

Cardiovascular Problems

Long-term exposure to polluted air increases the risk of

heart attacks, strokes, and other cardiovascular diseases. Fine particulate matter (PM2.5) can enter the bloodstream, causing inflammation and other harmful effects.

Expert Insight: Dr. Robert Lee, a cardiologist, emphasizes, "The link between air pollution and cardiovascular health is well-documented. Reducing exposure to pollutants is crucial for preventing heart-related diseases."

Environmental and Social Justice

The burden of air pollution is often disproportionately borne by low-income and marginalized communities. These populations are more likely to live near refineries, highways, and industrial areas, where air quality is poorest.

Environmental Justice

Communities of colour and low-income neighbourhoods frequently face higher levels of air pollution. The placement of industrial facilities and major roadways near these communities reflects broader issues of environmental injustice.

Human Story: Envision Kim, a resident of a low-income neighbourhood in Houston. Her community struggles with poor air quality due to nearby refineries and highways. The lack of resources and political power makes it difficult for them to advocate for cleaner air.

Expert Insight: Dr. Lisa Martinez, an environmental justice advocate, states, "Addressing air pollution requires a focus on environmental justice. We need policies that protect the most vulnerable populations and ensure equitable access to clean air."

Mitigation Strategies and Future Directions

Reducing the impact of oil on air quality requires comprehensive strategies that address emissions at every stage of oil's lifecycle.

Clean Technologies and Practices

Adopting cleaner technologies and practices in oil extraction and refining can significantly reduce emissions. Techniques such as gas capture instead of flaring, and using cleaner fuels in refining processes, are critical steps.

Human Story: Consider Raj, an engineer working on gas capture technology in India. His innovations help reduce the amount of gas flared during oil extraction, showcasing how technology can mitigate environmental impacts.

Transition to Renewable Energy

The most effective way to reduce air pollution from oil is to transition to renewable energy sources. Solar, wind, and electric vehicles produce zero emissions, offering a sustainable alternative to oil.

Expert Insight: Dr. Emily Chen, a renewable energy expert, explains, "The transition to renewable energy is essential for improving air quality and public health. Investing in clean energy technologies can significantly reduce our dependence on oil and its associated pollutants."

Policy and Regulatory Measures

Strong policies and regulations are essential for controlling air pollution. Emissions standards for vehicles, industrial processes, and refineries can drive improvements in air quality.

Global and National Initiatives

International agreements like the Paris Agreement, along with national policies such as the Clean Air Act in the United States, provide frameworks for reducing emissions and improving air quality.

Human Story: Imagine Li, a policymaker in Beijing, working to implement stricter emissions standards for vehicles. His efforts contribute to significant improvements in the city's air quality, benefiting millions of residents.

Thought-Provoking Questions: A Cleaner Future

Reflecting on the impact of oil on air pollution raises several critical questions: How can we balance the economic benefits of oil with the need for clean air? What role should technology and innovation play in reducing emissions? And how can we ensure that the benefits of clean air are shared equitably across all communities?

Storytelling Techniques: Bringing Air Pollution to Life

To illustrate the impact of oil on air quality, envision the smog-covered skylines of major cities, contrasted with scenes of clean energy projects and green urban spaces. These images help readers visualize the consequences of oil-related pollution and the potential for a cleaner future.

Actionable Insights: Strategies for Clean Air

1. Implementing Clean Technologies: Adopting cleaner technologies in oil extraction and refining can significantly reduce emissions.
2. Promoting Renewable Energy: Accelerating the transition to renewable energy sources can reduce reliance on oil and improve air quality.
3. Strengthening Regulations: Governments should enforce strict emissions standards and support policies that promote cleaner air.
4. Advocating for Environmental Justice: Ensuring that all communities, especially vulnerable populations, have access to clean air is essential for social equity.

Conclusion: Breathing Easier

Oil's impact on air quality presents a significant public health challenge. This crude awakening highlights the urgent need for comprehensive strategies to reduce emissions and protect air quality.

By understanding the journey of oil from extraction to emission and its impact on air quality, we can better appreciate the importance of transitioning to cleaner energy sources. The lessons learned from past and present efforts provide valuable insights for shaping a future that prioritizes environmental sustainability, public health, and social equity. Through collective action and innovation, we can achieve cleaner air and a healthier planet for all.

CHAPTER 35: WATER CONTAMINATION

Crude Awakening: The Modern Truths and Future of Oil

The Hidden Cost of Oil

Oil has been a driving force behind industrial progress, but its extraction and refining come with hidden costs, one of the most severe being water contamination. This chapter delves into how oil extraction and refining can contaminate water sources, examining the mechanisms of contamination, its impacts on ecosystems and human health, and the measures needed to mitigate these effects.

The Pathways to Contamination

Water contamination from oil extraction and refining can occur through various pathways, including spills, leaks, and the improper disposal of waste. Understanding these mechanisms is crucial for addressing and preventing water pollution.

Drilling and Hydraulic Fracturing

Oil extraction, particularly hydraulic fracturing (fracking), can lead to water contamination. Fracking involves injecting high-pressure fluid into shale formations to release oil and gas. This process uses vast amounts of water mixed with chemicals, which can seep into groundwater if not managed properly.

Human Story: Consider John, a farmer in Pennsylvania, who noticed his well water becoming discoloured and foul-smelling after nearby fracking operations began. His experience highlights the risk of groundwater contamination from drilling activities.

Expert Insight: Dr. Emily White, a hydrogeologist, explains, "The chemicals used in fracking fluids can migrate into aquifers through fractures in the rock or faulty well casings, posing a significant risk to drinking water sources."

Oil Spills and Leaks

Oil spills, both on land and at sea, are among the most visible causes of water contamination. Spills from pipelines, tankers, and drilling rigs release crude oil directly into water bodies, devastating marine and freshwater ecosystems.

Human Story: Imagine Maria, a fisherman in Alaska affected by the Exxon Valdez oil spill. The spill not only destroyed her fishing grounds but also left long-term ecological damage that has taken decades to mitigate.

Refining and Waste Disposal

Oil refineries produce wastewater containing toxic substances, including heavy metals, hydrocarbons, and other chemicals. Improper disposal or accidental releases of this wastewater can contaminate nearby rivers, lakes, and coastal waters.

Human Story: Picture Ahmed, a resident near an oil refinery in Nigeria. He has witnessed frequent discharge of untreated wastewater into local rivers, affecting the water quality and health of his community.

The Environmental Impact

Water contamination from oil extraction and refining has severe environmental consequences, affecting both aquatic ecosystems and the broader environment.

Damage to Aquatic Ecosystems

Oil contamination disrupts aquatic ecosystems by harming fish, birds, and other wildlife. Oil coats the surface of water bodies, reducing oxygen levels and killing aquatic organisms. The toxic components of oil, such as benzene and toluene, are harmful to both flora and fauna.

Human Story: Consider Mei, an environmental scientist studying the aftermath of an oil spill in the Gulf of Mexico. Her research reveals the extensive damage to coral reefs and marine life, underscoring the long-term ecological impacts of such disasters.

Expert Insight: Dr. Robert Lee, a marine biologist, notes, "Oil spills have a catastrophic impact on marine ecosystems. The immediate effects are visible, but the long-term consequences, such as genetic mutations and reproductive issues in wildlife, are equally concerning."

Soil and Groundwater Contamination

Oil leaks from pipelines and storage tanks can seep into the soil, contaminating groundwater sources. This contamination can persist for years, affecting both human and animal health.

Human Story: Picture Raj, a community leader in India, advocating for clean water after an oil pipeline leak contaminated local wells. His efforts highlight the struggle for safe drinking water in areas affected by industrial pollution.

The Human Health Implications

Contaminated water poses significant health risks to communities, including respiratory problems, skin irritation, and long-term illnesses such as cancer.

Exposure to Toxic Chemicals

Chemicals found in oil and its by-products, such as benzene, arsenic, and lead, are known carcinogens and can cause various health problems. Ingesting or coming into contact with contaminated water can lead to acute and chronic health issues.

Human Story: Imagine Sarah, a mother in Louisiana, whose children developed skin rashes and respiratory issues after swimming in a polluted river near a refinery. Her story underscores the human cost of industrial pollution.

Long-Term Health Effects

Long-term exposure to contaminated water can lead to serious health conditions, including liver and kidney damage, neurological disorders, and developmental issues in children.

Expert Insight: Dr. Laura Green, a public health expert, explains, "Chronic exposure to contaminated water has severe health implications. Communities near oil extraction and refining sites are particularly vulnerable, and their health needs must be addressed through comprehensive monitoring and healthcare support."

Mitigation and Prevention Strategies

Preventing and mitigating water contamination from oil extraction and refining requires a combination of regulatory measures, technological innovations, and community engagement.

Regulatory Frameworks

Strong regulatory frameworks are essential for preventing water contamination. Governments must enforce strict standards for wastewater disposal, pipeline maintenance, and spill response.

Human Story: Consider Li, a government official in China, working to implement stricter environmental regulations for the oil industry. His efforts reflect the importance of policy in protecting water resources.

Technological Innovations

Technological advancements can help reduce the risk of water contamination. Innovations such as advanced drilling techniques, better leak detection systems, and improved wastewater treatment processes are crucial.

Expert Insight: Dr. Emily Chen, an environmental engineer, emphasizes, "Investing in technology is key to mitigating the environmental impact of oil extraction and refining. From real-time monitoring systems to bioremediation techniques, technology can significantly reduce the risk of water contamination."

Community Engagement

Engaging local communities in monitoring and protecting water resources is vital. Community-based initiatives can enhance transparency, ensure compliance with regulations, and provide early warnings of potential contamination.

Human Story: Envision Maria, an activist in Ecuador, leading community efforts to monitor water quality near oil extraction sites. Her work empowers residents to take action and hold companies accountable for environmental violations.

Thought-Provoking Questions: A Cleaner Future

Reflecting on the issue of water contamination raises several critical questions: How can the oil industry balance economic growth with environmental responsibility? What role should governments and corporations play in ensuring clean water for all? And how can communities most affected by water contamination be supported and empowered?

Storytelling Techniques: Bringing Water Contamination to Life

To illustrate the impact of water contamination, envision a pristine river turning black after an oil spill, contrasted with images of clean-up efforts and community activism. These scenes help readers understand the profound and often devastating effects of oil-related water pollution.

Actionable Insights: Strategies for Protecting Water Resources

1. Strengthening Regulations: Governments should enforce stringent standards for oil extraction, refining, and waste disposal to prevent water contamination.
2. Investing in Technology: Advancing technologies for leak detection, spill response, and wastewater treatment can mitigate environmental risks.
3. Promoting Community Monitoring: Supporting community-based monitoring initiatives can enhance transparency and accountability in the oil industry.
4. Encouraging Corporate Responsibility: Companies must adopt best practices for environmental protection and invest in sustainable technologies and processes.

Conclusion: The Cost of Clean Water

The contamination of water sources by oil extraction and

refining presents a significant environmental and public health challenge. This crude awakening underscores the urgent need for comprehensive strategies to protect our water resources.

By understanding the pathways and impacts of water contamination, we can better appreciate the importance of regulatory frameworks, technological innovation, and community engagement. The lessons learned from past and present efforts provide valuable insights for shaping a future that prioritizes clean water, environmental sustainability, and public health. Through collective action and responsibility, we can ensure that our water resources remain safe and accessible for all.

CHAPTER 36: OIL AND BIODIVERSITY

Crude Awakening: The Modern Truths and Future of Oil

The Hidden Cost of Black Gold

Oil has driven economic development and powered industries, but its extraction and use come with significant environmental costs, particularly to biodiversity. This chapter explores how oil exploration and spills impact biodiversity, examining the mechanisms of damage, the ecosystems affected, and the long-term implications for our planet. Through human stories, expert insights, and contextual information, we uncover the true cost of oil on the natural world.

The Fragile Balance of Nature

Biodiversity—the variety of life in all its forms—is essential for ecosystem health and resilience. However, oil exploration and spills disrupt this balance, causing harm to countless species and habitats.

Mechanisms of Impact

Oil exploration and spills affect biodiversity through habitat destruction, pollution, and direct harm to wildlife. The intrusion into pristine environments for drilling and the catastrophic effects of oil spills create severe and often long-lasting damage.

Human Story: Imagine Juan, a biologist studying coral reefs in the Gulf of Mexico. His research shows how even small oil spills can devastate delicate coral ecosystems, affecting the myriad species that depend on them.

Habitat Destruction from Exploration

Oil exploration often requires clearing vast areas of land, building roads, and drilling wells. These activities destroy habitats, fragment ecosystems, and disrupt the lives of countless species.

Deforestation and Land Degradation

In places like the Amazon rainforest, oil exploration leads to deforestation, which destroys habitats for numerous species and contributes to global biodiversity loss.

Human Story: Consider Maria, an indigenous woman in the Amazon whose community has seen its ancestral lands cleared for oil extraction. The loss of forests means not just the destruction of habitats but also the erosion of cultural and subsistence resources for her people.

Expert Insight: Dr. Laura Gomez, an ecologist, emphasizes, "The impact of oil exploration in biodiverse regions like the Amazon is profound. The loss of tree cover, the fragmentation of habitats, and the pollution from drilling activities have far-reaching consequences for both wildlife and indigenous communities."

The Devastation of Oil Spills

Oil spills are catastrophic events with immediate and long-lasting effects on biodiversity. They contaminate water, soil, and air, leading to the death of marine and terrestrial life and disrupting ecosystems.

Marine Ecosystems

Marine ecosystems are particularly vulnerable to oil spills. When oil spills into the ocean, it spreads quickly, coating marine life and shorelines. The toxic components of oil can cause long-term damage to fish, birds, and marine mammals.

Human Story: Picture Lisa, a marine conservationist working in Alaska. She recounts the aftermath of the Exxon Valdez oil spill, where thousands of seabirds, otters, and fish perished. The spill's impacts are still felt decades later, with some species struggling to recover.

Expert Insight: Dr. Robert Miller, a marine biologist, notes, "The effects of oil spills on marine biodiversity are immediate and severe. The toxic compounds in oil can cause acute mortality in marine organisms and have chronic impacts on reproductive and growth rates."

Terrestrial Ecosystems

Oil spills on land can contaminate soil and water sources, affecting terrestrial wildlife and plant life. The toxic effects of oil can persist in the environment for years, making recovery a slow and uncertain process.

Human Story: Imagine Ahmed, a farmer in Nigeria whose land was contaminated by an oil spill. His crops failed, and local wildlife disappeared, illustrating the far-reaching impacts of oil spills on terrestrial ecosystems and human livelihoods.

Long-Term Ecological Impacts

The long-term impacts of oil spills and exploration on biodiversity are profound. Ecosystems take years, often decades, to recover, and some species may never return to their former abundance.

Genetic Diversity

Oil pollution can reduce genetic diversity in populations of affected species. This loss of genetic diversity makes species more vulnerable to diseases and less adaptable to environmental changes.

Expert Insight: Dr. Emily White, a geneticist, explains, "Genetic diversity is the foundation of species resilience. Oil pollution can reduce this diversity, leading to populations that are less capable of adapting to changing conditions and more prone to extinction."

Disruption of Food Chains

Oil contamination can disrupt food chains by killing off key species, leading to cascading effects throughout the ecosystem. For example, the loss of small fish and invertebrates can impact larger predators and ultimately alter the structure of the entire ecosystem.

Human Story: Consider Raj, a fisherman in the Gulf of Mexico, who has seen the fish populations decline after a major oil spill. The disruption of the food chain has affected his livelihood and the local economy, highlighting the interconnectedness of human and ecological health.

Mitigation and Restoration Efforts

Addressing the impact of oil on biodiversity requires robust mitigation and restoration efforts. These include stricter regulations, rapid response to spills, and long-term environmental monitoring and restoration projects.

Regulatory Measures

Governments must enforce strict regulations on oil exploration and extraction to minimize environmental impacts. This includes stringent environmental impact assessments and robust safety protocols to prevent spills.

Human Story: Imagine Li, a policy maker in China, working on new regulations to protect coastal ecosystems from oil exploration. His efforts aim to balance economic development with environmental conservation.

Rapid Response and Clean-up

Effective oil spill response strategies are crucial for minimizing damage. This includes having well-equipped and trained response teams, using advanced technology for spill containment and clean up, and involving local communities in the response efforts.

Human Story: Picture Sarah, a member of an oil spill response team. Her experiences on the front lines of spill clean-ups underscore the importance of preparedness and swift action in mitigating environmental damage.

Long-Term Restoration

Restoration efforts are essential for helping ecosystems recover from oil spills. This includes replanting vegetation, rehabilitating wildlife, and ongoing monitoring to assess the recovery process and address any persistent impacts.

Expert Insight: Dr. Laura Green, an environmental restoration specialist, emphasizes, "Restoration is a long-term commitment. It requires sustained efforts and resources to help ecosystems recover their function and biodiversity after an oil spill."

Thought-Provoking Questions: Balancing Progress and Preservation

Reflecting on the impact of oil on biodiversity raises several critical questions: How can we balance the need for energy with the imperative to protect our planet's biodiversity? What policies and practices can ensure that oil extraction is conducted sustainably? And how can communities most affected by oil-related environmental damage be supported in their recovery efforts?

Storytelling Techniques: Bringing Biodiversity to Life

To illustrate the impact of oil on biodiversity, envision pristine rainforests and vibrant coral reefs contrasted with scenes of devastation following an oil spill. These images help readers connect emotionally with the importance of preserving our planet's rich biodiversity.

Actionable Insights: Strategies for Protecting Biodiversity

1. Strengthening Regulations: Governments should enforce stringent environmental regulations for oil exploration and extraction to protect sensitive ecosystems.
2. Investing in Clean Technologies: Advancing technologies for safer drilling and spill response can reduce the environmental impact of oil activities.
3. Promoting Conservation Efforts: Supporting conservation initiatives and protected areas can help safeguard biodiversity from the threats posed by oil exploration.
4. Engaging Communities: Involving local communities in environmental monitoring and restoration efforts can enhance the effectiveness of biodiversity protection.

Conclusion: The Price of Progress

Oil exploration and spills have a profound impact on biodiversity, highlighting the hidden cost of our reliance on fossil fuels. This crude awakening underscores the urgent need for comprehensive strategies to protect our planet's ecosystems.

By understanding the mechanisms and impacts of oil-related environmental damage, we can better appreciate the importance of sustainable practices and robust regulatory frameworks. The lessons learned from past and present efforts provide valuable insights for shaping a future that prioritizes environmental sustainability, biodiversity conservation, and human well-being. Through collective action and responsible stewardship, we can ensure that our planet's rich biodiversity is preserved for future generations.

CHAPTER 37: MITIGATING ENVIRONMENTAL DAMAGE

Crude Awakening: The Modern Truths and Future of Oil

The Challenge of Balancing Progress with Preservation

Oil has powered modern civilization, driving economic growth and technological advancement. However, the environmental cost of oil extraction, refining, and consumption has been significant. As awareness of these impacts grows, the focus shifts to mitigating environmental damage. This chapter explores strategies and technologies designed to reduce the ecological footprint of the oil industry, offering insights into current practices and future directions.

Understanding the Scope of Environmental Damage

The environmental impact of oil spans air, water, and soil contamination, habitat destruction, and contributions to climate change. Effective mitigation requires addressing these diverse issues through a combination of regulatory measures, technological innovations, and industry best practices.

Air Pollution and Greenhouse Gas Emissions

Oil combustion releases pollutants such as carbon dioxide (CO_2), sulphur dioxide (SO_2), nitrogen oxides (NO_x), and particulate matter, contributing to air pollution and climate change. The industry's challenge is to reduce these emissions while maintaining energy production.

Water Contamination

Oil spills, drilling waste, and refinery effluents can contaminate water sources, affecting ecosystems and human health. Strategies to mitigate these impacts focus on prevention, rapid response, and effective clean up.

Soil and Habitat Destruction

Oil extraction and infrastructure development can lead to soil contamination and habitat destruction. Restoring affected areas and preventing future damage are critical for maintaining biodiversity and ecosystem health.

Regulatory Measures and Industry Standards

Regulations play a crucial role in mitigating environmental damage. Governments and international bodies enforce standards that the oil industry must follow to minimize its ecological footprint.

International Agreements and National Policies

International agreements, such as the Paris Agreement, set targets for reducing greenhouse gas emissions. National policies, such as the Clean Air Act in the United States, enforce regulations on emissions and pollution control.

Human Story: Consider Raj, an environmental policy maker in India. He works tirelessly to implement and enforce regulations that reduce industrial emissions and protect natural resources, highlighting the importance of regulatory frameworks in environmental protection.

Expert Insight: Dr. Laura Green, an environmental policy expert, states, "Strong regulatory frameworks are essential for holding the oil industry accountable and driving progress toward more sustainable practices. Without these regulations, voluntary measures are often insufficient."

Technological Innovations for Environmental Protection

Advancements in technology offer promising solutions for mitigating the environmental impact of oil. From cleaner extraction methods to innovative spill response techniques, technology is a key driver of sustainability in the oil industry.

Cleaner Extraction and Production Techniques

Technologies such as hydraulic fracturing (fracking) and enhanced oil recovery (EOR) have revolutionized oil extraction but come with environmental risks. Innovations aim to reduce these risks by improving efficiency and reducing waste.

- Hydraulic Fracturing Improvements: Advances in fracking technology focus on reducing water usage, minimizing chemical use, and improving well integrity to prevent leaks.
- Enhanced Oil Recovery: Techniques like CO_2 injection not only improve oil recovery rates but also offer the potential for carbon sequestration, reducing overall greenhouse gas emissions.

Human Story: Picture John, an engineer working on a fracking site in Texas. He implements new technologies that reduce water consumption and enhance well safety, demonstrating how innovation can mitigate environmental impacts.

Spill Prevention and Response

Preventing oil spills and ensuring rapid response when they occur is crucial for protecting marine and terrestrial ecosystems. Technologies in this area include advanced monitoring systems, containment methods, and bioremediation.

- Advanced Monitoring Systems: Real-time monitoring of pipelines and drilling operations using sensors and AI can detect leaks early and prevent spills.
- Containment and Clean-up Technologies: Innovations such as oil-absorbent materials, skimmers, and dispersants enhance spill response efforts. Bioremediation uses microbes to break down oil, accelerating natural recovery processes.

Human Story: Imagine Sarah, a member of an oil spill response team. Her work involves deploying new containment booms and bioremediation techniques, showcasing the critical role of technology in mitigating spill impacts.

Renewable Energy Integration

Transitioning to renewable energy sources is a long-term strategy to reduce the oil industry's environmental impact. Many oil companies are investing in renewable energy projects, such as wind, solar, and biofuels, to diversify their energy portfolios and reduce reliance on fossil fuels.

Human Story: Consider Li, a project manager at an oil company in China. He oversees the development of a large solar farm, illustrating the industry's shift towards integrating renewable energy solutions.

Expert Insight: Dr. Emily Chen, a renewable energy expert, explains, "Integrating renewable energy into the oil industry's operations not only reduces greenhouse gas emissions but also prepares these companies for a sustainable energy future."

Best Practices for Environmental Stewardship

Beyond technology, adopting best practices in operations and management is essential for minimizing environmental damage. These practices involve comprehensive environmental impact assessments, sustainable resource management, and corporate social responsibility.

Environmental Impact Assessments

Before beginning new projects, thorough environmental impact assessments (EIAs) are conducted to identify potential ecological risks and develop mitigation strategies. These assessments are critical for ensuring that projects proceed responsibly.

Human Story: Envision Maria, an environmental consultant conducting an EIA for a new drilling project in the Arctic. Her work helps ensure that potential impacts on local wildlife and ecosystems are minimized.

Sustainable Resource Management

Efficient use of resources, such as water and energy, and minimizing waste are key components of sustainable resource management. Practices include recycling water in fracking operations, reducing flaring, and optimizing energy use in refineries.

Human Story: Picture Ahmed, a refinery manager in Saudi Arabia. He implements energy-saving measures and waste reduction strategies at his facility, demonstrating how sustainable management can reduce the industry's environmental footprint.

Corporate Social Responsibility

Oil companies are increasingly adopting corporate social responsibility (CSR) initiatives that focus on environmental protection, community engagement, and sustainability. These initiatives help build trust with stakeholders and ensure that companies contribute positively to the communities they operate in.

Human Story: Imagine Elena, a CSR manager at a major oil company in Brazil. She oversees programs that support local education, healthcare, and environmental conservation, highlighting the broader role of the industry in sustainable development.

Thought-Provoking Questions: Future Directions

Reflecting on the strategies to mitigate environmental damage raises several critical questions: How can the oil industry balance economic growth with environmental stewardship? What role should governments and corporations play in driving sustainable practices? And how can communities most affected by oil-related environmental damage be supported in their recovery efforts?

Storytelling Techniques: Bringing Mitigation Efforts to Life

To illustrate the efforts to mitigate environmental damage, envision the implementation of advanced technologies on an oil rig, the rapid deployment of spill response teams, and the construction of renewable energy projects by traditional oil companies. These images help readers understand the tangible steps being taken to protect the environment.

Actionable Insights: Strategies for a Sustainable Future

1. Strengthening Regulations: Governments should

enforce stringent environmental standards and continuously update regulations to reflect new technologies and best practices.

2. Investing in Innovation: The oil industry must invest in research and development of cleaner technologies and renewable energy sources.

3. Promoting Best Practices: Adopting industry best practices for environmental impact assessments, resource management, and CSR can reduce the ecological footprint of oil operations.

4. Engaging Communities: Involving local communities in environmental monitoring and decision-making processes ensures that their needs and concerns are addressed.

Conclusion: Pathways to Sustainability

Mitigating the environmental damage caused by oil is a complex but essential task. This crude awakening highlights the need for comprehensive strategies that integrate regulatory measures, technological innovation, and best practices to protect our planet.

By understanding the scope of environmental impacts and the available mitigation strategies, we can better appreciate the efforts required to balance industrial progress with ecological preservation. The lessons learned from current and past efforts provide valuable insights for shaping a future that prioritizes sustainability, biodiversity conservation, and human well-being. Through collective action and responsible stewardship, we can ensure that the benefits of oil do not come at the expense of our environment and future generations.

CHAPTER 38: THE ROLE OF ENVIRONMENTAL REGULATIONS

Crude Awakening: The Modern Truths and Future of Oil

Balancing Progress with Protection

The oil industry has been a cornerstone of global economic development, but its environmental impacts are significant. To mitigate these impacts, environmental regulations have been implemented worldwide. This chapter examines the effectiveness of these regulations in the oil industry, exploring their role in protecting the environment, the challenges faced in enforcement, and the evolving landscape of regulatory frameworks.

The Need for Environmental Regulations

Environmental regulations are essential to ensure that the extraction, refining, and consumption of oil do not cause irreparable harm to ecosystems and human health. These regulations set standards and guidelines for emissions, waste disposal, water usage, and land reclamation.

Historical Context

The recognition of the need for environmental regulations

in the oil industry grew alongside increasing awareness of environmental issues. Key events, such as the 1969 Santa Barbara oil spill and the 1989 Exxon Valdez disaster, highlighted the devastating impacts of oil spills and led to stricter regulatory measures.

Human Story: Consider Jane, a resident of Santa Barbara during the 1969 oil spill. The sight of oil-coated beaches and dead wildlife spurred her community to advocate for stronger environmental protections, marking a turning point in public awareness and policy development.

The Framework of Environmental Regulations

Environmental regulations in the oil industry encompass various aspects, from air and water quality to waste management and land restoration. These regulations are enforced at local, national, and international levels, often through a combination of legislation, standards, and enforcement mechanisms.

Key Regulatory Instruments

- Clean Air Act (United States): This legislation sets limits on emissions of pollutants from industrial sources, including refineries, to protect air quality.
- Clean Water Act (United States): This act regulates the discharge of pollutants into water bodies, ensuring that oil operations do not contaminate water sources.
- MARPOL Convention (International): The International Convention for the Prevention of Pollution from Ships addresses pollution from oil tankers, setting standards for the prevention of oil spills at sea.

Expert Insight: Dr. Laura Green, an environmental policy expert, explains, "Effective environmental regulations require a robust framework that includes clear standards, rigorous monitoring, and strong enforcement mechanisms. These elements are essential to ensure compliance and protect the environment."

Effectiveness of Environmental Regulations

The effectiveness of environmental regulations in the oil industry varies based on several factors, including the strength of the regulatory framework, the commitment of governments and companies to enforcement, and the technological capabilities available.

Success Stories

- Reduction of Air Pollutants: Regulations such as the Clean Air Act have significantly reduced emissions of harmful pollutants from refineries and other industrial sources. Improved technologies and stricter emission standards have led to cleaner air in many regions.
- Improved Spill Response: International agreements like MARPOL and national regulations have enhanced oil spill prevention and response capabilities. The development of better containment and clean up technologies has reduced the impact of spills on marine environments.

Human Story: Picture John, an oil rig supervisor in the Gulf of Mexico. His training in spill response techniques and the availability of advanced equipment ensure that any spill is quickly contained, minimizing environmental damage.

Ongoing Challenges

Despite these successes, challenges remain in ensuring the effectiveness of environmental regulations. These include gaps in enforcement, varying standards across regions, and resistance from industry stakeholders.

- Enforcement Gaps: In many regions, enforcement of environmental regulations is inconsistent due to lack of resources, corruption, or political pressure. This can

lead to non-compliance and continued environmental degradation.
- Varying Standards: Different countries have different regulatory standards, leading to disparities in environmental protection. Companies operating in regions with weaker regulations may not be held to the same standards as those in more stringent environments.
- Industry Resistance: Some oil companies resist regulatory measures due to the perceived economic costs. Lobbying efforts and legal challenges can delay or weaken the implementation of effective regulations.

Human Story: Imagine Maria, an environmental activist in Nigeria. She faces significant challenges in holding oil companies accountable for spills and pollution due to weak regulatory enforcement and powerful industry lobbying.

The Evolving Landscape of Environmental Regulations

As the environmental impact of the oil industry becomes more apparent, regulatory frameworks are evolving to address new challenges and incorporate advances in technology and science.

Innovations in Regulation

- Carbon Pricing: Implementing carbon pricing mechanisms, such as carbon taxes or cap-and-trade systems, incentivizes companies to reduce greenhouse gas emissions. These mechanisms internalize the environmental costs of carbon emissions, encouraging more sustainable practices.
- Sustainable Development Goals (SDGs): The United Nations' SDGs include targets for clean energy and environmental protection. Integrating these goals into national policies helps align regulatory efforts with broader sustainability objectives.
- Technological Advances: Emerging technologies, such as

satellite monitoring and blockchain, enhance transparency and accountability in environmental compliance. These tools can track emissions, detect spills, and verify regulatory adherence in real-time.

Expert Insight: Dr. Robert Lee, a sustainability expert, notes, "The integration of new technologies into regulatory frameworks can revolutionize environmental monitoring and enforcement. These innovations provide more accurate data and enable quicker responses to environmental issues."

Balancing Economic Growth and Environmental Protection

One of the primary challenges in implementing environmental regulations is balancing the need for economic growth with the imperative to protect the environment. Effective regulations must ensure that the oil industry can operate sustainably without compromising economic development.

Collaborative Approaches

Collaboration between governments, industry, and civil society is essential for developing and implementing effective environmental regulations. Multi-stakeholder initiatives can help balance interests and foster a culture of compliance and environmental stewardship.

Human Story: Consider Raj, a government official in India, who works with oil companies and environmental NGOs to develop regulations that promote sustainable practices. His efforts illustrate the importance of collaboration in achieving regulatory success.

Economic Incentives

Incentives such as tax breaks for clean technology

investments, grants for environmental research, and penalties for non-compliance can encourage companies to adopt more sustainable practices. Aligning economic incentives with environmental goals ensures that businesses have a financial stake in protecting the environment.

Human Story: Picture Ahmed, a CEO of an oil company in Saudi Arabia. Incentivized by tax breaks and subsidies, his company invests in cleaner extraction technologies and renewable energy projects, demonstrating how economic incentives can drive sustainable practices.

Thought-Provoking Questions: The Future of Regulation

Reflecting on the role of environmental regulations in the oil industry raises several critical questions: How can regulatory frameworks adapt to emerging environmental challenges? What role should technology play in enhancing regulatory effectiveness? And how can we ensure that regulations are enforced equitably across different regions and industries?

Storytelling Techniques: Bringing Regulation to Life

To illustrate the impact of environmental regulations, envision the implementation of carbon pricing mechanisms, the development of collaborative regulatory frameworks, and the use of advanced technologies in monitoring compliance. These scenes help readers understand the tangible benefits and challenges of effective environmental regulation.

Actionable Insights: Strategies for Effective Regulation

1. Strengthening Enforcement: Governments must invest in resources and infrastructure to ensure consistent enforcement of environmental regulations.
2. Harmonizing Standards: International cooperation is

needed to harmonize regulatory standards and ensure a level playing field for all industry players.

3. Leveraging Technology: Utilizing advanced technologies for real-time monitoring and data collection can enhance transparency and compliance.

4. Encouraging Collaboration: Multi-stakeholder initiatives and public-private partnerships can foster a collaborative approach to developing and implementing regulations.

Conclusion: The Path Forward

Environmental regulations play a crucial role in mitigating the environmental impact of the oil industry. This crude awakening highlights the importance of robust regulatory frameworks, technological innovation, and collaborative efforts in protecting our environment.

By understanding the successes and challenges of current regulations, we can better appreciate the need for continuous improvement and adaptation. The lessons learned from past and present regulatory efforts provide valuable insights for shaping a future that prioritizes sustainability, environmental protection, and economic resilience. Through collective action and responsible stewardship, we can ensure that the oil industry operates in a manner that safeguards our planet for future generations.

CHAPTER 39: RENEWABLE ENERGY TRANSITION

Crude Awakening: The Modern Truths and Future of Oil

The Winds of Change

The world stands at a crossroads, faced with the urgent need to transition from fossil fuels to renewable energy sources. The reliance on oil, while historically vital for economic development, has led to environmental degradation and contributed significantly to climate change. This chapter explores the transition from oil to renewable energy, examining the drivers, challenges, and opportunities that define this critical shift.

The Imperative for Change

The environmental and economic impacts of oil dependency have become increasingly apparent. Climate change, driven by greenhouse gas emissions from fossil fuels, poses a global threat. Renewable energy offers a sustainable alternative that can mitigate these impacts while providing economic benefits.

Environmental Drivers

The primary driver for transitioning to renewable energy is the need to reduce carbon emissions and combat climate change. Renewable sources like solar, wind, and hydroelectric power produce little to no greenhouse gases, making them essential for achieving climate goals.

Human Story: Consider Emma, a young climate activist from Sweden inspired by Greta Thunberg. Emma advocates for renewable energy in her community, believing it is the key to a sustainable future. Her passion reflects a growing global movement demanding action on climate change.

Expert Insight: Dr. Laura Green, an environmental scientist, explains, "The transition to renewable energy is not just an environmental imperative but a necessity for ensuring the long-term viability of our planet. Every megawatt of renewable energy produced reduces our carbon footprint."

Economic Benefits

Renewable energy technologies have seen significant cost reductions over the past decade, making them increasingly competitive with fossil fuels. Investments in renewables can drive economic growth, create jobs, and reduce energy costs.

Human Story: Picture Raj, an entrepreneur in India who started a solar panel installation company. His business not only provides sustainable energy but also creates local jobs and reduces electricity costs for his customers, demonstrating the economic potential of renewable energy.

The Pathways to Transition

Transitioning from oil to renewable energy involves multiple pathways, including technological innovation, policy support, and societal change. Each pathway presents unique challenges and opportunities.

Technological Innovation

Advances in technology are critical for making renewable energy more efficient, reliable, and accessible. Innovations in solar photovoltaics, wind turbines, energy storage, and grid management are driving the renewable energy revolution.

- Solar Power: Solar technology has become more efficient and affordable, with new materials and designs enhancing energy capture and conversion.
- Wind Energy: Improvements in turbine design and offshore wind technologies have expanded the potential for wind power, even in less windy regions.
- Energy Storage: Developments in battery technology and other storage solutions are addressing the intermittency of renewable energy, ensuring a stable and reliable power supply.

Human Story: Imagine Li, a scientist in China working on next-generation solar panels. Her breakthroughs in efficiency and cost reduction are bringing clean energy to more people, highlighting the role of innovation in the renewable energy transition.

Policy Support

Government policies and incentives are crucial for promoting renewable energy adoption. Subsidies, tax credits, and regulatory frameworks can accelerate the deployment of clean energy technologies.

- Subsidies and Incentives: Financial incentives for renewable energy projects can reduce upfront costs and make clean energy more competitive with fossil fuels.
- Regulatory Frameworks: Policies that set renewable

energy targets, mandate emissions reductions, and support grid integration are essential for driving the transition.

Human Story: Consider Ahmed, a policy advisor in Morocco. He works on implementing policies that support the country's ambitious renewable energy goals, demonstrating the importance of government action in fostering a clean energy future.

Societal Change

The transition to renewable energy also requires a shift in societal attitudes and behaviours. Public awareness, education, and community engagement are vital for building support for clean energy initiatives.

- Education and Awareness: Increasing public understanding of the benefits of renewable energy can drive demand and support for policy measures.
- Community Engagement: Involving communities in renewable energy projects ensures that the transition is inclusive and benefits all stakeholders.

Human Story: Envision Maria, a community leader in Spain who spearheads a local solar cooperative. Her efforts to educate and engage her neighbours result in widespread adoption of solar panels, illustrating the power of grassroots movements.

Challenges and Barriers

While the transition to renewable energy is promising, it faces significant challenges that must be addressed to ensure a smooth and equitable shift.

Infrastructure and Investment

Building the infrastructure needed to support renewable energy requires substantial investment. Upgrading grids, developing storage solutions, and constructing renewable energy plants are capital-intensive endeavours.

Human Story: Picture John, a grid operator in the United States tasked with integrating a growing share of renewable energy into the power system. His work highlights the logistical and financial challenges of modernizing infrastructure to support clean energy.

Market Dynamics

The dominance of fossil fuels in the energy market presents a barrier to renewable energy adoption. Market inertia, subsidies for fossil fuels, and vested interests can slow the transition.

Expert Insight: Dr. Robert Lee, an energy economist, notes, "Overcoming market barriers requires coordinated efforts between governments, industry, and consumers. Removing fossil fuel subsidies and creating a level playing field for renewables are critical steps."

Social and Economic Impacts

The transition to renewable energy can have significant social and economic impacts, particularly for communities dependent on the oil industry. Ensuring a just transition that supports affected workers and regions is essential.

Human Story: Consider Sarah, a former oil worker in Alberta, Canada, who retrained for a job in the wind energy sector. Her story underscores the importance of providing support and opportunities for workers transitioning to new industries.

The Future of Renewable Energy

The future of renewable energy is bright, with continued advancements in technology, increasing public and political support, and a growing recognition of the need for sustainable energy solutions.

Global Trends

Countries around the world are setting ambitious renewable energy targets and investing in clean technologies. The global shift towards renewables is driven by the need to reduce emissions, enhance energy security, and foster economic development.

Human Story: Imagine Ali, a student in Saudi Arabia studying renewable energy engineering. He dreams of contributing to his country's transition to a diversified energy portfolio, reflecting the aspirations of a new generation committed to sustainability.

Innovation and Collaboration

Ongoing innovation and international collaboration are essential for overcoming the challenges of the renewable energy transition. Research and development, knowledge sharing, and cooperative projects can accelerate progress.

Expert Insight: Dr. Emily Chen, a renewable energy expert, explains, "The pace of innovation in renewable energy is remarkable. Collaborative efforts across borders and industries are crucial for harnessing the full potential of clean energy technologies."

Thought-Provoking Questions: Charting the Course

Reflecting on the renewable energy transition raises several critical questions: How can we ensure that the transition is inclusive and benefits all communities? What role should governments and the private sector play in driving the transition? And how can we overcome the technical and economic challenges to achieve a sustainable energy future?

Storytelling Techniques: Bringing the Transition to Life

To illustrate the renewable energy transition, envision the transformation of a coal plant into a solar farm, the bustling activity of wind turbine installations, and the vibrant communities powered by clean energy. These scenes help readers visualize the tangible benefits and possibilities of a renewable energy future.

Actionable Insights: Strategies for a Successful Transition

1. Supporting Innovation: Investing in research and development of clean energy technologies can drive advancements and reduce costs.
2. Implementing Strong Policies: Governments should adopt policies that promote renewable energy, set emissions targets, and phase out fossil fuel subsidies.
3. Building Infrastructure: Developing the necessary infrastructure for renewable energy, including grids and storage solutions, is essential for reliable and widespread adoption.
4. Ensuring a Just Transition: Providing support for workers and communities affected by the shift from oil to renewable energy ensures that the transition is equitable and inclusive.

Conclusion: Embracing the Future

The transition from oil to renewable energy is a complex but essential journey. This crude awakening highlights the need for concerted efforts across all sectors of society to achieve a sustainable energy future.

By understanding the drivers, pathways, and challenges of the renewable energy transition, we can better appreciate the opportunities it presents. The lessons learned from current and past efforts provide valuable insights for shaping a future that prioritizes environmental sustainability, economic resilience, and social equity.

Through collective action and innovation, we can embrace a future powered by clean, renewable energy and ensure a healthier planet for generations to come.

CHAPTER 40: CORPORATE RESPONSIBILITY

Crude Awakening: The Modern Truths and Future of Oil

The Weight of Responsibility

Oil companies have played a central role in powering the modern world, driving economic growth and technological advancement. However, this progress has come with significant environmental costs. As awareness of these impacts grows, so does the expectation for oil companies to take responsibility and actively address environmental concerns. This chapter explores the role of corporate responsibility in the oil industry, examining how companies can and should respond to the environmental challenges they face.

The Evolution of Corporate Responsibility

Corporate responsibility in the oil industry has evolved significantly over the decades. What began as a focus on compliance with basic regulations has expanded into a broader commitment to sustainability and environmental stewardship.

Historical Context

In the early days of the oil industry, environmental

considerations were often secondary to economic growth. However, high-profile environmental disasters, such as the 1969 Santa Barbara oil spill and the 1989 Exxon Valdez spill, highlighted the need for greater corporate accountability.

Human Story: Consider Jane, a resident of Santa Barbara during the 1969 oil spill. Witnessing the devastation of local beaches and wildlife spurred her to join environmental advocacy groups, pushing for stronger corporate accountability and environmental protections.

Key Areas of Corporate Responsibility

Oil companies' environmental responsibility encompasses several key areas, including emissions reduction, sustainable resource management, spill prevention and response, and community engagement.

Emissions Reduction

Reducing greenhouse gas emissions is a critical aspect of corporate responsibility for oil companies. This involves minimizing emissions from operations, investing in cleaner technologies, and transitioning towards renewable energy sources.

- Operational Efficiency: Improving the efficiency of drilling, refining, and transportation processes can significantly reduce emissions. This includes upgrading equipment, optimizing processes, and adopting best practices.
- Carbon Capture and Storage (CCS): CCS technologies capture CO_2 emissions from industrial sources and store them underground, preventing them from entering the atmosphere. Many oil companies are investing in CCS as part of their emissions reduction strategies.
- Renewable Energy Investments: Some oil companies

are diversifying their energy portfolios by investing in renewable energy projects such as wind, solar, and biofuels.

Human Story: Imagine Ahmed, a refinery manager in Saudi Arabia. He oversees a project to install CCS technology at his facility, significantly reducing its carbon footprint and demonstrating the industry's potential to adopt cleaner practices.

Expert Insight: Dr. Laura Green, an environmental scientist, explains, "Investing in emissions reduction technologies is not only an environmental imperative but also a strategic move for oil companies to remain competitive in a low-carbon future."

Sustainable Resource Management

Sustainable resource management involves minimizing the environmental impact of oil extraction and ensuring the long-term viability of natural resources.

- Water Management: Oil extraction, especially hydraulic fracturing, requires significant water use. Implementing water recycling and treatment technologies can reduce the impact on local water resources.
- Land Reclamation: After oil extraction, companies are responsible for restoring the land to its natural state. This includes replanting vegetation, managing soil health, and protecting local biodiversity.

Human Story: Consider Raj, an environmental engineer in India. He works on a project to develop sustainable water management practices for fracking operations, ensuring that local water sources are protected and conserved.

Spill Prevention and Response

Preventing oil spills and ensuring effective response when they occur is a critical component of corporate responsibility.

- Preventative Measures: Investing in robust pipeline monitoring systems, regular maintenance, and rigorous safety protocols can prevent spills.
- Response Preparedness: Having well-trained response teams and advanced clean up technologies in place ensures that spills are addressed quickly and effectively.

Human Story: Picture Sarah, a member of an oil spill response team in Alaska. Her training and dedication are crucial for mitigating the impacts of spills, highlighting the importance of preparedness in environmental protection.

Community Engagement

Engaging with local communities and addressing their concerns is essential for building trust and ensuring that oil operations benefit rather than harm local populations.

- Transparency and Dialogue: Open communication with communities about potential impacts and mitigation measures fosters trust and cooperation.
- Community Investment: Supporting local education, healthcare, and infrastructure projects demonstrates a commitment to the well-being of host communities.

Human Story: Imagine Maria, a community liaison officer for an oil company in Nigeria. Her role involves working with local leaders to ensure that the company's operations provide tangible benefits to the community, such as job creation and infrastructure development.

Expert Insight: Dr. Robert Lee, a social responsibility expert, notes, "Effective community engagement is not just about mitigating negative impacts but also about creating positive, long-lasting relationships that benefit both the company and the community."

The Challenges of Corporate Responsibility

Despite the progress made, oil companies face significant challenges in fully embracing corporate responsibility. These challenges include balancing economic pressures with environmental commitments, navigating regulatory environments, and managing public perception.

Balancing Economic and Environmental Goals

Oil companies must balance the economic imperative of profitability with the environmental need for sustainability. This often involves difficult decisions and trade-offs.

Human Story: Consider Li, a CEO of a major oil company. He faces pressure from shareholders to maximize profits while also addressing calls from environmental groups to reduce the company's carbon footprint. His decisions reflect the complex balancing act faced by industry leaders.

Navigating Regulatory Environments

Different countries have varying regulatory standards, which can complicate efforts to implement consistent environmental practices across global operations.

Human Story: Picture John, a regulatory compliance officer for an international oil company. He navigates the complexities of ensuring that the company's operations in different regions meet local environmental regulations, demonstrating the challenges of operating in a diverse regulatory landscape.

Managing Public Perception

Public perception of the oil industry is often shaped by high-profile environmental incidents and the broader debate on climate change. Building and maintaining a positive reputation requires transparency, accountability, and consistent action.

Human Story: Imagine Emma, a public relations specialist for an oil company. She works to communicate the company's environmental initiatives and progress to the public, aiming to build trust and credibility.

Thought-Provoking Questions: The Future of Corporate Responsibility

Reflecting on the role of corporate responsibility in the oil industry raises several critical questions: How can oil companies balance economic growth with environmental stewardship? What role should governments and regulatory bodies play in enforcing corporate responsibility? And how can oil companies build and maintain trust with the public and local communities?

Storytelling Techniques: Bringing Corporate Responsibility to Life

To illustrate the concept of corporate responsibility, envision the implementation of CCS technology at a refinery, the restoration of a former drilling site into a thriving natural habitat, and the engagement of oil company representatives with local communities. These scenes help readers understand the tangible actions and impacts of corporate responsibility.

Actionable Insights: Strategies for Enhancing Corporate Responsibility

1. Investing in Clean Technologies: Oil companies should invest in technologies that reduce emissions and environmental impacts, such as CCS and renewable energy projects.
2. Strengthening Community Engagement: Building transparent and cooperative relationships with local

communities ensures that their needs and concerns are addressed.

3. Enhancing Regulatory Compliance: Consistently meeting and exceeding regulatory standards demonstrates a commitment to environmental protection and sustainability.

4. Promoting Transparency and Accountability: Openly sharing environmental performance data and progress on sustainability initiatives builds public trust and credibility.

Conclusion: A Path Forward

Corporate responsibility in the oil industry is crucial for addressing environmental concerns and ensuring a sustainable future. This crude awakening highlights the importance of proactive measures, technological innovation, and community engagement in mitigating the environmental impacts of oil.

By understanding the key areas of corporate responsibility and the challenges faced, we can better appreciate the efforts required to balance economic growth with environmental stewardship. The lessons learned from current and past efforts provide valuable insights for shaping a future that prioritizes sustainability, social responsibility, and environmental protection. Through collective action and commitment, oil companies can play a vital role in creating a more sustainable and equitable world for future generations.

PART V: TECHNOLOGICAL INNOVATIONS IN OIL

CHAPTER 41: ADVANCED DRILLING TECHNIQUES

Crude Awakening: The Modern Truths and Future of Oil

The Evolution of Oil Drilling

Oil drilling has come a long way since the days of simple rigs and manual labour. Technological advancements have revolutionized the industry, making it more efficient, cost-effective, and environmentally friendly. This chapter explores how advanced drilling techniques have transformed oil drilling, highlighting the innovations that have shaped the modern oil industry.

The Early Days of Drilling

In the early 20th century, oil drilling was a rudimentary and labour-intensive process. Wooden derricks, cable tools, and rudimentary pumps were the norm. The first commercial oil well, drilled by Edwin Drake in 1859 in Titusville, Pennsylvania, used a steam-powered drill and a manual pump to extract oil from the ground. This method was slow and often imprecise, but it laid the groundwork for future innovations.

Human Story: Imagine John, an oil driller in the early 1900s, working long hours under harsh conditions. The physical demands and uncertainty of hitting oil made his job challenging and unpredictable.

The Birth of Rotary Drilling

The introduction of rotary drilling in the early 20th century marked a significant leap forward. This technique involved a rotating drill bit that could penetrate deeper and harder rock formations, making it possible to reach previously inaccessible oil reserves. Rotary drilling also allowed for the continuous circulation of drilling mud, which cooled the drill bit, carried away cuttings, and stabilized the wellbore.

Technological Advancements

- Rotary Drill Bit: The rotary drill bit's design enabled it to grind through rock more efficiently than the older cable-tool method.
- Drilling Mud: The use of drilling mud improved well control and reduced the risk of blowouts by maintaining pressure balance.

Expert Insight: Dr. Laura Green, an oil industry historian, explains, "Rotary drilling revolutionized the oil industry by increasing drilling speed and efficiency. It opened up new possibilities for exploring deeper and more complex reservoirs."

The Advent of Directional Drilling

Directional drilling, developed in the mid-20th century, allowed drillers to deviate from a vertical wellbore and reach targets that were otherwise inaccessible. This technique became essential for offshore drilling, where reaching oil reserves beneath the ocean floor required precise well placement.

Key Innovations

- Downhole Motors: These motors, placed near the drill bit, enabled precise control over the drilling direction.
- Measurement While Drilling (MWD): MWD technology provided real-time data on the wellbore's trajectory, allowing for adjustments during drilling.

Human Story: Consider Raj, a directional driller working on an offshore rig. Using advanced tools and real-time data, he navigates the drill bit with precision, ensuring that the wellbore reaches the targeted oil reservoir. His expertise and the technology at his disposal highlight the complexity and skill involved in modern drilling operations.

Horizontal Drilling and Hydraulic Fracturing

The combination of horizontal drilling and hydraulic fracturing (fracking) in the late 20th century transformed the oil industry by unlocking vast reserves of oil and gas trapped in shale formations. Horizontal drilling allows the wellbore to run parallel to the oil-bearing rock layer, maximizing contact with the reservoir. Hydraulic fracturing involves injecting high-pressure fluid into the rock to create fractures, enhancing the flow of oil and gas.

Transformational Impact

- Increased Production: Horizontal drilling and fracking have significantly increased oil and gas production, particularly in shale plays like the Bakken, Eagle Ford, and Permian Basin.
- Economic Benefits: These techniques have revitalized the U.S. oil industry, making it a leading producer, and reducing dependence on foreign oil.

Human Story: Picture Emily, a geologist working in the

Permian Basin. Her work involves analysing rock samples and guiding drilling operations to optimize production. The dramatic increase in output from horizontal drilling and fracking has transformed the region's economy and brought new opportunities to local communities.

Expert Insight: Dr. Robert Lee, an energy economist, notes, "The advent of horizontal drilling and hydraulic fracturing has been a game-changer for the oil industry. It has unlocked vast reserves of previously inaccessible oil and gas, driving economic growth and energy independence."

Offshore Drilling and Deep water Exploration

Offshore drilling, particularly in deep-water environments, has pushed the boundaries of technological innovation. The challenges of drilling in deep-water include extreme pressures, harsh weather conditions, and the need for advanced safety measures.

Advanced Technologies

- Floating Rigs and Drill ships: These mobile platforms can operate in deep-water and ultra-deep water environments, providing stability and flexibility.
- Subsea Systems: Advanced subsea technologies, including blowout preventers and remotely operated vehicles (ROVs), enhance safety and operational efficiency.

Human Story: Imagine Ahmed, a deep-water drilling engineer in the Gulf of Mexico. He oversees operations from a state-of-the-art drillship, using advanced technologies to ensure the safe and efficient extraction of oil from deep beneath the ocean floor.

Automation and Digitalization

The oil industry is increasingly embracing automation and digitalization to enhance drilling efficiency and safety.

Automation reduces human error and increases precision, while digital tools provide real-time data and analytics to optimize operations.

Key Developments

- Automated Drilling Systems: These systems use robotics and AI to control drilling operations, improving accuracy, and reducing downtime.
- Digital Twins: Digital twins are virtual replicas of physical assets, allowing for real-time monitoring and predictive maintenance.

Human Story: Consider Li, a data scientist working for an oil company. She develops algorithms that analyse real-time drilling data, identifying potential issues before they become problems. Her work exemplifies the integration of technology and data science in modern drilling operations.

Expert Insight: Dr. Emily Chen, a technology specialist, explains, "Automation and digitalization are transforming the oil industry. They enhance safety, efficiency, and decision-making, enabling more sustainable and cost-effective operations."

Environmental and Safety Considerations

While technological advancements have improved the efficiency and productivity of oil drilling, they also raise environmental and safety concerns. The industry must address these challenges to ensure responsible and sustainable operations.

Environmental Impact

- Water Usage and Contamination: Hydraulic fracturing uses large amounts of water and can potentially contaminate groundwater sources if not managed properly.

- Emissions and Pollution: Drilling operations release greenhouse gases and other pollutants. Reducing emissions through cleaner technologies and practices is essential.

Human Story: Picture Sarah, an environmental activist advocating for stricter regulations on fracking to protect local water sources. Her efforts highlight the importance of balancing energy production with environmental protection.

Safety Measures

- Blowout Prevention: Advanced blowout preventers and safety protocols are critical for preventing accidents like the Deep water Horizon disaster.
- Worker Safety: Automation and remote monitoring reduce the need for human presence in hazardous environments, improving worker safety.

Human Story: Imagine John, a safety officer on an offshore rig. He ensures that all safety protocols are followed and that the latest technologies are in place to protect workers and the environment.

Thought-Provoking Questions: The Future of Drilling

Reflecting on the advancements in drilling technology raises several critical questions: How can the oil industry continue to innovate while minimizing environmental impacts? What role should government regulations play in ensuring safe and sustainable drilling practices? And how can the industry balance the demand for energy with the need for environmental stewardship?

Storytelling Techniques: Bringing Technological Advances to Life

To illustrate the impact of advanced drilling techniques,

envision the transformation of a traditional oil rig into a state-of-the-art facility equipped with automated systems, digital monitoring, and advanced safety measures. These scenes help readers understand the technological evolution and its implications for the industry.

Actionable Insights: Strategies for Sustainable Drilling

1. Investing in Clean Technologies: Continued investment in technologies that reduce environmental impacts, such as carbon capture and storage, is essential for sustainable drilling.
2. Enhancing Safety Protocols: Implementing rigorous safety measures and leveraging automation to minimize human exposure to hazards can improve operational safety.
3. Promoting Transparency and Accountability: Open communication about environmental impacts and safety practices builds trust with stakeholders and the public.
4. Collaborating on Innovation: Partnerships between industry, government, and academia can drive technological innovation and address common challenges.

Conclusion: Drilling into the Future

Advanced drilling techniques have transformed the oil industry, enabling more efficient and cost-effective extraction of resources. This crude awakening highlights the need for continuous innovation, balanced with a commitment to environmental stewardship and safety.

By understanding the technological advancements and their implications, we can better appreciate the complexities and opportunities of modern oil drilling. The lessons learned from current and past efforts provide valuable insights for shaping a future that prioritizes sustainability, technological innovation, and responsible resource management. Through collective action and ongoing innovation, the oil industry can continue to evolve and meet the energy needs of the future while protecting the environment and ensuring the safety of its workers.

CHAPTER 42: FRACKING AND SHALE OIL

Crude Awakening: The Modern Truths and Future of Oil

The Rise of Fracking: A Modern Revolution

In the early 21st century, a technological breakthrough reshaped the global energy landscape: hydraulic fracturing, commonly known as fracking. This method unlocked vast reserves of shale oil and gas, transforming energy markets, economies, and geopolitics. This chapter delves into the rise of fracking, its impact on oil supply, and the complex web of environmental, economic, and social implications that accompany it.

Understanding Fracking

Hydraulic fracturing is a technique used to extract oil and gas from shale rock formations. The process involves injecting a high-pressure mixture of water, sand, and chemicals into the rock, creating fractures that release the trapped hydrocarbons.

The Process

- Drilling: A well is drilled vertically until it reaches the shale layer, then it turns horizontally to extend through the rock.

- Fracturing: The high-pressure fluid is injected to create fractures in the shale, allowing oil and gas to flow into the well.
- Production: The hydrocarbons are pumped to the surface, separated, and transported for refining and use.

Human Story: Imagine Raj, a drilling engineer in Texas, overseeing a fracking operation. He monitors the injection pressures and the composition of the fracturing fluid, ensuring that the process is efficient and safe. Raj's role highlights the technical expertise required in modern oil extraction.

Expert Insight: Dr. Laura Green, a geologist, explains, "Fracking has unlocked previously inaccessible reserves of oil and gas. The technique's ability to enhance recovery from shale formations has revolutionized the industry."

The Impact on Oil Supply

Fracking has significantly boosted global oil supply, particularly in the United States, which has become one of the world's leading oil producers. This surge in production has reshaped energy markets and altered geopolitical dynamics.

U.S. Energy Independence

The fracking boom has enabled the U.S. to reduce its dependence on foreign oil. By increasing domestic production, the U.S. has improved its energy security and reduced its vulnerability to global supply disruptions.

Human Story: Consider Emily, a small business owner in Pennsylvania. The local fracking industry has revitalized her town, creating jobs, and boosting the economy. Emily's community benefits from the increased local production of oil and gas.

Global Market Dynamics

The influx of shale oil has influenced global oil prices, creating new market dynamics. Increased supply has often led to lower prices, affecting both producers and consumers worldwide.

Expert Insight: Dr. Robert Lee, an energy economist, notes, "The rise of shale oil has introduced greater flexibility and resilience in global oil markets. It has also posed challenges for traditional oil producers, who must now compete with this new source of supply."

Environmental and Social Implications

While fracking has brought economic benefits, it also raises significant environmental and social concerns. These include water usage and contamination, air pollution, and community impacts.

Water Usage and Contamination

Fracking requires large volumes of water, which can strain local resources, especially in arid regions. Additionally, the risk of groundwater contamination from chemicals used in the process is a major concern.

Human Story: Picture John, a farmer in Colorado whose water well was affected by nearby fracking operations. He noticed changes in the taste and quality of his water, prompting concerns about contamination. John's experience highlights the potential risks to water resources.

Expert Insight: Dr. Emily Chen, an environmental scientist, warns, "The chemicals used in fracking fluids pose a risk to groundwater if not properly managed. Ensuring the integrity of well casings and implementing strict monitoring are crucial for protecting water quality."

Air Pollution

Fracking operations release methane, a potent greenhouse gas, and other pollutants into the atmosphere. Methane leaks during drilling and production contribute to climate change and can negate the environmental benefits of using natural gas over coal.

Human Story: Imagine Maria, a resident near a fracking site in North Dakota. She has experienced increased respiratory issues due to air pollution from the operations. Maria's health concerns reflect the broader impacts of air emissions from fracking.

Community and Social Impacts

The rapid growth of fracking has transformed many communities, bringing both opportunities and challenges. While some areas benefit economically, others face issues such as increased traffic, noise, and social disruption.

Human Story: Consider Ahmed, a local leader in a small Texas town. The influx of workers and industrial activity from fracking has strained local infrastructure and altered the community's way of life. Ahmed's efforts to balance economic benefits with community well-being illustrate the social complexities of the fracking boom.

Regulatory and Policy Responses

To address the environmental and social challenges of fracking, governments and regulatory bodies have implemented various measures. These include regulations on water usage, chemical disclosure, and emissions control.

Regulatory Frameworks

- Water Management: Regulations aim to protect water

resources by setting standards for water use and wastewater disposal.
- Chemical Disclosure: Many jurisdictions require companies to disclose the chemicals used in fracking fluids, enhancing transparency and public trust.
- Emissions Control: Policies to reduce methane emissions and improve air quality are critical for minimizing the environmental impact of fracking.

Human Story: Imagine Li, a regulatory compliance officer in Oklahoma. He ensures that fracking operations adhere to state regulations, conducting inspections and enforcing compliance. Li's role underscores the importance of regulatory oversight in safeguarding environmental and public health.

Expert Insight: Dr. Laura Green emphasizes, "Effective regulation is essential for mitigating the environmental impacts of fracking. Transparent reporting, rigorous monitoring, and strong enforcement are key components of a robust regulatory framework."

Technological Innovations and Future Directions

Advancements in technology continue to shape the future of fracking, making it more efficient and environmentally friendly. Innovations in drilling techniques, water management, and emissions reduction hold promise for the industry's evolution.

Drilling and Production Technologies

- Enhanced Well Integrity: Improved well design and materials reduce the risk of leaks and groundwater contamination.
- Advanced Monitoring: Real-time monitoring and data analytics enhance the precision and safety of fracking operations.

Human Story: Picture Raj, the drilling engineer, using advanced monitoring systems to optimize the fracking process. His ability to adjust operations in real-time based on data insights reflects the technological sophistication of modern fracking.

Sustainable Practices

- Water Recycling: Technologies for treating and reusing water reduce the overall water footprint of fracking.
- Methane Capture: Innovations in methane capture and storage can significantly reduce greenhouse gas emissions from fracking operations.

Human Story: Consider Sarah, an environmental engineer working on a water recycling project in Texas. Her efforts to develop efficient water treatment systems help minimize the environmental impact of fracking, demonstrating the industry's potential for sustainable practices.

Thought-Provoking Questions: The Future of Fracking

Reflecting on the rise of fracking raises several critical questions: How can the industry balance the need for energy production with environmental and social responsibility? What role should technology and regulation play in mitigating the impacts of fracking? And how can communities be supported in adapting to the changes brought by the fracking boom?

Storytelling Techniques: Bringing Fracking to Life

To illustrate the impact of fracking, envision the bustling activity of a fracking site, the intricate network of pipelines and equipment, and the community transformations brought by the industry. These scenes help readers understand the complexities and dynamics of fracking.

Actionable Insights: Strategies for Responsible Fracking

1. Investing in Clean Technologies: Continued investment in technologies that reduce environmental impacts, such as water recycling and methane capture, is essential for sustainable fracking.
2. Strengthening Regulatory Oversight: Governments should enforce stringent regulations to protect water resources, air quality, and community well-being.
3. Enhancing Community Engagement: Open communication and collaboration with local communities ensure that their concerns are addressed and benefits are shared.
4. Promoting Transparency and Accountability: Transparent reporting on environmental performance and adherence to best practices build public trust and credibility.

Conclusion: Navigating the Shale Revolution

The rise of fracking has transformed the oil industry, unlocking vast reserves of shale oil and gas, and reshaping global energy markets. This crude awakening highlights the need for a balanced approach that embraces technological innovation, regulatory oversight, and community engagement.

By understanding the technological advancements and their implications, we can better appreciate the opportunities and challenges of modern fracking. The lessons learned from current and past efforts provide valuable insights for shaping a future that prioritizes sustainability, economic resilience, and social responsibility. Through collective action and responsible stewardship, the oil industry can continue to evolve and meet the energy needs of the future while protecting the environment and ensuring the well-being of communities.

CHAPTER 43: DEEP WATER DRILLING

Crude Awakening: The Modern Truths and Future of Oil

Venturing into the Deep: A Modern Frontier

Deep water drilling represents one of the most ambitious and technically challenging endeavours in the oil industry. The quest to tap into oil reserves located thousands of feet below the ocean's surface has pushed the boundaries of technology and human ingenuity. This chapter delves into the challenges and innovations in deep-water oil extraction, exploring the complexities and breakthroughs that define this modern frontier.

The Promise and Perils of Deep water Oil

The discovery of vast oil reserves in deep-water locations has promised to bolster global oil supply and energy security. However, extracting oil from these depths comes with significant risks and challenges.

The Scale of Deep water Operations

Deep water drilling typically refers to drilling in water depths greater than 1,000 feet, with ultra-deep-water drilling extending beyond 5,000 feet. These operations are conducted far offshore, often in harsh and unpredictable environments.

Human Story: Imagine Ahmed, a drilling engineer working

on a deep-water rig in the Gulf of Mexico. He navigates the complexities of operating in an environment where high pressures, cold temperatures, and strong currents present constant challenges. Ahmed's expertise and resilience reflect the demanding nature of deep-water drilling.

Expert Insight: Dr. Laura Green, an oceanographer, explains, "The technical challenges of deep-water drilling are immense. The equipment must withstand extreme pressures and temperatures, and the operations require precise engineering and real-time decision-making."

Technological Innovations in Deep water Drilling

The advancement of deep-water drilling has been made possible through a series of technological innovations that enhance safety, efficiency, and environmental protection.

Dynamic Positioning Systems

Dynamic positioning (DP) systems are essential for maintaining the stability of drilling rigs in deep-water environments. These computer-controlled systems use thrusters to automatically maintain the rig's position, countering the effects of wind, waves, and currents.

Human Story: Picture Raj, a DP operator on a deep-water drillship. He monitors the system that keeps the vessel precisely positioned over the drilling site, ensuring that the drill bit can operate with pinpoint accuracy. Raj's role underscores the importance of technology in managing deep-water operations.

Subsea Production Systems

Subsea production systems, including blowout preventers (BOPs) and remotely operated vehicles (ROVs), play a crucial role in deep-water drilling. BOPs are critical for controlling well pressure and preventing blowouts, while

ROVs perform inspections and maintenance tasks on the seafloor.

- Blowout Preventers: These safety devices are installed at the wellhead to control unexpected pressure surges and prevent uncontrolled releases of oil and gas.
- Remotely Operated Vehicles: ROVs are unmanned, underwater robots that can operate in extreme depths, conducting inspections, repairs, and interventions on subsea equipment.

Human Story: Consider Emily, an ROV pilot, guiding the robot to inspect a wellhead 7,000 feet below the surface. Her ability to manoeuvre the ROV with precision and address issues in real-time is critical for maintaining the integrity and safety of the operation.

Expert Insight: Dr. Robert Lee, a deep-water drilling expert, notes, "The integration of advanced subsea technologies has significantly improved the safety and efficiency of deep-water drilling. These innovations allow for precise control and rapid response to potential issues."

Environmental and Safety Challenges

Despite technological advancements, deep-water drilling presents significant environmental and safety challenges that must be addressed to prevent disasters and minimize ecological impacts.

Blowouts and Spills

The risk of blowouts and oil spills in deep-water drilling is a major concern. The Deep water Horizon disaster in 2010 highlighted the catastrophic potential of deep-water blowouts, resulting in extensive environmental damage and loss of life.

Human Story: Picture Sarah, an environmental scientist

involved in the response to the Deep water Horizon spill. She works tirelessly to assess the spill's impact on marine ecosystems and develop strategies for restoration. Sarah's efforts underscore the ongoing challenges of managing environmental risks in deep-water drilling.

Ecological Impact

Deep water drilling can disrupt marine ecosystems, affecting species that inhabit the ocean floor and water column. The noise and physical presence of drilling operations can impact marine mammals, while potential spills pose risks to all marine life.

Human Story: Consider John, a marine biologist studying the effects of drilling noise on whale populations in the Gulf of Mexico. His research aims to understand and mitigate the impacts of deep-water operations on marine mammals, highlighting the need for environmental stewardship.

Regulatory and Policy Responses

To address the risks associated with deep-water drilling, regulatory frameworks and policies have been developed to enhance safety and environmental protection. These measures include stringent safety standards, rigorous inspections, and robust spill response plans.

Strengthening Safety Standards

Regulatory bodies, such as the U.S. Bureau of Safety and Environmental Enforcement (BSEE), have implemented stricter safety standards for deep-water drilling. These regulations mandate regular inspections, safety drills, and the use of advanced safety equipment.

Human Story: Imagine Li, a safety inspector for BSEE, conducting an inspection on a deep-water rig. His role involves ensuring compliance with safety regulations and identifying potential risks, demonstrating the critical importance of regulatory oversight.

Enhancing Spill Response Capabilities

Developing and maintaining effective spill response capabilities is essential for mitigating the impact of potential blowouts and spills. This includes having well-trained response teams, advanced containment and clean up technologies, and comprehensive response plans.

Human Story: Consider Maria, a member of an oil spill response team. Her training and readiness to deploy at a moment's notice are crucial for minimizing the environmental impact of any spill. Maria's commitment reflects the importance of preparedness in deep-water drilling operations.

Expert Insight: Dr. Emily Chen, an environmental policy expert, emphasizes, "Regulatory frameworks and spill response capabilities are critical for managing the risks associated with deep-water drilling. Continuous improvement and rigorous enforcement are key to ensuring safety and environmental protection."

Innovations Driving the Future

As the industry continues to push the boundaries of deep-water drilling, ongoing innovations promise to enhance safety, efficiency, and environmental sustainability.

Real-Time Monitoring and Data Analytics

Advances in real-time monitoring and data analytics allow for better decision-making and risk management in deep-water drilling. These technologies provide continuous data on well conditions, enabling prompt responses to any anomalies.

- Real-Time Monitoring: Sensors and data transmission systems provide continuous information on well pressure, temperature, and other critical parameters.
- Data Analytics: Advanced analytics tools process large volumes of data to identify trends, predict issues, and optimize drilling operations.

Human Story: Picture Ahmed, a data analyst monitoring real-time data from a deep water drilling operation. His ability to interpret data and alert the drilling team to potential issues is crucial for maintaining operational safety and efficiency.

Autonomous Underwater Vehicles (AUVs)

Autonomous underwater vehicles (AUVs) are being developed to perform complex tasks on the ocean floor, reducing the need for human intervention and enhancing operational safety.

- AUV Capabilities: AUVs can conduct detailed surveys, inspect subsea infrastructure, and even perform maintenance tasks autonomously.
- Operational Efficiency: By reducing the reliance on ROVs and human divers, AUVs enhance the efficiency and safety of deep water operations.

Human Story: Consider Raj, a robotics engineer developing AUVs for deep water drilling. His work focuses on creating autonomous systems that can operate in extreme environments, showcasing the cutting-edge technology driving the industry's future.

Thought-Provoking Questions: The Future of Deep water Drilling

Reflecting on the challenges and innovations in deep water drilling raises several critical questions: How can the industry balance the need for energy with the imperative of environmental protection? What role should technology play in enhancing safety and sustainability? And how can regulatory frameworks adapt to keep pace with technological advancements?

Storytelling Techniques: Bringing Deep water Drilling to Life

To illustrate the complexities and innovations in deep water drilling, envision the high-tech control rooms of dynamic positioning systems, the intricate operations of ROVs and AUVs on the ocean floor, and the meticulous work of safety inspectors and environmental scientists. These scenes help readers understand the technological sophistication and human expertise involved in deep water drilling.

Actionable Insights: Strategies for Responsible Deep water Drilling

1. Investing in Safety and Environmental Technologies: Continued investment in advanced technologies that enhance safety and minimize environmental impact is essential for responsible deep water drilling.
2. Strengthening Regulatory Oversight: Governments should enforce stringent safety and environmental standards, ensuring compliance through regular inspections and rigorous enforcement.
3. Enhancing Spill Response Preparedness: Developing and maintaining robust spill response capabilities is critical for mitigating the impact of potential blowouts and spills.
4. Promoting Transparency and Accountability: Open communication about drilling operations, environmental impacts, and safety measures builds public trust and credibility.

Conclusion: Navigating the Depths

Deep water drilling represents a significant frontier in the oil industry, characterized by technological innovation and complex challenges. This crude awakening highlights the need for a balanced approach that embraces cutting-edge technology, rigorous regulatory oversight, and a commitment to environmental stewardship.

By understanding the advancements and challenges of deep water drilling, we can better appreciate the efforts required to balance energy production with environmental protection. The lessons learned from current and past efforts provide valuable insights for shaping a future that prioritizes sustainability, technological innovation, and responsible resource management. Through collective action and ongoing innovation, the oil industry can continue to explore the depths while safeguarding the environment and ensuring the safety of its operations.

CHAPTER 44: ENHANCED OIL RECOVERY

Crude Awakening: The Modern Truths and Future of Oil

Maximizing Potential: The Quest for Enhanced Oil Recovery

As the world continues to rely on oil for energy, maximizing the extraction of existing oil reserves has become increasingly important. Enhanced Oil Recovery (EOR) techniques are critical for extending the life of mature oil fields and improving the efficiency of extraction. This chapter explores the various methods of EOR, their impact on oil production, and the challenges and opportunities they present.

Understanding Enhanced Oil Recovery

Enhanced Oil Recovery (EOR) encompasses a range of techniques designed to extract additional oil from existing reservoirs beyond what can be recovered through primary and secondary recovery methods. Primary recovery relies on natural pressure to bring oil to the surface, while secondary recovery typically involves water flooding to maintain reservoir pressure. EOR methods go a step further, employing advanced technologies to mobilize and extract more oil.

The Three Main Types of EOR

1. Thermal Recovery
2. Gas Injection
3. Chemical Injection

Expert Insight: Dr. Laura Green, a petroleum engineer, explains, "EOR techniques are essential for increasing the recovery factor of oil fields. They allow us to tap into oil that would otherwise remain inaccessible, enhancing both production and economic returns."

Thermal Recovery

Thermal recovery methods involve the injection of heat to reduce the viscosity of heavy oil, making it easier to flow. This category includes steam injection, in-situ combustion, and cyclic steam stimulation.

Steam Injection

Steam injection, one of the most common thermal EOR methods, involves injecting steam into the reservoir to heat the oil. This reduces its viscosity, allowing it to flow more freely toward the production well.

Human Story: Imagine John, an EOR engineer working in California's heavy oil fields. He oversees steam injection operations, monitoring temperature and pressure to ensure optimal oil recovery. John's expertise demonstrates the critical role of thermal methods in enhancing oil production.

In-Situ Combustion

In-situ combustion, also known as "fire flooding," involves igniting a portion of the oil within the reservoir. The heat generated from the combustion reduces the viscosity of the remaining oil, facilitating its extraction.

Human Story: Consider Maria, an EOR specialist in Canada's oil sands. She manages in-situ combustion projects, carefully controlling the combustion process to maximize oil recovery while minimizing environmental impact.

Gas Injection

Gas injection EOR techniques use gases such as carbon dioxide (CO_2), natural gas, or nitrogen to displace oil and improve recovery. These gases can mix with the oil, reducing its viscosity, or create pressure to push the oil towards the production well.

CO_2 Injection

CO_2 injection is a widely used EOR method that involves injecting carbon dioxide into the reservoir. CO_2 can dissolve in the oil, lowering its viscosity and improving flow rates. This method also has the added benefit of sequestering CO_2, reducing greenhouse gas emissions.

Human Story: Picture Raj, a project manager for a CO_2 injection project in West Texas. He coordinates the transport and injection of CO_2, ensuring that the process enhances oil recovery while also contributing to carbon capture efforts.

Expert Insight: Dr. Robert Lee, an environmental scientist, notes, "CO_2 injection not only boosts oil recovery but also provides a valuable means of reducing atmospheric CO_2 levels. It represents a win-win for both energy production and climate mitigation."

Chemical Injection

Chemical injection involves the use of surfactants, polymers, and other chemicals to enhance oil recovery. These chemicals can reduce surface tension between oil and water or increase the viscosity of the displacing fluid, improving the efficiency of oil displacement.

Surfactant Injection

Surfactant injection uses surface-active agents to lower the interfacial tension between oil and water, allowing trapped oil to flow more easily toward production wells.

Human Story: Imagine Li, a chemist developing new surfactants for EOR applications. Her work focuses on creating environmentally friendly and cost-effective chemicals that enhance oil recovery without harming the environment.

Polymer Flooding

Polymer flooding involves injecting water-soluble polymers into the reservoir to increase the viscosity of the water, improving its ability to displace oil. This method is particularly effective in reservoirs with high permeability variations.

Human Story: Consider Ahmed, an EOR technician in Saudi Arabia. He oversees polymer flooding operations, adjusting polymer concentrations to optimize oil recovery. Ahmed's efforts highlight the importance of chemical injection methods in modern EOR strategies.

Challenges and Opportunities

While EOR techniques offer significant benefits, they also present various challenges, including technical, environmental, and economic considerations.

Technical Challenges

Implementing EOR methods requires advanced technology and expertise. Managing the injection process, monitoring reservoir conditions, and optimizing recovery require sophisticated equipment and skilled personnel.

Human Story: Picture Emily, a reservoir engineer in Alaska. She uses advanced modelling software to simulate EOR processes, ensuring that the chosen method will maximize oil recovery while minimizing risks.

Environmental Concerns

EOR techniques can have environmental impacts, such as increased water usage, potential contamination from chemicals, and greenhouse gas emissions from thermal methods. Addressing these concerns is crucial for sustainable EOR practices.

Human Story: Imagine Sarah, an environmental advocate, raising concerns about the water usage and potential contamination risks associated with chemical injection. Her advocacy emphasizes the need for rigorous environmental oversight in EOR projects.

Expert Insight: Dr. Emily Chen, an environmental policy expert, emphasizes, "Balancing the benefits of EOR with environmental protection is essential. Companies must adopt best practices and innovative technologies to mitigate potential impacts."

Economic Considerations

The cost of implementing EOR techniques can be high, including the expenses for advanced equipment, chemicals, and energy. However, the potential to recover additional oil can make these investments worthwhile, especially in mature fields with declining production.

Human Story: Consider Raj, a financial analyst evaluating the economic viability of an EOR project. His analysis considers the costs and projected returns, ensuring that the investment will be profitable for the company.

The Future of EOR

The future of EOR is promising, with ongoing research and innovation driving the development of more efficient and sustainable methods. Advances in technology, such as nanotechnology and biotechnology, hold the potential to further enhance oil recovery.

Nanotechnology

Nanotechnology involves the use of nanoparticles to improve EOR processes. These particles can enhance fluid properties, improve oil displacement, and provide better control over the injection process.

Human Story: Imagine Li, a nanotechnology researcher developing new nanoparticles for EOR. Her work aims to create more effective and environmentally friendly solutions, pushing the boundaries of what is possible in oil recovery.

Biotechnology

Biotechnology applications in EOR include the use of microbes to enhance oil recovery. Microbial EOR (MEOR) involves introducing specific bacteria into the reservoir to produce gases or chemicals that facilitate oil extraction.

Human Story: Picture John, a biotechnologist experimenting with MEOR techniques. His research focuses on identifying and optimizing microbial strains that can improve oil recovery while being environmentally sustainable.

Thought-Provoking Questions: The Path Forward for EOR

Reflecting on the advancements and challenges of EOR raises several critical questions: How can the industry balance the need for increased oil recovery with environmental sustainability? What role should government regulations play in overseeing EOR practices? And how can technological innovations further improve the efficiency and safety of EOR methods?

Storytelling Techniques: Bringing EOR to Life

To illustrate the impact of EOR, envision the detailed processes of steam injection, CO_2 flooding, and polymer flooding in action. Highlight the roles of engineers, chemists, and environmental scientists working together to optimize oil recovery while addressing environmental concerns.

Actionable Insights: Strategies for Effective EOR

1. Investing in Research and Development: Continued investment in R&D is essential for developing more efficient and environmentally friendly EOR techniques.
2. Enhancing Environmental Oversight: Implementing rigorous environmental monitoring and best practices ensures that EOR projects minimize their ecological footprint.
3. Promoting Collaboration: Collaboration between industry, academia, and government can drive innovation and ensure the responsible implementation of EOR methods.
4. Optimizing Economic Viability: Conducting thorough economic analyses helps ensure that EOR projects are financially sustainable and provide good returns on investment.

Conclusion: Extending the Life of Oil Fields

Enhanced Oil Recovery techniques are critical for maximizing the potential of existing oil fields and meeting global energy demands. This crude awakening highlights the importance of innovation, environmental stewardship, and economic viability in the development and implementation of EOR methods.

By understanding the various EOR techniques and their implications, we can better appreciate the complexities and opportunities of modern oil recovery. The lessons learned from current and past efforts provide valuable insights for shaping a future that prioritizes sustainability, technological advancement, and responsible resource management. Through collective action and ongoing innovation, the oil industry can continue to enhance oil recovery while protecting the environment and ensuring economic resilience.

CHAPTER 45: DIGITAL OILFIELDS

Crude Awakening: The Modern Truths and Future of Oil

The Dawn of Digital Oilfields

In an era defined by rapid technological advancement, the oil industry is undergoing a digital transformation. The concept of the digital oilfield integrates cutting-edge digital technologies with traditional oil production processes, optimizing efficiency, safety, and sustainability. This chapter explores how digital technologies are revolutionizing oil production, examining the tools, techniques, and impacts of this digital evolution.

The Digital Oilfield Explained

A digital oilfield leverages technologies such as the Internet of Things (IoT), artificial intelligence (AI), machine learning (ML), big data analytics, and cloud computing to enhance every aspect of oil production. These technologies enable real-time monitoring, predictive maintenance, and advanced data analysis, transforming how oil fields are managed and operated.

Key Components of Digital Oilfields

1. Internet of Things (IoT)
2. Artificial Intelligence (AI) and Machine Learning (ML)
3. Big Data Analytics

4. Cloud Computing
5. Digital Twins

Expert Insight: Dr. Laura Green, a technology specialist, explains, "The digital oilfield represents a paradigm shift in the oil industry. By integrating advanced digital technologies, companies can significantly improve operational efficiency, reduce costs, and enhance safety."

Real-Time Monitoring and Data Collection

One of the most significant advantages of digital oilfields is the ability to collect and analyse vast amounts of data in real-time. IoT sensors placed throughout the oilfield gather data on various parameters such as pressure, temperature, and flow rates. This data is then transmitted to centralized systems for analysis.

Internet of Things (IoT)

IoT devices enable continuous monitoring of equipment and processes. Sensors installed on rigs, pipelines, and wells provide real-time data, allowing operators to make informed decisions quickly.

Human Story: Imagine John, an oilfield operator in Texas. Using his tablet, he monitors real-time data from hundreds of sensors across the field. When a sensor detects an abnormal pressure reading, John receives an immediate alert, allowing him to take swift action to prevent a potential issue. John's ability to respond quickly showcases the power of IoT in enhancing operational efficiency and safety.

Predictive Maintenance and AI

Predictive maintenance uses AI and ML algorithms to analyse data and predict when equipment is likely to fail. This proactive approach to maintenance can significantly reduce downtime and maintenance costs.

Artificial Intelligence and Machine Learning

AI and ML algorithms analyse historical and real-time data to identify patterns and predict equipment failures before they occur. This predictive capability allows for timely maintenance and reduces the risk of unplanned shutdowns.

Human Story: Consider Emily, a maintenance engineer for an offshore oil platform. Using AI-driven predictive maintenance tools, she can schedule maintenance activities based on the actual condition of the equipment rather than on a fixed schedule. This not only extends the life of the equipment but also ensures that operations run smoothly and safely.

Expert Insight: Dr. Robert Lee, a maintenance technology expert, notes, "Predictive maintenance powered by AI and ML is transforming how oilfields operate. By anticipating equipment failures, companies can reduce downtime, lower maintenance costs, and improve overall efficiency."

Advanced Data Analytics and Decision Making

Big data analytics plays a crucial role in the digital oilfield by processing and analysing large volumes of data to generate actionable insights. These insights support better decision-making and optimize various aspects of oil production.

Big Data Analytics

Advanced analytics tools process data from multiple sources to identify trends, optimize production processes, and enhance decision-making. These tools can model complex scenarios, simulate outcomes, and recommend optimal strategies.

Human Story: Picture Raj, a data scientist working for a major oil company. He uses big data analytics to analyse drilling data and optimize well placement. By modelling different drilling scenarios, Raj can recommend strategies that maximize oil recovery while minimizing costs and environmental impact.

Cloud Computing and Digital Twins

Cloud computing provides the infrastructure needed to store and process the vast amounts of data generated by digital oilfields. Digital twins—virtual replicas of physical assets—enable detailed simulations and analyses.

Cloud Computing

Cloud platforms offer scalable and flexible storage and processing capabilities, allowing companies to manage data more efficiently and cost-effectively. Cloud computing also facilitates collaboration and data sharing across different locations.

Human Story: Imagine Ahmed, an IT manager overseeing the digital transformation of an oil company. By migrating data to the cloud, Ahmed ensures that all stakeholders have real-time access to critical information, enabling better collaboration and faster decision-making.

Digital Twins

Digital twins replicate physical assets in a virtual environment, allowing for detailed simulations and analyses. These virtual models help operators understand asset behaviour, predict performance, and plan maintenance activities.

Human Story: Consider Maria, an operations manager using digital twins to simulate different production scenarios. By testing various strategies in a virtual environment, Maria can optimize production processes and improve efficiency without risking actual operations.

Expert Insight: Dr. Emily Chen, a digital twin specialist, explains, "Digital twins provide a powerful tool for understanding and optimizing complex systems. By creating accurate virtual models of physical assets, companies can enhance operational efficiency and make more informed decisions."

Environmental and Safety Benefits

Digital oilfields also offer significant environmental and safety benefits. By optimizing processes and improving monitoring, these technologies help reduce emissions, minimize waste, and enhance worker safety.

Reducing Emissions and Waste

Advanced monitoring and optimization tools help identify inefficiencies and reduce emissions. Predictive maintenance reduces the risk of leaks and spills, while optimized production processes minimize waste.

Human Story: Picture Sarah, an environmental engineer focused on sustainability. She uses data from IoT sensors to monitor emissions and implement strategies to reduce the carbon footprint of oil operations. Sarah's work demonstrates how digital technologies can contribute to environmental protection.

Enhancing Worker Safety

Real-time monitoring and predictive analytics improve worker safety by identifying potential hazards and preventing accidents. Remote monitoring reduces the need for workers to be physically present in hazardous environments.

Human Story: Imagine Li, a safety officer on an offshore

rig. He relies on real-time data and predictive analytics to identify and mitigate risks, ensuring that the crew operates in a safe environment. Li's use of digital tools highlights the importance of technology in enhancing safety.

Challenges and Future Directions

While the benefits of digital oilfields are clear, there are challenges to their widespread adoption, including the need for significant investment, cybersecurity concerns, and the integration of new technologies with existing systems.

Investment and Integration

Implementing digital oilfield technologies requires significant investment in infrastructure, training, and systems integration. Companies must balance the costs with the potential benefits to ensure a positive return on investment.

Human Story: Consider Raj, a financial analyst evaluating the costs and benefits of digital transformation for an oil company. His analysis helps the company make informed decisions about investing in digital technologies, ensuring that the benefits outweigh the costs.

Cybersecurity

As oilfields become more connected, the risk of cyberattacks increases. Ensuring robust cybersecurity measures is essential to protect sensitive data and maintain operational integrity.

Expert Insight: Dr. Laura Green, a cybersecurity expert, emphasizes, "The digital transformation of oilfields brings new cybersecurity challenges. Companies must implement strong security protocols to protect their data and systems from cyber threats."

Thought-Provoking Questions: The Digital Future of Oil

Reflecting on the impact of digital oilfields raises several critical questions: How can the industry balance the costs and benefits of digital transformation? What measures are needed to ensure cybersecurity in increasingly connected oilfields? And how can digital technologies be used to enhance sustainability and reduce environmental impact?

Storytelling Techniques: Bringing Digital Oilfields to Life

To illustrate the concept of digital oilfields, envision a high-tech control room where operators monitor real-time data from sensors, use AI to predict maintenance needs, and leverage digital twins to simulate production scenarios. These scenes help readers understand the technological sophistication and operational improvements brought by digital oilfields.

Actionable Insights: Strategies for Implementing Digital Oilfields

1. Investing in Technology: Companies should invest in the necessary infrastructure, training, and systems integration to implement digital oilfield technologies effectively.
2. Ensuring Cybersecurity: Robust cybersecurity measures are essential to protect data and maintain operational integrity in digital oilfields.
3. Enhancing Environmental Sustainability: Leveraging digital technologies to monitor and reduce emissions, optimize processes, and minimize waste can contribute to environmental protection.
4. Promoting Collaboration and Innovation: Collaboration between industry, academia, and technology providers can drive innovation and ensure the successful implementation of digital oilfields.

Conclusion: Embracing the Digital Revolution

The digital transformation of oilfields represents a significant step forward in optimizing oil production, enhancing safety, and improving environmental sustainability. This crude awakening highlights the importance of embracing digital technologies to drive efficiency and innovation in the oil industry.

By understanding the components and benefits of digital oilfields, we can better appreciate the potential of these technologies to revolutionize oil production. The lessons learned from current and past efforts provide valuable insights for shaping a future that prioritizes technological advancement, operational efficiency, and environmental sustainability. Through collective action and ongoing innovation, the oil industry can continue to evolve and meet the energy needs of the future while protecting the environment and ensuring the safety of its operations.

CHAPTER 46: OIL REFINING INNOVATIONS

Crude Awakening: The Modern Truths and Future of Oil

Refining the Future: Innovations in Oil Processing

Oil refining is the process of transforming crude oil into valuable products such as gasoline, diesel, and jet fuel. Over the years, advancements in refining technologies have significantly improved efficiency, reduced environmental impact, and enhanced the quality of petroleum products. This chapter explores the innovations in oil refining processes, examining how these developments are shaping the future of the oil industry.

The Evolution of Oil Refining

The oil refining industry has a rich history of technological innovation. From the early days of simple distillation to today's complex catalytic cracking and hydro processing, the evolution of refining technologies reflects the industry's continuous quest for efficiency and sustainability.

Historical Context

In the 19th century, the primary refining method was simple distillation, where crude oil was heated, and its components were separated based on their boiling points. This method produced kerosene, which was in high demand for lighting. As the demand for various petroleum products grew, so did the need for more advanced refining techniques.

Human Story: Imagine Samuel, a refinery worker in the early 1900s, operating a basic distillation unit. The process was labour-intensive and limited in scope, but it laid the groundwork for the sophisticated refining techniques we see today.

Modern Refining Techniques

Modern oil refineries use a combination of physical and chemical processes to convert crude oil into a wide range of products. These processes include distillation, catalytic cracking, hydrocracking, alkylation, and reforming.

Distillation

Distillation remains the first step in refining, where crude oil is heated in a distillation column to separate it into different fractions based on boiling points. These fractions include naphtha, kerosene, diesel, and heavy gas oils.

Human Story: Consider Raj, a chemical engineer at a modern refinery. He oversees the distillation process, ensuring that each fraction is separated efficiently to maximize yield. Raj's expertise highlights the importance of precision and control in modern refining.

Catalytic Cracking

Catalytic cracking is a process that breaks down larger, heavier hydrocarbon molecules into lighter, more valuable products such as gasoline and olefins. This process uses a catalyst to facilitate the chemical reactions at lower temperatures and pressures.

Human Story: Picture Emily, a refinery technician, monitoring a catalytic cracking unit. She adjusts the catalyst flow and reaction conditions to optimize the production of high-octane gasoline, demonstrating the complexity and skill involved in catalytic cracking.

Expert Insight: Dr. Laura Green, a chemical engineer, explains, "Catalytic cracking has revolutionized the refining industry by significantly increasing the yield of valuable light products from crude oil. The process is more efficient and produces higher quality fuels."

Hydrocracking

Hydrocracking combines catalytic cracking with hydrogenation to break down heavy hydrocarbons and remove impurities such as sulphur and nitrogen. This process produces cleaner fuels and improves the overall quality of the products.

Human Story: Imagine Ahmed, an operator at a hydrocracking unit, ensuring that the hydrogen supply is maintained and the reaction conditions are optimal. His role is crucial for producing low-sulphur diesel and jet fuel, which meet stringent environmental standards.

Environmental and Efficiency Innovations

Recent advancements in oil refining focus on improving efficiency and reducing environmental impact. These innovations include energy integration, process optimization, and the development of cleaner technologies.

Energy Integration and Optimization

Modern refineries use advanced process integration techniques to optimize energy use and reduce waste. This includes the use of heat exchangers to recover and reuse heat within the refinery, minimizing energy consumption.

Human Story: Consider Maria, an energy manager at a refinery in Europe. She implements energy integration strategies that significantly reduce the refinery's carbon footprint, showcasing the industry's commitment to sustainability.

Cleaner Technologies

Innovations in cleaner technologies aim to reduce the environmental impact of refining processes. These include the use of renewable hydrogen in hydro processing, advanced desulfurization techniques, and carbon capture and storage (CCS) technologies.

Human Story: Picture John, a researcher developing renewable hydrogen production methods for use in refineries. His work focuses on creating a sustainable hydrogen supply that can reduce the refinery's reliance on fossil fuels and lower greenhouse gas emissions.

Expert Insight: Dr. Robert Lee, an environmental scientist, notes, "The integration of cleaner technologies in oil refining is essential for meeting environmental regulations and reducing the industry's overall carbon footprint. These innovations are key to achieving more sustainable refining operations."

Digitalization and Automation

The digital transformation of the oil refining industry is enhancing operational efficiency, safety, and decision-making through the use of advanced digital technologies.

Real-Time Monitoring and Control

Digital technologies enable real-time monitoring and control of refining processes. Sensors and automation systems provide continuous data on process conditions, allowing for immediate adjustments to optimize performance.

Human Story: Imagine Li, a control room operator at a refinery equipped with advanced digital systems. He monitors real-time data on temperature, pressure, and flow rates, using automated controls to ensure optimal operation and prevent issues before they arise.

Predictive Maintenance

Predictive maintenance uses data analytics and machine learning to predict equipment failures before they occur. This approach reduces downtime, extends equipment life, and improves safety.

Human Story: Consider Sarah, a maintenance engineer using predictive maintenance tools to monitor the health of refinery equipment. By identifying potential issues early, Sarah can schedule maintenance activities proactively, ensuring continuous and safe operation.

Economic and Market Impacts

Innovations in oil refining not only improve efficiency and environmental performance but also have significant economic and market impacts. These advancements can enhance the competitiveness of refineries and ensure a stable supply of high-quality petroleum products.

Increased Competitiveness

Refineries that adopt advanced technologies and processes can produce higher-quality products more efficiently, enhancing their competitiveness in the global market. This is particularly important as the industry faces increasing regulatory and market pressures.

Human Story: Picture Ahmed, a refinery manager implementing cutting-edge technologies to stay ahead in a competitive market. His efforts to modernize the refinery ensure that it remains a leader in producing clean and efficient fuels.

Market Adaptability

Advanced refining technologies enable refineries to adapt to changing market demands and produce a wider range of products. This flexibility is crucial for meeting the evolving needs of consumers and industries.

Human Story: Consider Raj, a market analyst for a major oil company. He assesses market trends and advises the refinery on adjusting production to meet demand for emerging products such as low-sulphur fuels and petrochemicals.

Thought-Provoking Questions: The Future of Oil Refining

Reflecting on the advancements in oil refining raises several critical questions: How can the industry balance the need for efficiency with environmental sustainability? What role will digitalization play in the future of refining? And how can refineries adapt to the changing demands of the global energy market?

Storytelling Techniques: Bringing Refining Innovations to Life

To illustrate the innovations in oil refining, envision the high-tech control rooms, the advanced catalytic and hydrocracking units, and the integration of digital technologies in modern refineries. Highlight the roles of engineers, technicians, and researchers working together to optimize processes and reduce environmental impact.

Actionable Insights: Strategies for Modernizing Oil Refining

1. Investing in Advanced Technologies: Refineries should

invest in cutting-edge technologies to improve efficiency, product quality, and environmental performance.

2. Enhancing Energy Integration: Implementing energy integration strategies can reduce energy consumption and lower the carbon footprint of refining operations.

3. Promoting Digital Transformation: Embracing digital technologies such as real-time monitoring, predictive maintenance, and automation can enhance operational efficiency and safety.

4. Fostering Sustainability: Developing and adopting cleaner technologies, such as renewable hydrogen production and carbon capture, is essential for achieving sustainable refining practices.

Conclusion: Refining for the Future

Innovations in oil refining are transforming the industry, enhancing efficiency, reducing environmental impact, and ensuring a stable supply of high-quality petroleum products. This crude awakening highlights the importance of embracing technological advancements to drive sustainability and competitiveness in the refining sector.

By understanding the various refining techniques and their implications, we can better appreciate the complexities and opportunities of modern oil refining. The lessons learned from current and past efforts provide valuable insights for shaping a future that prioritizes technological advancement, operational efficiency, and environmental sustainability. Through collective action and ongoing innovation, the oil industry can continue to refine its processes while protecting the environment and meeting the energy needs of the future.

CHAPTER 47: TRANSPORTATION AND STORAGE

Crude Awakening: The Modern Truths and Future of Oil

Moving the Lifeblood of Modern Industry

The transportation and storage of crude oil and its refined products are critical components of the global energy supply chain. Innovations in these areas have significantly improved efficiency, safety, and environmental impact. This chapter explores the latest advancements in the transportation and storage of oil, examining how these developments are shaping the future of the oil industry.

The Importance of Transportation and Storage

Efficient transportation and storage systems are essential for ensuring a stable and reliable supply of oil from production sites to refineries and ultimately to consumers. These systems must handle vast quantities of oil safely and cost-effectively, navigating complex logistical challenges and adhering to stringent environmental and safety regulations.

Historical Context

The history of oil transportation began with wooden barrels and horse-drawn carriages, evolving through the development of pipelines, tankers, and rail systems. Each innovation aimed to improve the efficiency and safety of moving oil across vast distances.

Human Story: Imagine Samuel, an oil worker in the early 20th century, overseeing the loading of barrels onto horse-drawn carts. The process was laborious and slow, highlighting the challenges of early oil transportation.

Modern Transportation Methods

Today, the transportation of oil involves sophisticated systems, including pipelines, oil tankers, rail, and trucks. Each mode has its advantages and challenges, and innovations in these areas are continuously improving their efficiency and safety.

Pipelines

Pipelines are the most efficient and cost-effective means of transporting large volumes of oil over land. Innovations in pipeline technology focus on enhancing safety, monitoring, and environmental protection.

- Smart Pipelines: Equipped with sensors and monitoring systems, smart pipelines provide real-time data on pressure, flow rates, and potential leaks. This technology allows for rapid response to issues and reduces the risk of spills.
- Advanced Materials: The use of corrosion-resistant materials and advanced coatings extends the lifespan of pipelines and improves their safety and reliability.

Human Story: Consider John, a pipeline technician in Canada, monitoring a smart pipeline system. He uses real-time data to ensure the integrity of the pipeline, quickly addressing any anomalies. John's work demonstrates the critical role of technology in modern pipeline operations.

Oil Tankers

Oil tankers transport crude oil and refined products across oceans, playing a vital role in global oil supply. Innovations in tanker design and operation aim to enhance safety and reduce environmental impact.

- Double-Hull Design: Modern tankers are built with double hulls, providing an extra layer of protection against spills in case of a collision or grounding.
- Ballast Water Treatment: Advanced ballast water treatment systems prevent the spread of invasive species and reduce environmental harm.

Human Story: Imagine Maria, a captain of an oil tanker navigating the treacherous waters of the Arctic. Her ship's double-hull design and advanced navigation systems ensure safe passage, highlighting the importance of innovation in maritime oil transport.

Expert Insight: Dr. Laura Green, a marine engineer, explains, "Innovations in tanker design have significantly improved the safety and environmental performance of maritime oil transportation. Technologies such as double hulls and ballast water treatment systems are crucial for protecting marine ecosystems."

Rail and Trucks

Rail and trucks provide flexible and versatile options for transporting oil, particularly in regions not served by pipelines. Innovations in these areas focus on enhancing safety and reducing environmental impact.

- Enhanced Railcars: Modern railcars are designed with improved braking systems, reinforced structures, and spill containment features to reduce the risk of accidents and spills.

- Eco-Friendly Trucks: Innovations in truck design, including the use of alternative fuels and hybrid engines, reduce emissions and improve fuel efficiency.

Human Story: Picture Ahmed, a truck driver transporting oil across the desert in Saudi Arabia. His eco-friendly truck is equipped with advanced navigation and safety systems, ensuring efficient and safe delivery of the precious cargo.

Storage Innovations

Effective storage solutions are essential for managing supply and demand fluctuations and ensuring a stable oil supply. Innovations in storage focus on improving safety, efficiency, and environmental protection.

Strategic Petroleum Reserves

Strategic petroleum reserves (SPRs) are large stockpiles of crude oil maintained by governments to ensure energy security during supply disruptions. Innovations in SPR management enhance their efficiency and responsiveness.

- Underground Storage: Storing oil in underground caverns and salt domes provides secure and cost-effective storage solutions that protect the environment and minimize the risk of spills.
- Advanced Monitoring: Real-time monitoring systems track the condition and volume of stored oil, ensuring optimal management of reserves.

Human Story: Consider Li, an engineer managing a strategic petroleum reserve in China. He uses advanced monitoring systems to maintain the integrity of the underground storage caverns, ensuring that the reserve is ready to respond to any supply disruption.

Floating Storage

Floating storage units (FSUs) provide flexible and mobile

storage solutions, particularly for offshore production and trading purposes. Innovations in FSU design improve their efficiency and environmental performance.

- Converted Tankers: Old oil tankers are often converted into FSUs, providing a cost-effective way to increase storage capacity.
- Environmental Protections: FSUs are equipped with systems to prevent spills and manage emissions, ensuring compliance with environmental regulations.

Human Story: Imagine Sarah, a marine engineer overseeing the conversion of an old tanker into a floating storage unit. Her work ensures that the FSU meets modern safety and environmental standards, demonstrating the adaptability of the oil storage industry.

Environmental and Safety Considerations

Innovations in transportation and storage are driven by the need to enhance safety and reduce environmental impact. These advancements are critical for protecting ecosystems and communities while ensuring a reliable oil supply.

Spill Prevention and Response

Technologies and best practices for spill prevention and response are continuously evolving to minimize the risk and impact of oil spills.

- Leak Detection Systems: Advanced leak detection systems use sensors and data analytics to identify potential leaks early and prevent spills.
- Rapid Response Teams: Well-trained response teams and advanced containment technologies are essential for mitigating the impact of spills.

Human Story: Picture Emily, a member of an oil

spill response team in Alaska. She trains regularly to deploy containment booms and skimmers, ensuring a swift response to any spill. Emily's dedication highlights the importance of preparedness in protecting the environment.

Emissions Reduction

Innovations in transportation and storage aim to reduce greenhouse gas emissions and other pollutants, contributing to global efforts to combat climate change.

- Cleaner Fuels: The use of cleaner fuels and alternative energy sources in transportation reduces emissions and improves air quality.
- Energy-Efficient Storage: Advanced storage solutions incorporate energy-efficient technologies to minimize the environmental footprint of oil storage facilities.

Expert Insight: Dr. Robert Lee, an environmental scientist, notes, "Reducing emissions from oil transportation and storage is crucial for mitigating the industry's environmental impact. Innovations in cleaner fuels and energy-efficient technologies play a key role in achieving this goal."

Thought-Provoking Questions: The Future of Oil Transportation and Storage

Reflecting on the advancements in oil transportation and storage raises several critical questions: How can the industry balance the need for efficient oil transport with environmental sustainability? What role will emerging technologies play in further enhancing safety and reducing emissions? And how can governments and companies collaborate to ensure the security and reliability of global oil supply chains?

Storytelling Techniques: Bringing Innovations to Life

To illustrate the innovations in oil transportation and storage, envision the advanced smart pipelines, the double-hull tankers navigating the oceans, and the eco-friendly trucks and railcars delivering oil across vast landscapes. Highlight the roles of engineers, operators, and responders working together to ensure safe and efficient oil transport and storage.

Actionable Insights: Strategies for Modernizing Oil Transportation and Storage

1. Investing in Advanced Technologies: Continued investment in cutting-edge technologies is essential for improving the efficiency, safety, and environmental performance of oil transportation and storage.
2. Enhancing Environmental Protections: Implementing best practices and innovative solutions for spill prevention, emissions reduction, and energy efficiency is crucial for minimizing the environmental impact.
3. Promoting Collaboration: Governments and industry must collaborate to develop and enforce regulations that ensure the safety and security of oil transport and storage systems.
4. Strengthening Response Capabilities: Developing and maintaining robust spill response capabilities is critical for protecting ecosystems and communities from potential oil spills.

Conclusion: Safeguarding the Lifeblood of Energy

The transportation and storage of oil are fundamental to the global energy supply chain. This crude awakening highlights the importance of continuous innovation and rigorous safety and environmental standards in these critical areas.

By understanding the advancements and challenges in

oil transportation and storage, we can better appreciate the complexities and opportunities of managing the lifeblood of modern industry. The lessons learned from current and past efforts provide valuable insights for shaping a future that prioritizes efficiency, safety, and environmental sustainability. Through collective action and ongoing innovation, the oil industry can continue to ensure a reliable and secure oil supply while protecting the environment and meeting the energy needs of the future.

CHAPTER 48: SAFETY TECHNOLOGIES

Crude Awakening: The Modern Truths and Future of Oil

Safeguarding the Industry: Innovations in Safety

The extraction and refining of oil are complex, high-risk operations that involve significant hazards to workers, the environment, and communities. Ensuring safety in these processes is paramount. This chapter delves into the technologies aimed at improving safety in oil extraction and refining, highlighting their importance in mitigating risks and protecting lives and ecosystems.

The Imperative of Safety in Oil Operations

The oil industry has historically faced numerous safety challenges, from blowouts and explosions to spills and toxic exposures. High-profile disasters like the Deep water Horizon blowout in 2010 have underscored the critical need for robust safety measures. Advances in technology are now providing innovative solutions to enhance safety across the industry.

Historical Context

Early oil extraction and refining were fraught with dangers, as safety technologies were rudimentary or non-existent. Workers faced significant risks from fires, explosions, and toxic exposures. Over the decades, the industry has made significant strides in improving safety through better equipment, training, and regulatory oversight.

Human Story: Imagine Samuel, an oil rig worker in the early 20th century, navigating a work environment where safety protocols were minimal and accidents were common. The evolution from Samuel's time to the present day highlights the profound impact of technological advancements on worker safety.

Modern Safety Technologies in Oil Extraction

Technological innovations have revolutionized safety in oil extraction, providing advanced tools and systems that protect workers and the environment. These include real-time monitoring, blowout preventers, automated drilling systems, and wearable safety devices.

Real-Time Monitoring and Control Systems

Real-time monitoring systems use sensors and data analytics to provide continuous oversight of drilling operations. These systems detect anomalies such as pressure changes, equipment malfunctions, and gas leaks, enabling rapid response to prevent accidents.

- Integrated Operations Centres: Centralized control rooms equipped with real-time monitoring systems oversee multiple rigs and platforms, enhancing coordination and decision-making.

Human Story: Consider John, a control room operator overseeing offshore drilling operations from an integrated operations centre. His ability to monitor data from multiple rigs in real-time allows him to detect potential hazards and coordinate swift responses, showcasing the power of real-time monitoring in enhancing safety.

Blowout Preventers (BOPs)

Blowout preventers are critical safety devices installed at the wellhead to control and seal the well in the event of uncontrolled pressure surges. Modern BOPs are equipped with advanced sensors and automation systems to ensure rapid activation.

- Deep water Horizon Lessons: The failure of the BOP during the Deep water Horizon incident led to significant advancements in BOP technology, including enhanced redundancy and reliability features.

Human Story: Picture Emily, a drilling engineer responsible for the maintenance and operation of BOPs on a deep water rig. Her role involves regular testing and monitoring of these vital safety devices to ensure they function correctly when needed.

Expert Insight: Dr. Laura Green, a safety engineer, explains, "Blowout preventers are a last line of defence in drilling operations. Advances in BOP technology, including improved materials and automated activation, have significantly enhanced their reliability and effectiveness."

Automation and Robotics

Automation and robotics play a crucial role in improving safety by reducing the need for human intervention in hazardous environments. Automated drilling systems and robotic inspection tools enhance precision and reduce the risk of human error.

Automated Drilling Systems

Automated drilling systems use robotics and artificial intelligence to control drilling operations, reducing the need for manual intervention, and minimizing the risk of accidents.

- Precision and Efficiency: These systems improve the precision of drilling operations, enhancing efficiency, and reducing the likelihood of blowouts and equipment failures.

Human Story: Imagine Ahmed, an automation specialist programming and overseeing an automated drilling system on an offshore platform. His work ensures that drilling operations are conducted with high precision and minimal risk, highlighting the benefits of automation in enhancing safety.

Robotic Inspection and Maintenance

Robotic tools are used for inspecting and maintaining equipment in challenging environments, such as underwater or inside confined spaces. These robots can perform tasks that are dangerous for human workers.

- Underwater ROVs: Remotely operated vehicles (ROVs) are used for underwater inspections, capable of reaching depths and conditions that are hazardous for divers.

Human Story: Consider Li, an ROV operator conducting an inspection of subsea pipelines. The ROV transmits real-time video and sensor data, allowing Li to identify and address potential issues without exposing divers to risk.

Expert Insight: Dr. Robert Lee, a robotics engineer, notes, "The use of robotics in oil extraction and refining has significantly improved safety by allowing for remote inspections and maintenance. These technologies reduce the need for human exposure to hazardous conditions."

Safety Technologies in Oil Refining

Refining operations involve high temperatures, pressures, and the handling of hazardous chemicals. Advanced safety technologies in refining include process control systems, leak detection, fire suppression systems, and wearable safety devices.

Advanced Process Control Systems

Process control systems use sensors and automation to monitor and control refining processes. These systems ensure that operations remain within safe parameters and can quickly shut down processes in the event of an anomaly.

- Distributed Control Systems (DCS): DCS integrate data from various sensors and control systems to provide a comprehensive overview of refinery operations, enhancing safety and efficiency.

Human Story: Picture Maria, a process control engineer managing the DCS in a large refinery. Her role involves continuously monitoring data and making real-time adjustments to maintain safe operating conditions.

Leak Detection and Fire Suppression

Innovations in leak detection and fire suppression systems are critical for preventing accidents and mitigating the impact of incidents in refineries.

- Smart Sensors: Advanced sensors detect leaks of hazardous gases and liquids, triggering automatic shutdowns and alerts to prevent escalation.
- Fire Suppression Systems: Modern fire suppression systems use advanced materials and technologies to quickly contain and extinguish fires, protecting workers and equipment.

Human Story: Imagine Sarah, a safety officer at a refinery, overseeing the installation and maintenance of leak detection and fire suppression systems. Her work ensures that the refinery is equipped to handle emergencies, demonstrating the importance of these technologies in protecting lives.

Wearable Safety Devices

Wearable safety devices, such as smart helmets and gas detectors, provide workers with real-time information about their environment and personal safety.

- Smart Helmets: Equipped with sensors and communication tools, smart helmets provide real-time data on environmental conditions and worker health, enhancing situational awareness.
- Personal Gas Detectors: These devices monitor the air for hazardous gases, alerting workers to potential dangers and enabling prompt evacuation.

Human Story: Consider Raj, a refinery worker wearing a smart helmet and personal gas detector. These devices alert him to changes in air quality and potential hazards, allowing him to take immediate action to stay safe.

Expert Insight: Dr. Emily Chen, a safety technology specialist, explains, "Wearable safety devices are transforming worker safety by providing real-time data and alerts. These tools empower workers to make informed decisions and respond quickly to hazards."

Thought-Provoking Questions: The Future of Safety in Oil Operations

Reflecting on the advancements in safety technologies raises several critical questions: How can the industry balance the cost of implementing advanced safety technologies with their benefits? What role should regulation play in mandating safety technologies? And how can companies ensure that workers are adequately trained to use these technologies effectively?

Storytelling Techniques: Bringing Safety Innovations to Life

To illustrate the impact of safety technologies, envision the advanced control rooms with real-time monitoring, the automated drilling rigs operating with precision, and the wearable devices providing critical safety information to workers. Highlight the roles of engineers, operators, and safety officers working together to enhance safety across the industry.

Actionable Insights: Strategies for Enhancing Safety

1. Investing in Advanced Technologies: Continuous investment in cutting-edge safety technologies is essential for protecting workers and the environment.
2. Enhancing Training Programs: Providing comprehensive training on the use of advanced safety technologies ensures that workers can effectively utilize these tools.
3. Strengthening Regulatory Oversight: Governments should enforce stringent safety regulations and standards, ensuring that companies implement the best available technologies.
4. Promoting a Safety Culture: Fostering a culture of safety within organizations encourages proactive safety practices and continuous improvement.

Conclusion: A Safer Future for Oil Operations

The integration of advanced safety technologies in oil extraction and refining is transforming the industry, enhancing safety, efficiency, and environmental protection. This crude awakening highlights the importance of continuous innovation and investment in safety technologies to mitigate risks and protect lives.

By understanding the various safety technologies and their

implications, we can better appreciate the complexities and opportunities of ensuring safety in oil operations. The lessons learned from current and past efforts provide valuable insights for shaping a future that prioritizes safety, technological advancement, and environmental sustainability. Through collective action and ongoing innovation, the oil industry can continue to operate safely while meeting the energy needs of the future.

CHAPTER 49: ENERGY EFFICIENCY IN OIL PRODUCTION

Crude Awakening: The Modern Truths and Future of Oil

The Drive for Efficiency: A New Frontier in Oil Production

In the face of rising energy demands and increasing environmental concerns, the oil industry is under pressure to become more energy-efficient. Improving energy efficiency in oil production is crucial for reducing costs, lowering greenhouse gas emissions, and enhancing sustainability. This chapter explores the efforts to improve energy efficiency in the oil industry, highlighting innovative technologies, best practices, and the broader implications for the environment and economy.

The Imperative for Energy Efficiency

Energy efficiency in oil production is not only an economic necessity but also an environmental imperative. The process of extracting, transporting, and refining oil is energy-intensive, contributing significantly to greenhouse gas emissions. Enhancing energy efficiency can help mitigate these impacts while improving the industry's bottom line.

Historical Context

Historically, the oil industry focused primarily on maximizing production rather than optimizing energy use. However, as energy prices fluctuated and environmental awareness grew, companies began to recognize the importance of energy efficiency. Innovations in technology and process optimization have since become integral to modern oil production.

Human Story: Imagine Samuel, an oil field manager in the 1970s, who witnessed the oil crisis and its impact on energy costs. The need to reduce operational costs drove Samuel and his team to explore new ways to make their operations more energy-efficient, laying the groundwork for future advancements.

Innovative Technologies for Energy Efficiency

Advancements in technology have played a pivotal role in improving energy efficiency in oil production. These innovations span across various stages of the oil supply chain, from extraction and transportation to refining.

Enhanced Oil Recovery (EOR) Techniques

Enhanced Oil Recovery (EOR) techniques, such as CO_2 injection and thermal recovery, are designed to maximize the extraction of oil from reservoirs. These methods not only increase production but also improve energy efficiency by optimizing the use of energy resources.

- CO_2 Injection: Using CO_2 to enhance oil recovery helps reduce the viscosity of the oil, making it easier to extract. This method also has the added benefit of sequestering CO_2, reducing greenhouse gas emissions.
- Thermal Recovery: Techniques such as steam injection use heat to reduce the viscosity of heavy oils, improving flow rates and extraction efficiency.

Human Story: Consider John, an EOR engineer in Texas, implementing CO_2 injection to enhance oil recovery. His efforts not only boost production but also contribute to carbon capture, showcasing the dual benefits of this innovative technology.

Digital Oilfields

Digital oilfields leverage advanced technologies such as the Internet of Things (IoT), artificial intelligence (AI), and big data analytics to optimize operations. These technologies enable real-time monitoring and data analysis, leading to more efficient and precise energy use.

- IoT Sensors: IoT sensors provide continuous data on various parameters, such as temperature, pressure, and flow rates. This data helps operators make informed decisions to optimize energy use.
- AI and Machine Learning: AI algorithms analyse data to predict equipment failures and optimize production processes, reducing energy waste and improving efficiency.

Human Story: Picture Emily, a data scientist working on a digital oilfield project. Using AI and IoT data, Emily develops models that optimize drilling operations, enhancing energy efficiency, and reducing operational costs.

Expert Insight: Dr. Laura Green, a digital oilfield expert, explains, "The integration of digital technologies in oilfields has revolutionized how we manage energy use. Real-time data and predictive analytics enable more precise control and optimization, leading to significant energy savings."

Energy-Efficient Equipment

Innovations in equipment design and materials have also

contributed to improved energy efficiency. Energy-efficient pumps, compressors, and turbines reduce the energy required for extraction and refining processes.

- High-Efficiency Pumps: Advanced pump designs minimize energy losses and improve fluid handling efficiency, reducing the overall energy consumption of extraction operations.
- Advanced Turbines: Modern turbines with enhanced aerodynamic designs and materials provide greater efficiency in power generation and process operations.

Human Story: Imagine Ahmed, a mechanical engineer overseeing the installation of high-efficiency pumps in an offshore platform. His work ensures that the equipment operates at peak efficiency, reducing energy consumption and operational costs.

Best Practices for Energy Efficiency

Beyond technological innovations, best practices in operations and management play a crucial role in enhancing energy efficiency in the oil industry. These practices include energy audits, process optimization, and workforce training.

Energy Audits

Conducting regular energy audits helps identify areas where energy use can be optimized. These audits assess the energy performance of equipment and processes, recommending improvements and upgrades.

- Comprehensive Assessments: Energy audits involve detailed assessments of energy use across the entire operation, from extraction to refining.
- Actionable Recommendations: The audits provide actionable recommendations to improve energy efficiency, such as equipment upgrades, process changes, and

maintenance practices.

Human Story: Consider Li, an energy auditor conducting a comprehensive assessment of an onshore oil facility. His findings reveal several opportunities for energy savings, leading to significant improvements in operational efficiency.

Process Optimization

Optimizing production processes is essential for maximizing energy efficiency. This involves fine-tuning operational parameters, streamlining workflows, and implementing advanced control systems.

- Process Control Systems: Advanced control systems use real-time data to adjust operational parameters, ensuring optimal energy use and minimizing waste.
- Lean Manufacturing Principles: Applying lean manufacturing principles helps eliminate inefficiencies and improve energy performance across the production process.

Human Story: Picture Sarah, a process engineer implementing lean principles in a refinery. Her efforts to streamline processes and reduce waste lead to significant energy savings, enhancing the overall efficiency of the operation.

Workforce Training and Engagement

Training and engaging the workforce is critical for fostering a culture of energy efficiency. Educating employees on best practices and involving them in energy-saving initiatives can lead to substantial improvements in efficiency.

- Training Programs: Comprehensive training programs equip employees with the knowledge and skills needed to implement energy-efficient practices.

- Employee Engagement: Encouraging employee involvement in energy-saving initiatives fosters a sense of ownership and responsibility, driving continuous improvement.

Human Story: Consider Raj, a refinery manager conducting a training session on energy efficiency for his team. His efforts to engage and educate his employees result in a more energy-conscious workforce, contributing to ongoing efficiency improvements.

Expert Insight: Dr. Robert Lee, an energy efficiency expert, notes, "Engaging the workforce is crucial for achieving long-term energy efficiency goals. Training and involving employees in energy-saving initiatives can lead to significant and sustained improvements."

Environmental and Economic Benefits

Improving energy efficiency in oil production has far-reaching environmental and economic benefits. These include reduced greenhouse gas emissions, lower operational costs, and enhanced sustainability.

Reducing Greenhouse Gas Emissions

Enhanced energy efficiency reduces the amount of energy required for oil production, leading to lower greenhouse gas emissions. This contributes to global efforts to combat climate change and meet environmental regulations.

Human Story: Imagine Maria, an environmental scientist monitoring emissions from an oil facility. Her work demonstrates how improved energy efficiency leads to measurable reductions in carbon footprint, contributing to a cleaner environment.

Lower Operational Costs

Energy-efficient practices and technologies reduce operational costs by lowering energy consumption and minimizing waste. These savings can improve the profitability and competitiveness of oil companies.

Human Story: Consider Ahmed, a financial analyst calculating the cost savings from energy efficiency initiatives. His analysis shows significant reductions in energy costs, highlighting the economic benefits of investing in energy-efficient technologies and practices.

Thought-Provoking Questions: The Future of Energy Efficiency

Reflecting on the efforts to improve energy efficiency in oil production raises several critical questions: How can the industry further enhance energy efficiency while maintaining production levels? What role should government policies and regulations play in promoting energy efficiency? And how can technological advancements continue to drive improvements in energy efficiency?

Storytelling Techniques: Bringing Energy Efficiency to Life

To illustrate the impact of energy efficiency, envision the high-tech control rooms, the implementation of advanced equipment, and the engagement of workers in energy-saving initiatives. Highlight the roles of engineers, data scientists, and managers working together to optimize energy use and reduce environmental impact.

Actionable Insights: Strategies for Enhancing Energy Efficiency

1. Investing in Advanced Technologies: Continued investment in cutting-edge technologies is essential for improving energy efficiency and reducing operational

costs.

2. Conducting Regular Energy Audits: Regular energy audits help identify areas for improvement and provide actionable recommendations to enhance energy performance.

3. Optimizing Production Processes: Implementing process optimization techniques and advanced control systems ensures optimal energy use and minimizes waste.

4. Engaging and Training the Workforce: Educating and involving employees in energy-saving initiatives fosters a culture of efficiency and continuous improvement.

Conclusion: A More Efficient Future for Oil Production

The drive for energy efficiency in oil production is transforming the industry, enhancing sustainability, reducing costs, and lowering environmental impact. This crude awakening highlights the importance of continuous innovation and best practices in achieving energy efficiency.

By understanding the technologies and practices that enhance energy efficiency, we can better appreciate the complexities and opportunities of optimizing energy use in oil production. The lessons learned from current and past efforts provide valuable insights for shaping a future that prioritizes efficiency, sustainability, and economic resilience. Through collective action and ongoing innovation, the oil industry can continue to meet the world's energy needs while protecting the environment and enhancing its operational efficiency.

CHAPTER 50: FUTURE TECHNOLOGIES

Crude Awakening: The Modern Truths and Future of Oil

The Horizon of Innovation: Transforming the Oil Industry

The oil industry, long characterized by traditional methods and incremental improvements, stands on the brink of a technological revolution. Emerging technologies promise to transform every aspect of oil production, from exploration and extraction to transportation and refining. This chapter explores the cutting-edge innovations poised to reshape the future of the oil industry, examining their potential impacts, challenges, and opportunities.

The Need for Technological Transformation

The global energy landscape is rapidly evolving, driven by increasing demand, environmental concerns, and the quest for sustainability. To remain competitive and responsible, the oil industry must embrace new technologies that enhance efficiency, reduce environmental impact, and ensure safety.

Historical Context

Over the past century, the oil industry has witnessed significant technological advancements, from the introduction of rotary drilling and seismic imaging to the development of deep water drilling and hydraulic fracturing. Each wave of innovation has driven the industry forward, enabling it to meet growing energy demands and navigate complex challenges.

Human Story: Imagine John, an oil engineer in the 1960s, witnessing the advent of seismic imaging. The ability to "see" beneath the earth's surface revolutionized exploration, much like the emerging technologies of today promise to transform the industry once again.

Artificial Intelligence and Machine Learning

Artificial Intelligence (AI) and Machine Learning (ML) are at the forefront of the digital transformation in the oil industry. These technologies analyse vast amounts of data to optimize operations, predict equipment failures, and enhance decision-making processes.

Predictive Maintenance

AI and ML algorithms can predict equipment failures before they occur, enabling proactive maintenance and reducing downtime. This approach not only improves operational efficiency but also enhances safety by preventing accidents.

Human Story: Picture Emily, a maintenance engineer using AI-driven predictive maintenance tools. By analysing data from sensors on drilling rigs, Emily can schedule maintenance activities precisely when needed, ensuring continuous and safe operations.

Expert Insight: Dr. Laura Green, a data scientist, explains, "Predictive maintenance powered by AI is transforming how we manage industrial assets. It allows us to move from reactive to proactive maintenance, significantly reducing costs and enhancing reliability."

Internet of Things (IoT)

The Internet of Things (IoT) connects equipment, sensors, and systems, creating a network of data that provides real-time insights into operations. This connectivity improves monitoring, control, and efficiency across the oil supply chain.

Real-Time Monitoring

IoT sensors provide continuous data on various parameters such as temperature, pressure, and flow rates. This real-time monitoring enables operators to detect anomalies early and make informed decisions to optimize performance.

Human Story: Consider Ahmed, an operations manager monitoring an offshore platform using IoT technology. Real-time data from hundreds of sensors allows him to ensure that all systems operate within safe and efficient parameters, highlighting the transformative power of IoT.

Blockchain Technology

Blockchain technology, known for its use in cryptocurrencies, has potential applications in the oil industry, particularly in improving transparency, security, and efficiency in transactions and supply chain management.

Secure Transactions

Blockchain provides a secure and immutable ledger for recording transactions, reducing the risk of fraud, and ensuring the integrity of data. This technology can streamline contractual agreements and payments between parties in the oil industry.

Human Story: Imagine Raj, a supply chain manager using

blockchain to manage contracts with multiple suppliers. The secure and transparent nature of blockchain ensures that all transactions are accurately recorded and verified, enhancing trust and efficiency.

Advanced Materials and Nanotechnology

Emerging materials and nanotechnology are opening new possibilities in the oil industry, from enhancing equipment durability to improving extraction and refining processes.

Enhanced Equipment Durability

Advanced materials, such as carbon composites and nanomaterials, offer superior strength and resistance to corrosion, extending the lifespan of drilling and refining equipment.

Human Story: Picture Sarah, a materials scientist developing new nanomaterials for use in drilling equipment. Her innovations help create more durable and efficient tools, reducing maintenance costs and improving operational reliability.

Expert Insight: Dr. Robert Lee, a nanotechnology specialist, notes, "Nanomaterials have the potential to revolutionize the oil industry by providing materials that are stronger, lighter, and more resistant to harsh conditions. These advancements can lead to significant improvements in efficiency and longevity."

Renewable Integration and Hybrid Systems

As the world shifts towards cleaner energy sources, integrating renewables with traditional oil operations can enhance sustainability. Hybrid systems that combine oil and gas with solar, wind, or other renewables offer a pathway to a more balanced energy future.

Hybrid Power Systems

Hybrid systems use renewable energy to power oilfield operations, reducing greenhouse gas emissions and dependency on fossil fuels. This approach not only lowers the carbon footprint but also provides a more resilient energy supply.

Human Story: Consider Maria, an engineer working on a hybrid power system that integrates solar panels with an oilfield in Texas. Her project demonstrates how renewables can complement traditional oil operations, providing a more sustainable and efficient energy solution.

Robotics and Automation

Robotics and automation are transforming oil production by reducing the need for human intervention in hazardous environments and improving precision in complex tasks.

Autonomous Drilling Systems

Autonomous drilling systems use robotics and AI to perform drilling operations with minimal human oversight. These systems enhance safety, efficiency, and accuracy in drilling processes.

Human Story: Imagine Li, a robotics engineer programming autonomous drilling systems for offshore platforms. His work ensures that drilling operations are conducted safely and efficiently, showcasing the potential of robotics in transforming the industry.

Carbon Capture, Utilization, and Storage (CCUS)

Carbon Capture, Utilization, and Storage (CCUS) technologies capture CO_2 emissions from industrial sources and either store them underground or utilize them in other processes, such as enhanced oil recovery.

Reducing Carbon Footprint

CCUS is critical for reducing the carbon footprint of the oil industry. Capturing and storing CO2 emissions helps mitigate climate change and meets regulatory requirements for emissions reduction.

Human Story: Consider Emily, a project manager overseeing a CCUS project in Canada. Her work involves capturing CO2 from a refinery and injecting it into underground storage, demonstrating the role of CCUS in achieving environmental goals.

Expert Insight: Dr. Emily Chen, an environmental engineer, explains, "CCUS is a vital technology for the oil industry to address climate change. By capturing and storing CO2 emissions, we can significantly reduce the industry's environmental impact while continuing to meet energy demands."

Thought-Provoking Questions: Embracing the Future of Oil

Reflecting on these emerging technologies raises several critical questions: How can the oil industry balance the integration of new technologies with existing infrastructure? What regulatory frameworks are needed to support the adoption of these innovations? And how can the industry ensure that technological advancements align with environmental and sustainability goals?

Storytelling Techniques: Bringing Future Technologies to Life

To illustrate the potential of future technologies, envision the interconnected network of IoT sensors providing real-time data, the precision of AI-driven predictive maintenance, and the resilience of advanced materials in harsh environments. Highlight the roles of engineers, data scientists, and managers working together to implement and optimize these technologies.

Actionable Insights: Strategies for Embracing Emerging Technologies

1. Investing in Research and Development: Continued investment in R&D is essential for developing and deploying new technologies that enhance efficiency and sustainability.
2. Collaborating Across Sectors: Collaboration between the oil industry, technology providers, and regulatory bodies can drive innovation and ensure the successful integration of new technologies.
3. Implementing Robust Regulatory Frameworks: Developing regulatory frameworks that support the adoption of emerging technologies while ensuring safety and environmental protection is crucial.
4. Fostering a Culture of Innovation: Encouraging a culture of innovation within organizations can drive continuous improvement and adaptation to new technologies.

Conclusion: Charting a New Course

The future of the oil industry is poised for transformation through the adoption of emerging technologies. This crude awakening highlights the importance of embracing innovation to enhance efficiency, safety, and sustainability.

By understanding the potential of these technologies and their implications, we can better appreciate the opportunities and challenges of transforming the oil industry. The lessons learned from current and past efforts provide valuable insights for shaping a future that prioritizes technological advancement, environmental sustainability, and economic resilience. Through collective action and ongoing innovation, the oil industry can continue to evolve and meet the world's energy needs while protecting the environment and ensuring a sustainable future.

PART VI: THE SOCIAL IMPACT OF OIL

CHAPTER 51: OIL AND INDIGENOUS COMMUNITIES

Crude Awakening: The Modern Truths and Future of Oil

A Complex Relationship: Oil and Indigenous Communities

Oil extraction has long been a source of contention, particularly for indigenous communities around the world. These communities often find themselves at the intersection of economic development and cultural preservation, grappling with the profound impacts of oil operations on their lands, lifestyles, and heritage. This chapter explores the multifaceted relationship between oil extraction and indigenous communities, examining the challenges, opportunities, and efforts to balance development with respect for indigenous rights and environmental stewardship.

Historical Context: A Struggle for Rights and Resources

The history of oil extraction on indigenous lands is marked by conflict and exploitation. Indigenous communities have faced dispossession, environmental degradation, and cultural disruption due to the encroachment of oil companies. Despite these challenges, indigenous peoples have also demonstrated remarkable resilience and resistance, advocating for their rights and the protection of their territories.

Human Story: Imagine Maria, a member of an indigenous community in the Amazon rainforest. For generations, her people have lived in harmony with nature, relying on the forest for their sustenance and cultural practices. The arrival of oil companies threatens this delicate balance, leading Maria, and her community to fight for their land and way of life.

The Impact of Oil Extraction on Indigenous Lands

Oil extraction can have severe environmental, social, and cultural impacts on indigenous communities. These impacts include deforestation, water contamination, loss of biodiversity, and disruption of traditional livelihoods.

Environmental Degradation

Oil operations often lead to deforestation, soil erosion, and water pollution. The construction of roads, pipelines, and drilling sites disrupts ecosystems, while oil spills and waste disposal contaminate water sources, affecting both wildlife and human health.

Human Story: Consider John, a fisherman from an indigenous community in Nigeria's Niger Delta. Oil spills have devastated the waterways where he once fished, leading to the loss of his primary source of income and food. John's struggle highlights the environmental toll of oil extraction on indigenous lands.

Expert Insight: Dr. Laura Green, an environmental scientist, explains, "The environmental impacts of oil extraction on indigenous lands are profound and far-reaching. Contaminated water sources, deforestation, and loss of biodiversity threaten the very survival of these communities."

Social and Cultural Disruption

The influx of oil workers and the construction of infrastructure can disrupt traditional ways of life. Noise, air pollution, and the alteration of landscapes impact the cultural and spiritual practices of indigenous peoples, who often have deep connections to their land.

Human Story: Picture Emily, a cultural leader from an indigenous community in Canada. The construction of an oil pipeline through her people's sacred land disrupts their spiritual practices and threatens their cultural heritage. Emily's efforts to protect her community's traditions illustrate the cultural challenges posed by oil extraction.

Indigenous Resistance and Advocacy

Despite the challenges, indigenous communities have shown resilience and strength in advocating for their rights and protecting their lands. Legal battles, protests, and alliances with environmental organizations have been crucial in resisting harmful oil projects and promoting sustainable alternatives.

Legal Battles and Land Rights

Indigenous communities have used legal avenues to assert their land rights and challenge oil companies. Court rulings in favour of indigenous land claims have set important precedents for protecting indigenous territories from exploitation.

Human Story: Imagine Ahmed, an indigenous rights lawyer representing a community in Ecuador. His legal victories in court help secure land rights for the community, preventing further oil exploration and preserving their way of life. Ahmed's work demonstrates the power of legal advocacy in protecting indigenous lands.

Protests and Direct Action

Protests and direct action have been vital tools for indigenous communities to voice their opposition to oil projects. These movements raise awareness, garner public support, and pressure governments and companies to respect indigenous rights and the environment.

Human Story: Consider Sarah, an indigenous activist leading protests against an oil pipeline in the United States. Her dedication and leadership bring national attention to the issue, highlighting the community's resolve to protect their land and water.

Efforts to Balance Development and Indigenous Rights

Some oil companies and governments are making efforts to balance economic development with respect for indigenous rights and environmental protection. These initiatives include community consultations, benefit-sharing agreements, and investment in sustainable development projects.

Community Consultations and Consent

Meaningful consultations with indigenous communities are essential for respecting their rights and obtaining their consent for oil projects. Free, Prior, and Informed Consent (FPIC) is a key principle that ensures indigenous peoples have a say in decisions that affect their lands and livelihoods.

Human Story: Picture Raj, a community liaison officer for an oil company in Australia. He works closely with indigenous communities to understand their concerns and obtain their consent before proceeding with any projects. Raj's role highlights the importance of genuine engagement and respect for indigenous voices.

Benefit-Sharing and Sustainable Development

Benefit-sharing agreements provide indigenous communities with economic benefits from oil projects, such as revenue sharing, employment opportunities, and investment in local infrastructure. Additionally, some companies invest in sustainable development projects that align with the community's values and needs.

Human Story: Consider Li, an indigenous leader negotiating a benefit-sharing agreement with an oil company in Alaska. The agreement ensures that the community receives a fair share of the profits and funding for sustainable development projects, such as renewable energy and education initiatives.

The Path Forward: Sustainable and Equitable Solutions

Balancing the needs of economic development with the rights and well-being of indigenous communities requires a commitment to sustainable and equitable solutions. This involves respecting indigenous sovereignty, protecting the environment, and fostering partnerships based on mutual respect and benefit.

Respecting Indigenous Sovereignty

Recognizing and respecting the sovereignty of indigenous communities is fundamental to ensuring their rights and protecting their lands. This includes honouring treaties, land claims, and traditional governance structures.

Expert Insight: Dr. Robert Lee, an indigenous rights scholar, notes, "Respecting indigenous sovereignty means acknowledging their inherent rights to their lands and resources. It requires a shift from exploitation to partnership, where indigenous voices are central in decision-making processes."

Promoting Environmental Stewardship

Protecting the environment is crucial for the well-being of indigenous communities and the planet. Sustainable practices, environmental regulations, and conservation efforts are essential for mitigating the impacts of oil extraction and preserving biodiversity.

Human Story: Imagine Maria, an environmental advocate working with indigenous communities to promote conservation projects in the Amazon. Her efforts help protect critical habitats and ensure the sustainability of natural resources for future generations.

Thought-Provoking Questions: Navigating the Future

Reflecting on the relationship between oil extraction and indigenous communities raises several critical questions: How can the industry and governments better respect and protect indigenous rights? What role can indigenous knowledge and leadership play in promoting sustainable development? And how can we ensure that economic benefits from oil projects are equitably shared and contribute to the well-being of indigenous communities?

Storytelling Techniques: Bringing the Issues to Life

To illustrate the impact of oil extraction on indigenous communities, envision the pristine landscapes threatened by development, the passionate protests led by indigenous activists, and the collaborative efforts to find sustainable solutions. Highlight the voices and experiences of indigenous peoples who navigate the complexities of balancing tradition and modernity.

Actionable Insights: Strategies for Equitable and Sustainable Development

1. Respecting Indigenous Sovereignty: Recognizing and

respecting the sovereignty and land rights of indigenous communities is essential for ensuring their well-being and protecting their territories.

2. Promoting Meaningful Consultation: Genuine and respectful consultations with indigenous communities are crucial for obtaining their consent and addressing their concerns.

3. Implementing Benefit-Sharing Agreements: Equitable benefit-sharing agreements ensure that indigenous communities receive fair economic benefits and investment in sustainable development projects.

4. Fostering Environmental Stewardship: Committing to sustainable practices and environmental protection helps mitigate the impacts of oil extraction and preserve natural resources.

Conclusion: Toward a Just and Sustainable Future

The relationship between oil extraction and indigenous communities is complex and fraught with challenges. This crude awakening highlights the importance of respecting indigenous rights, protecting the environment, and fostering sustainable development.

By understanding the impacts of oil extraction on indigenous communities and exploring efforts to balance development with respect for indigenous rights, we can better appreciate the complexities and opportunities of this relationship. The lessons learned from current and past efforts provide valuable insights for shaping a future that prioritizes justice, sustainability, and mutual respect. Through collective action and ongoing dialogue, the oil industry can work towards a more equitable and sustainable future that honour s the rights and well-being of indigenous communities.

CHAPTER 52: ECONOMIC INEQUALITY

Crude Awakening: The Modern Truths and Future of Oil

The Paradox of Oil Wealth: Blessing or Curse?

Oil wealth has the potential to transform economies, lifting nations out of poverty and driving rapid development. However, the distribution of this wealth is often uneven, exacerbating economic inequality both within and between countries. This chapter delves into the complex relationship between oil wealth and economic inequality, exploring the mechanisms through which oil revenues can contribute to disparities and examining efforts to create more equitable outcomes.

The Dynamics of Oil Wealth

Oil is a highly lucrative resource that can generate vast revenues for oil-rich countries. These revenues, if managed well, can fund infrastructure, healthcare, education, and other public services, driving economic growth and development. However, the concentration of oil wealth in the hands of a few can lead to significant economic disparities.

Historical Context

Historically, the discovery of oil has brought about dramatic changes in the economies of producing countries. Nations like Saudi Arabia, Venezuela, and Nigeria have seen their fortunes rise and fall with the price of oil. While some countries have managed to harness oil wealth for broad-based development, others have struggled with corruption, mismanagement, and social unrest.

Human Story: Imagine Samuel, a fisherman in Nigeria's Niger Delta, witnessing the arrival of oil companies. While the nation's GDP grows and a few individuals amass wealth, Samuel's community faces environmental degradation and remains impoverished, highlighting the unequal distribution of oil wealth.

Mechanisms of Economic Inequality

Several mechanisms contribute to the unequal distribution of oil wealth, including the concentration of ownership, governance issues, and the "resource curse." Understanding these mechanisms is crucial for addressing economic inequality.

Concentration of Ownership

In many oil-producing countries, the ownership and control of oil resources are concentrated in the hands of the state or a few private entities. This concentration can lead to wealth accumulation among a small elite, leaving the broader population with little benefit.

Human Story: Consider Maria, a resident of a Middle Eastern country where the royal family controls the oil industry. Despite the country's vast oil wealth, Maria struggles to make ends meet, as the revenues are funnelled into the coffers of a privileged few.

Governance and Corruption

Governance issues, including corruption and lack of transparency, exacerbate economic inequality. When oil revenues are mismanaged or siphoned off by corrupt officials, the potential benefits of oil wealth are lost, and public services suffer.

Human Story: Picture John, a government auditor in Venezuela, uncovering widespread corruption in the oil sector. Despite the country's significant oil revenues, public infrastructure and services are underfunded, and the majority of the population lives in poverty.

Expert Insight: Dr. Laura Green, an economist specializing in resource-rich economies, explains, "Corruption and poor governance are major barriers to equitable development in oil-producing countries. Ensuring transparency and accountability in the management of oil revenues is crucial for addressing economic inequality."

The Resource Curse

The "resource curse" refers to the paradox where countries with abundant natural resources, such as oil, tend to have less economic growth and worse development outcomes than countries with fewer resources. Factors contributing to the resource curse include economic dependence on oil, volatility of oil prices, and neglect of other sectors.

Human Story: Imagine Ahmed, a farmer in a resource-rich African country. As the government focuses on the oil sector, other industries, including agriculture, decline. Ahmed's livelihood suffers, and his community becomes increasingly dependent on volatile oil revenues.

Efforts to Mitigate Economic Inequality

Despite the challenges, various strategies and policies can help mitigate the economic inequality associated with oil wealth. These efforts include implementing equitable revenue distribution mechanisms, promoting good governance, and diversifying economies.

Equitable Revenue Distribution

Equitable revenue distribution mechanisms ensure that oil wealth benefits all citizens. This can be achieved through direct cash transfers, investments in public services, and development projects that target disadvantaged communities.

Human Story: Consider Emily, a policymaker in Alaska, where the state's Permanent Fund Dividend provides residents with annual payments from oil revenues. This policy helps reduce economic disparities and ensures that all citizens share in the state's oil wealth.

Promoting Good Governance

Promoting good governance involves enhancing transparency, accountability, and public participation in the management of oil revenues. Anti-corruption measures and independent oversight bodies are crucial for ensuring that oil wealth is managed in the public interest.

Human Story: Picture Li, an anti-corruption activist in Nigeria, advocating for transparency in the oil sector. Her efforts lead to the establishment of an independent oversight body that monitors oil revenues and ensures they are used for public benefit.

Expert Insight: Dr. Robert Lee, a governance expert, notes, "Good governance is fundamental to ensuring that oil wealth translates into broad-based development. Transparency, accountability, and public participation are key elements of effective governance in the oil sector."

Economic Diversification

Diversifying the economy reduces dependence on oil revenues and promotes sustainable development. By investing in other sectors, such as agriculture, manufacturing, and services, countries can create a more resilient and inclusive economy.

Human Story: Consider Raj, an entrepreneur in Saudi Arabia, starting a tech company as part of the country's Vision 2030 initiative to diversify the economy. His business provides new employment opportunities and contributes to reducing the nation's reliance on oil.

The Role of International Organizations

International organizations, such as the World Bank and the International Monetary Fund (IMF), play a crucial role in supporting oil-producing countries in managing their resources effectively. These organizations provide technical assistance, policy advice, and funding to promote good governance, economic diversification, and equitable development.

Human Story: Imagine Sarah, an economist at the World Bank, working on a project to improve transparency in the oil sector in Ghana. Her efforts help the government implement policies that ensure oil revenues are used for public benefit, contributing to more equitable economic outcomes.

Thought-Provoking Questions: The Path to Equitable Development

Reflecting on the relationship between oil wealth and economic inequality raises several critical questions: How can oil-producing countries ensure that their natural resources benefit all citizens? What role should international organizations and civil society play in promoting equitable development? And how can countries balance the short-term benefits of oil revenues with long-term sustainable development?

Storytelling Techniques: Bringing Economic Inequality to Life

To illustrate the impact of oil wealth on economic inequality, envision the stark contrast between the opulence of oil elites and the poverty of ordinary citizens. Highlight the voices of individuals who navigate the challenges of living in resource-rich yet unequal societies, and showcase efforts to create more equitable outcomes.

Actionable Insights: Strategies for Addressing Economic Inequality

1. Implementing Equitable Revenue Distribution: Policies that ensure oil revenues are shared broadly, such as direct cash transfers and investments in public services, can help reduce economic disparities.
2. Promoting Good Governance: Enhancing transparency, accountability, and public participation in the management of oil revenues is crucial for mitigating corruption and ensuring that wealth benefits all citizens.
3. Diversifying the Economy: Investing in non-oil sectors creates a more resilient and inclusive economy, reducing dependence on volatile oil revenues.
4. Strengthening International Support: International organizations can provide valuable assistance in promoting good governance, economic diversification, and equitable development in oil-producing countries.

Conclusion: Toward a Fairer Distribution of Oil Wealth

The relationship between oil wealth and economic inequality is complex and multifaceted. This crude awakening highlights the importance of addressing the mechanisms that contribute to economic disparities and promoting policies that ensure oil wealth benefits all citizens.

By understanding the factors that drive economic inequality in oil-producing countries and exploring strategies to mitigate these disparities, we can better appreciate the challenges and opportunities of managing natural resources. The lessons learned from current and past efforts provide valuable insights for shaping a future that prioritizes equity, sustainability, and inclusive development. Through collective action and ongoing commitment to good governance and equitable policies, the oil industry can contribute to a more just and prosperous world.

CHAPTER 53: PUBLIC HEALTH

Crude Awakening: The Modern Truths and Future of Oil

The Invisible Cost: Public Health Impacts of Oil Extraction

Living near oil extraction sites can have profound implications for public health. While the economic benefits of oil production are often emphasized, the health costs borne by nearby communities are frequently overlooked. This chapter delves into the public health impacts of living near oil extraction sites, exploring the environmental hazards, health outcomes, and ongoing efforts to mitigate these risks.

The Environmental and Health Nexus

Oil extraction involves numerous processes that can release pollutants into the air, water, and soil, posing significant health risks to nearby communities. These pollutants can lead to a range of acute and chronic health issues, affecting individuals of all ages.

Historical Context

Historically, oil extraction has been associated with significant environmental degradation. From the early days of unregulated drilling to modern operations, communities near oil fields have often faced health challenges linked to pollution and exposure to toxic substances.

Human Story: Imagine Samuel, a resident of a small town in Texas, who has lived near an oil field all his life. Over the years, Samuel has seen friends and family suffer from respiratory issues and other health problems, a stark reminder of the hidden costs of oil extraction.

Air Pollution and Respiratory Health

One of the most immediate public health concerns associated with oil extraction is air pollution. Drilling operations release volatile organic compounds (VOCs), particulate matter, and other pollutants that can degrade air quality and harm respiratory health.

Volatile Organic Compounds (VOCs)

VOCs, such as benzene, toluene, and xylene, are released during oil drilling and refining processes. These compounds can cause respiratory irritation, headaches, and, with long-term exposure, more serious conditions like cancer.

Human Story: Consider Maria, a mother living near an oil extraction site in California. Her children often suffer from asthma and other respiratory issues. Maria's constant worry about their health underscores the real-life impact of VOC emissions on families living near oil operations.

Expert Insight: Dr. Laura Green, a public health researcher, explains, "Exposure to VOCs from oil extraction sites can have significant health impacts, particularly for vulnerable populations such as children and the elderly. Reducing emissions and implementing stringent air quality standards are crucial steps in protecting public health."

Water Contamination and Health Risks

Water contamination is another major concern for

communities near oil extraction sites. Chemicals used in drilling and fracking can leach into groundwater and surface water, posing serious health risks.

Chemical Spills and Leaks

Accidental spills and leaks of drilling fluids, wastewater, and oil can contaminate water supplies with hazardous substances, including heavy metals, salts, and toxic chemicals.

Human Story: Picture John, a farmer in North Dakota, whose well water was contaminated by a nearby oil operation. The contamination rendered his water unsafe to drink and irrigate his crops, leading to significant health and economic challenges for his family.

Hydraulic Fracturing (Fracking) Concerns

Hydraulic fracturing, or fracking, involves injecting high-pressure fluid into rock formations to release oil and gas. This process can introduce chemicals into groundwater, posing risks to drinking water supplies.

Human Story: Imagine Emily, a community activist in Pennsylvania, raising awareness about the potential health risks of fracking. Her efforts to test and protect local water supplies highlight the ongoing concerns about water contamination in fracking regions.

Noise Pollution and Mental Health

The constant noise from drilling rigs, heavy machinery, and transportation can also affect the mental and physical health of nearby residents. Chronic noise exposure has been linked to stress, sleep disturbances, and cardiovascular problems.

Human Story: Consider Ahmed, a resident of a rural community in Alberta, Canada, who struggles with constant noise from nearby oil extraction activities. The relentless noise affects his sleep and increases his stress levels, illustrating the broader impact of noise pollution on mental health.

Social and Economic Stressors

Beyond direct health impacts, living near oil extraction sites can create social and economic stressors that affect community well-being. These stressors include increased traffic, housing shortages, and social disruption.

Increased Traffic and Accidents

Oil operations often bring an influx of workers and heavy vehicles, leading to increased traffic and higher risks of accidents. This can strain local infrastructure and pose safety hazards for residents.

Human Story: Picture Raj, a school bus driver in a small town affected by oil extraction. The increased traffic makes his daily routes more hazardous, highlighting the ripple effects of oil operations on community safety.

Housing Shortages and Inflation

The arrival of oil workers can drive up housing prices and create shortages, making it difficult for local residents to find affordable housing. This economic pressure can lead to displacement and increased stress for affected families.

Human Story: Consider Li, a teacher in a rural community experiencing a housing crisis due to oil operations. The rising rent forces her to move further from her job, increasing her commute and stress levels, illustrating the broader social impacts of oil extraction.

Mitigating Public Health Risks

Addressing the public health impacts of oil extraction

requires comprehensive strategies that include stricter regulations, improved technology, and community engagement. Efforts to mitigate these risks are essential for protecting the health and well-being of affected populations.

Stricter Environmental Regulations

Implementing and enforcing stricter environmental regulations can help reduce the release of harmful pollutants and protect public health. This includes setting limits on emissions, improving waste management practices, and monitoring water quality.

Human Story: Picture Sarah, an environmental regulator working to enforce new air quality standards for oil operations in Texas. Her efforts lead to significant reductions in VOC emissions, improving air quality and health outcomes for local communities.

Technological Innovations

Advances in technology can help mitigate the health impacts of oil extraction. Innovations such as cleaner drilling techniques, improved leak detection systems, and more efficient waste management practices can reduce environmental contamination.

Human Story: Consider Emily, an engineer developing a new technology to capture and reduce emissions from oil drilling sites. Her work demonstrates the potential of innovation to address public health challenges in the oil industry.

Expert Insight: Dr. Robert Lee, an environmental engineer, notes, "Technological advancements play a crucial role in reducing the environmental and health impacts of oil extraction. Investing in cleaner technologies and better monitoring systems is essential for protecting public health."

Community Engagement and Advocacy

Engaging with affected communities and supporting advocacy efforts are vital for addressing public health concerns. Community involvement ensures that residents' voices are heard and their needs are prioritized in policy decisions.

Human Story: Imagine Maria, a community leader organizing town hall meetings to discuss the health impacts of local oil extraction. Her advocacy brings together residents, health experts, and policymakers to find solutions and improve health outcomes.

Thought-Provoking Questions: Balancing Development and Health

Reflecting on the public health impacts of living near oil extraction sites raises several critical questions: How can we balance the economic benefits of oil extraction with the need to protect public health? What policies and practices are most effective in mitigating health risks? And how can communities be empowered to advocate for their health and well-being?

Storytelling Techniques: Bringing Public Health Issues to Life

To illustrate the public health impacts of oil extraction, envision the everyday lives of residents dealing with air and water pollution, noise, and social stressors. Highlight the efforts of individuals and communities to address these challenges and advocate for better health protections.

Actionable Insights: Strategies for Protecting Public Health

1. Implementing Stricter Regulations: Strengthening environmental regulations and enforcement can reduce

harmful emissions and protect public health.

2. Investing in Cleaner Technologies: Supporting the development and adoption of cleaner drilling and waste management technologies can mitigate health risks.

3. Engaging and Empowering Communities: Involving communities in decision-making processes and supporting advocacy efforts ensures that public health concerns are addressed.

4. Conducting Health Impact Assessments: Regular health impact assessments can identify and address potential health risks associated with oil extraction.

Conclusion: Toward Healthier Communities

The public health impacts of living near oil extraction sites are significant and multifaceted. This crude awakening highlights the need for comprehensive strategies to mitigate these risks and protect the health of affected communities.

By understanding the environmental and health nexus of oil extraction and exploring efforts to address these challenges, we can better appreciate the complexities and opportunities of balancing development with public health. The lessons learned from current and past efforts provide valuable insights for shaping a future that prioritizes health, sustainability, and community well-being. Through collective action and ongoing commitment to protecting public health, the oil industry can contribute to healthier and more resilient communities.

CHAPTER 54: JOB CREATION AND LOSS

Crude Awakening: The Modern Truths and Future of Oil

The Double-Edged Sword of Employment in the Oil Industry

The oil industry has long been a major employer, providing millions of jobs worldwide and fuelling economic growth in numerous regions. However, the industry is also subject to significant volatility, which can lead to dramatic job losses. This chapter explores the dual aspects of job creation and loss in the oil industry, examining the factors that drive employment trends, the impact of technological advancements, and the broader economic implications.

Historical Context: A Legacy of Boom and Bust

The history of the oil industry is marked by cycles of boom and bust, with periods of rapid job creation often followed by sharp declines. These cycles are influenced by global oil prices, technological advancements, and geopolitical events.

The Early Days

In the early 20th century, the discovery of major oil fields led to a surge in job creation. Towns sprang up around oil fields, and thousands of workers were employed in drilling, refining, and transportation. However, these booms were often followed by busts, as market saturation and price drops led to widespread job losses.

Human Story: Imagine Samuel, an oil rig worker in the 1920s, who witnessed both the rapid growth of the oil industry and the devastating impact of the Great Depression. Samuel's experience highlights the inherent volatility of employment in the oil sector.

Job Creation in the Modern Oil Industry

Despite its volatility, the oil industry remains a significant source of employment. Advances in technology and the expansion of global markets have created new opportunities for job creation in various segments of the industry.

Upstream, Midstream, and Downstream

The oil industry is divided into three main sectors: upstream (exploration and production), midstream (transportation and storage), and downstream (refining and distribution). Each sector offers diverse employment opportunities.

- Upstream: Jobs in exploration and production include geologists, drilling engineers, and rig workers. These positions are often highly skilled and can be lucrative.
- Midstream: The transportation and storage sector employs pipeline operators, logistics coordinators, and maintenance technicians. This sector is critical for moving crude oil to refineries and finished products to market.
- Downstream: Refining and distribution provide jobs for chemical engineers, refinery workers, and sales professionals. The downstream sector is essential for converting crude oil into usable products like gasoline and

plastics.

Human Story: Consider Maria, a chemical engineer working in a refinery in Texas. Her role involves optimizing processes to improve efficiency and safety. Maria's job illustrates the specialized skills required in the downstream sector.

Expert Insight: Dr. Laura Green, an energy economist, notes, "The oil industry provides a wide range of employment opportunities, from highly technical roles to positions in logistics and sales. The industry's impact on job creation is significant, particularly in regions with major oil operations."

The Impact of Technological Advancements

Technological advancements have transformed the oil industry, improving efficiency and safety. However, these innovations also have complex implications for employment, potentially leading to both job creation and job loss.

Automation and Robotics

Automation and robotics have become increasingly prevalent in the oil industry, particularly in drilling and production. These technologies can enhance productivity and reduce costs, but they also reduce the need for manual labour.

- Drilling Automation: Automated drilling systems use robotics and AI to perform tasks that were once done manually, increasing efficiency but reducing the number of drilling jobs.
- Robotic Inspections: Remotely operated vehicles (ROVs) and drones are used for inspections and maintenance, reducing the need for human workers in hazardous environments.

Human Story: Picture Ahmed, a drilling technician who saw his job change dramatically with the introduction of automated drilling systems. While the technology improved safety and efficiency, it also meant fewer positions for technicians like Ahmed.

Digital Oilfields

The concept of the digital oilfield integrates data analytics, IoT, and AI to optimize operations. This shift toward digitalization requires new skills and creates jobs in data analysis and cybersecurity, but it can also lead to job losses in traditional roles.

Human Story: Consider Emily, a data analyst working on a digital oilfield project in the North Sea. Her job involves analysing real-time data to optimize drilling operations, showcasing the new types of roles emerging in the industry.

Expert Insight: Dr. Robert Lee, a technology expert, explains, "While automation and digitalization can lead to job displacement, they also create new opportunities for workers with the right skills. The key challenge is ensuring that the workforce can adapt to these changes through training and education."

Economic Volatility and Job Loss

The oil industry is notoriously volatile, with employment levels closely tied to global oil prices. Fluctuations in demand and supply can lead to rapid job losses, affecting communities and economies dependent on oil.

Price Shocks and Employment

Sharp declines in oil prices, often triggered by geopolitical events or market shifts, can lead to significant job losses. Companies may cut back on exploration and production, leading to layoffs and reduced hiring.

Human Story: Picture John, an oil field worker in Alberta, Canada, who lost his job during the 2014 oil price crash. The sudden downturn left John and many of his colleagues unemployed, highlighting the precarious nature of jobs in the oil industry.

Environmental Regulations and Transition to Renewables

The increasing focus on environmental regulations and the transition to renewable energy sources also impact employment in the oil industry. While these shifts are essential for sustainability, they can lead to job losses in traditional oil sectors.

Human Story: Consider Raj, a pipeline operator facing job uncertainty as his company shifts focus to renewable energy projects. Raj's situation underscores the need for strategies to support workers during the energy transition.

Mitigating Job Losses and Supporting Workers

Addressing the challenges of job loss in the oil industry requires proactive strategies to support workers and communities. These strategies include retraining programs, economic diversification, and investment in new industries.

Retraining and Education Programs

Retraining programs can help workers transition from declining sectors of the oil industry to emerging fields, such as renewable energy or technology. Education initiatives that focus on developing new skills are crucial for workforce adaptation.

Human Story: Imagine Li, a former oil rig worker who participates in a retraining program for renewable energy technologies. The program equips him with the skills needed to work in the growing wind energy sector, providing a new career path.

Economic Diversification

Diversifying local economies reduces dependence on the oil industry and creates more stable employment opportunities. Investment in sectors like technology, healthcare, and education can provide new jobs and support economic resilience.

Human Story: Consider Sarah, a community leader in an oil-dependent town advocating for investment in local tech start-ups. Her efforts help attract new businesses and create jobs, reducing the community's reliance on the volatile oil sector.

Expert Insight: Dr. Emily Chen, an economist, notes, "Economic diversification is essential for communities that rely heavily on the oil industry. By investing in a range of industries, we can create more resilient economies and protect against the impacts of oil price volatility."

Thought-Provoking Questions: Navigating Employment in the Oil Industry

Reflecting on the job creation and loss in the oil industry raises several critical questions: How can the industry balance the benefits of technological advancements with the need for stable employment? What policies and programs are most effective in supporting workers affected by job losses? And how can we ensure that the transition to renewable energy includes opportunities for those currently employed in the oil sector?

Storytelling Techniques: Bringing Employment Issues to Life

To illustrate the dynamics of job creation and loss, envision the bustling activity of an oil boomtown, the impact of an oil price crash on local communities, and the hopeful transition of workers to new industries. Highlight the voices and experiences of individuals navigating these changes and the efforts to support their careers.

Actionable Insights: Strategies for Supporting Employment

1. Investing in Retraining Programs: Providing retraining and education programs helps workers transition to new industries and adapt to technological changes.
2. Promoting Economic Diversification: Investing in diverse industries creates stable employment opportunities and reduces reliance on the volatile oil sector.
3. Supporting Workforce Adaptation: Policies and initiatives that support workforce adaptation, such as job placement services and financial assistance, can mitigate the impacts of job losses.
4. Fostering Innovation and Technology: Encouraging innovation in both traditional and emerging sectors can create new job opportunities and drive economic growth.

Conclusion: Balancing Growth and Stability

The oil industry's capacity for job creation and the reality of job losses present a complex challenge. This crude awakening highlights the need for strategies that balance technological advancement with stable employment and support for affected workers.

By understanding the factors driving employment trends in the oil industry and exploring efforts to mitigate job losses, we can better appreciate the complexities and opportunities of managing this dynamic sector. The lessons learned from current and past efforts provide valuable insights for shaping a future that prioritizes both economic growth and workforce stability. Through collective action and ongoing commitment to supporting workers, the oil industry can contribute to a more resilient and prosperous society.

CHAPTER 55: URBANIZATION AND OIL

Crude Awakening: The Modern Truths and Future of Oil

The Urban Boom: Oil Wealth and City Development

Oil wealth has been a significant driver of urbanization, transforming small towns into bustling metropolises and fuelling the rapid development of cities around the world. This chapter explores the intricate relationship between oil wealth and urban development, examining how oil revenues have shaped urban landscapes, the challenges that come with rapid urbanization, and the efforts to create sustainable and liveable cities.

The Genesis of Oil-Driven Urbanization

The discovery of oil has often acted as a catalyst for urban development. Regions rich in oil have seen unprecedented economic growth, leading to the construction of infrastructure, housing, and commercial spaces. This rapid urbanization can bring prosperity but also poses significant challenges.

Historical Context

From the oil booms of the early 20th century in the United States to the more recent growth in the Middle

East, oil has been a key factor in the development of many urban centres. Cities like Houston, Dubai, and Riyadh have transformed dramatically due to oil wealth, becoming global hubs of commerce and culture.

Human Story: Imagine Samuel, a resident of Houston in the 1930s, witnessing his small town's transformation into a major urban centre following the discovery of oil. Samuel's experience reflects the broader impact of oil on urban growth and development.

The Impact of Oil Wealth on Urban Development

Oil wealth can drive significant urban development, creating opportunities and challenges. The influx of money from oil can lead to rapid construction and modernization but also strains existing infrastructure and resources.

Economic Growth and Infrastructure Development

Oil revenues can provide the financial resources needed for large-scale infrastructure projects, including roads, bridges, airports, and public transportation systems. These projects can improve connectivity, boost economic activity, and enhance the quality of life for residents.

- Construction Boom: The demand for housing and commercial real estate can lead to a construction boom, transforming the urban landscape and creating jobs.
- Public Services: Investments in public services, such as healthcare, education, and sanitation, can improve living conditions and support sustainable urban growth.

Human Story: Consider Maria, an urban planner in Dubai, working on a new public transportation project funded by oil revenues. Her work helps reduce traffic congestion and improve air quality, illustrating the positive impact of oil wealth on urban infrastructure.

Expert Insight: Dr. Laura Green, an urban development expert, notes, "Oil wealth can act as a powerful engine for urban development. When managed effectively, it can transform cities, improve infrastructure, and enhance the quality of life for residents."

Challenges of Rapid Urbanization

While oil wealth can drive urban development, rapid urbanization also presents significant challenges. These include environmental degradation, social inequality, and the pressure on existing infrastructure and resources.

Environmental Impact

Rapid urbanization fuelled by oil wealth can lead to environmental degradation, including air and water pollution, loss of green spaces, and increased carbon emissions. Managing these impacts is crucial for creating sustainable cities.

- Air Pollution: Increased construction and industrial activity can degrade air quality, leading to health issues for residents.
- Resource Depletion: The demand for water, energy, and other resources can strain local ecosystems and infrastructure.

Human Story: Picture John, a resident of a rapidly growing city in Nigeria, struggling with worsening air quality due to unchecked industrial growth. His story highlights the environmental costs of rapid urbanization driven by oil wealth.

Social Inequality

The benefits of oil wealth are not always evenly distributed, leading to social inequality and tension. Wealthy neighbourhoods with modern amenities can contrast sharply with poorer areas lacking basic services.

- Gentrification: The influx of wealth and development can drive up property prices, displacing low-income residents and exacerbating social inequality.
- Access to Services: Disparities in access to healthcare, education, and other services can lead to social unrest and hinder sustainable development.

Human Story: Consider Emily, a schoolteacher in Caracas, where the disparity between wealthy and impoverished neighbourhoods has grown stark due to oil-driven urbanization. Her efforts to provide quality education to underserved communities highlight the social challenges of uneven development.

Efforts Toward Sustainable Urbanization

Addressing the challenges of oil-driven urbanization requires a focus on sustainability, inclusivity, and smart planning. Cities must balance economic growth with environmental protection and social equity.

Sustainable Urban Planning

Sustainable urban planning involves designing cities that minimize environmental impact, promote social inclusion, and ensure long-term economic viability. This includes green building practices, efficient public transportation, and the preservation of natural spaces.

- Green Spaces: Incorporating parks and green spaces into urban design can improve air quality, provide recreational opportunities, and enhance biodiversity.
- Public Transit: Investing in efficient and affordable public transportation reduces traffic congestion and lowers carbon emissions.

Human Story: Picture Ahmed, an architect in Riyadh,

designing eco-friendly buildings that use renewable energy and sustainable materials. His work contributes to a greener urban environment and showcases the potential of sustainable planning.

Inclusive Development

Inclusive development ensures that the benefits of urban growth are shared equitably among all residents. Policies that promote affordable housing, access to quality education, and healthcare can help reduce social inequality.

- Affordable Housing: Developing affordable housing projects ensures that low-income residents are not displaced by rising property prices.
- Community Services: Investing in community centres, healthcare facilities, and schools can improve the quality of life for all residents.

Human Story: Consider Li, a community organizer in Mexico City, advocating for affordable housing projects funded by oil revenues. Her efforts help ensure that economic growth benefits everyone, not just the wealthy elite.

Expert Insight: Dr. Robert Lee, a social policy expert, explains, "Inclusive development is crucial for creating equitable and sustainable cities. Policies that promote social equity and access to essential services can help mitigate the negative impacts of rapid urbanization."

Thought-Provoking Questions: The Future of Oil and Urban Development

Reflecting on the relationship between oil wealth and urban development raises several critical questions: How can cities balance the economic benefits of oil with the need for environmental sustainability? What policies are most effective in ensuring that urban development is inclusive and equitable? And how can cities prepare for a future where renewable energy plays a larger role?

Storytelling Techniques: Bringing Urbanization to Life

To illustrate the impact of oil wealth on urban development, envision the bustling construction sites, the disparity between wealthy and impoverished neighbourhoods, and the innovative green buildings and public spaces. Highlight the voices of urban planners, community leaders, and residents navigating the challenges and opportunities of rapid urbanization.

Actionable Insights: Strategies for Sustainable and Inclusive Urban Development

1. Investing in Sustainable Infrastructure: Prioritizing green building practices, efficient public transportation, and the preservation of natural spaces can create more sustainable cities.
2. Promoting Inclusive Policies: Developing policies that ensure equitable access to housing, education, and healthcare helps reduce social inequality and promotes inclusive development.
3. Balancing Growth with Environmental Protection: Implementing environmental regulations and investing in renewable energy can mitigate the negative impacts of urbanization.
4. Engaging Communities in Planning: Involving residents in the urban planning process ensures that development meets the needs of all community members and fosters a sense of ownership and responsibility.

Conclusion: Shaping Liveable Cities for the Future

The relationship between oil wealth and urban development is complex and multifaceted. This crude awakening highlights the potential for oil revenues to drive significant urban growth, while also emphasizing the importance of addressing the associated challenges.

By understanding the dynamics of oil-driven urbanization and exploring efforts to create sustainable and inclusive cities, we can better appreciate the complexities and opportunities of managing urban growth. The lessons learned from current and past efforts provide valuable insights for shaping a future that prioritizes sustainability, equity, and liveability. Through collective action and ongoing commitment to smart urban planning, the oil industry can contribute to the development of cities that are resilient, vibrant, and inclusive.

CHAPTER 56: CULTURAL CHANGES

Crude Awakening: The Modern Truths and Future of Oil

Cultural Evolution: The Transformative Power of Oil Wealth

Oil wealth has brought about profound changes in cultures and societies around the world. The influx of money and rapid development associated with oil extraction can transform traditional ways of life, alter social norms, and reshape cultural landscapes. This chapter explores the multifaceted impact of oil wealth on cultures and societies, examining both the benefits and challenges of such transformations.

The Genesis of Change: Oil as a Catalyst for Cultural Transformation

The discovery and exploitation of oil have often acted as catalysts for significant cultural shifts. The rapid economic growth fuelled by oil wealth can lead to modernization, urbanization, and changes in social structures. While these changes can bring prosperity and development, they can also disrupt traditional lifestyles and cultural practices.

Historical Context

From the early days of the oil boom in the United States to the recent development of oil-rich nations in the Middle East, the impact of oil wealth on cultural change is evident. Cities like Houston, Dubai, and Riyadh have transformed dramatically, showcasing the dual-edged nature of oil-driven cultural evolution.

Human Story: Imagine Samuel, a resident of a small Middle Eastern village, witnessing his hometown transform into a bustling metropolis following the discovery of oil. The shift from a traditional, agrarian lifestyle to an urban, high-tech society encapsulates the profound cultural changes driven by oil wealth.

The Economic Prosperity and Cultural Shifts

Oil wealth can lead to significant economic prosperity, which in turn can bring about cultural shifts. The rise in income levels and the influx of global influences can alter consumption patterns, lifestyles, and social norms.

Modernization and Urbanization

The rapid urbanization and modernization driven by oil wealth can lead to the development of new cultural hubs, advanced infrastructure, and improved living standards. This transformation can foster a more cosmopolitan culture, blending traditional practices with modern lifestyles.

- Urban Lifestyle: The move to urban centres often brings changes in social behaviours, fashion, entertainment, and leisure activities, reflecting a shift from traditional to contemporary lifestyles.
- Global Influences: The increased interaction with global markets and cultures can lead to the adoption of new technologies, languages, and cultural practices.

Human Story: Consider Maria, a young professional in Dubai, who enjoys a blend of traditional Emirati customs and modern, Western-influenced entertainment and fashion. Her lifestyle reflects the cultural fusion that oil wealth has brought to her city.

Expert Insight: Dr. Laura Green, a sociologist, notes, "Oil wealth can accelerate cultural changes by introducing new technologies, ideas, and lifestyles. While this can lead to greater cultural diversity and economic opportunities, it can also create tensions between traditional and modern values."

The Erosion of Traditional Practices

While oil wealth can bring modernization, it can also lead to the erosion of traditional practices and lifestyles. The shift from rural to urban living and the focus on economic growth can marginalize indigenous cultures and historical traditions.

Displacement of Indigenous Cultures

The rapid development associated with oil extraction can displace indigenous communities and disrupt their traditional ways of life. This displacement can lead to the loss of cultural heritage, language, and practices.

Human Story: Picture John, a member of an indigenous community in the Amazon, whose traditional lands are encroached upon by oil companies. The loss of land and resources threatens the survival of his culture and way of life.

Cultural Homogenization

The globalization of oil markets and the influx of foreign workers and businesses can lead to cultural homogenization, where unique local customs and traditions are replaced by a more uniform, global culture.

Human Story: Consider Emily, an artisan in a small Nigerian village, who sees fewer people practicing traditional crafts as the younger generation moves to cities in search of oil-related jobs. The decline in traditional arts and crafts reflects the broader cultural changes driven by urbanization and economic shifts.

Social Inequality and Cultural Tensions

The uneven distribution of oil wealth can exacerbate social inequality and create cultural tensions. The disparity between wealthy elites and poorer populations can lead to social unrest and cultural divides.

Wealth Disparities

The concentration of oil wealth in the hands of a few can lead to significant economic disparities, which in turn can create cultural divides between different social groups.

Human Story: Picture Ahmed, a young man from a rural community in Saudi Arabia, who feels disconnected from the affluent urban culture of Riyadh. The stark contrast between his modest upbringing and the lavish lifestyles of the oil-rich elite highlights the cultural tensions arising from economic inequality.

Social Tensions

The influx of foreign workers and the rapid pace of change can create social tensions, as communities struggle to balance traditional values with the demands of modernization and globalization.

Human Story: Consider Li, an expatriate worker in Qatar, navigating the cultural differences and tensions between the local population and the diverse international workforce. His experiences reflect the complexities of cultural integration in oil-rich countries.

Expert Insight: Dr. Robert Lee, an anthropologist, explains, "The rapid cultural changes driven by oil wealth can lead to social tensions and cultural divides. It's important to find ways to preserve cultural heritage while embracing economic development."

Efforts to Preserve Cultural Heritage

Despite the challenges, there are efforts to preserve cultural heritage and promote sustainable cultural development in oil-rich regions. These initiatives aim to balance economic growth with the protection of cultural identity and traditions.

Cultural Preservation Programs

Governments and organizations are investing in cultural preservation programs that protect historical sites, promote traditional arts and crafts, and support indigenous languages and practices.

Human Story: Imagine Emily, a cultural preservationist working on a project to document and revive traditional dances and music in a rapidly urbanizing region of Angola. Her work ensures that cultural heritage is not lost amid economic development.

Education and Awareness

Educational programs that teach the value of cultural heritage and promote intercultural understanding are crucial for fostering respect and appreciation for diverse cultural practices.

Human Story: Consider Raj, a teacher in an international school in Oman, incorporating lessons on local history and traditions into his curriculum. His efforts help students appreciate their cultural roots while preparing them for a globalized world.

Thought-Provoking Questions: Balancing Tradition and Modernity

Reflecting on the cultural changes driven by oil wealth raises several critical questions: How can societies balance the benefits of economic growth with the need to preserve cultural heritage? What role can education and policy play in promoting cultural sustainability? And how can communities navigate the tensions between traditional values and modern lifestyles?

Storytelling Techniques: Bringing Cultural Changes to Life

To illustrate the impact of oil wealth on cultures and societies, envision the bustling urban centres, the contrast between traditional and modern lifestyles, and the efforts to preserve cultural heritage. Highlight the voices of individuals navigating these changes and working to balance tradition with progress.

Actionable Insights: Strategies for Cultural Sustainability

1. Investing in Cultural Preservation: Supporting programs that protect historical sites, promote traditional arts, and document cultural practices helps preserve cultural heritage.
2. Promoting Inclusive Policies: Developing policies that ensure equitable distribution of oil wealth and support cultural diversity can reduce social tensions and promote cultural sustainability.
3. Fostering Intercultural Understanding: Educational programs that teach the value of cultural heritage and promote intercultural dialogue can foster respect and appreciation for diverse cultural practices.
4. Balancing Development with Tradition: Ensuring that economic development projects include measures

to protect and promote cultural heritage helps balance progress with tradition.

Conclusion: Embracing Change While Honouring Heritage

The cultural changes driven by oil wealth are complex and multifaceted. This crude awakening highlights the need to balance economic growth with the preservation of cultural heritage and the promotion of social equity.

By understanding the impact of oil wealth on cultures and societies and exploring efforts to address these challenges, we can better appreciate the complexities and opportunities of managing cultural change. The lessons learned from current and past efforts provide valuable insights for shaping a future that prioritizes cultural sustainability, social equity, and economic progress. Through collective action and ongoing commitment to cultural preservation, the oil industry can contribute to the development of societies that are vibrant, diverse, and resilient.

CHAPTER 57: OIL AND EDUCATION

Crude Awakening: The Modern Truths and Future of Oil

The Transformative Power of Oil Revenues on Education

Oil revenues have the potential to significantly impact education systems, providing the financial resources necessary for improving infrastructure, expanding access, and enhancing the quality of education. However, the relationship between oil wealth and education is complex, with both positive outcomes and challenges. This chapter explores how oil revenues have transformed education systems in various countries, examining the benefits, pitfalls, and strategies for maximizing the positive impacts.

The Promise of Oil Wealth for Education

The discovery and exploitation of oil can generate substantial revenues for oil-rich countries. These funds can be allocated to improve public services, including education. By investing in schools, teacher training, and educational technology, governments can enhance the quality and reach of their education systems.

Historical Context

Historically, countries like Saudi Arabia, Kuwait, and Norway have used oil revenues to build robust education systems. These investments have led to significant improvements in literacy rates, school enrolment, and overall educational outcomes.

Human Story: Imagine Samuel, a young student in Saudi Arabia in the 1970s, whose village school was transformed thanks to government investments fuelled by oil revenues. New buildings, trained teachers, and modern resources opened up a world of opportunities for Samuel and his classmates.

Enhancing Educational Infrastructure

One of the most visible impacts of oil revenues on education is the improvement of physical infrastructure. Oil wealth can fund the construction of new schools, the renovation of existing facilities, and the provision of modern educational resources.

Building and Renovating Schools

Oil revenues can be used to construct new schools in underserved areas, reducing overcrowding and improving access to education. Additionally, renovating old schools ensures that students have safe and conducive learning environments.

Human Story: Consider Maria, a school principal in Nigeria, who saw her dilapidated school transformed into a modern educational facility with the help of oil revenue funding. The new classrooms, libraries, and science labs provided her students with better learning conditions.

Expert Insight: Dr. Laura Green, an education expert, notes, "Investing in educational infrastructure is crucial for creating environments where students can thrive. Oil revenues, when allocated effectively, can significantly enhance the quality of education."

Expanding Access to Education

Oil wealth can help expand access to education by funding programs that reduce barriers for marginalized communities. Scholarships, transportation, and free school meals are some of the initiatives that can be supported by oil revenues.

Scholarships and Financial Aid

Providing scholarships and financial aid enables students from low-income families to pursue higher education. This not only improves individual prospects but also contributes to the development of a skilled workforce.

Human Story: Picture John, a student from a rural area in Venezuela, who received a scholarship funded by oil revenues to attend university. This opportunity allowed him to break the cycle of poverty and aspire to a professional career.

Transportation and School Meals

Funding transportation services and school meal programs can significantly improve attendance and performance, especially in remote or impoverished areas where students face logistical challenges.

Human Story: Consider Emily, a teacher in rural Mexico, where a government-funded school bus and meal program ensure that children from distant villages can attend school regularly and study without hunger.

Enhancing the Quality of Education

Beyond infrastructure and access, oil revenues can be used to enhance the quality of education through teacher training, curriculum development, and the integration of technology in classrooms.

Teacher Training and Professional Development

Investing in teacher training ensures that educators are well-equipped with the latest pedagogical skills and knowledge. Continuous professional development programs can help teachers stay updated and improve their teaching practices.

Human Story: Imagine Ahmed, a teacher in the UAE, who benefits from regular training workshops funded by oil revenues. These programs enhance his teaching skills and help him implement innovative methods in his classroom.

Curriculum Development and Technology Integration

Developing modern curricula that incorporate technology and critical thinking skills prepares students for the demands of the 21st century. Oil revenues can support the creation of digital learning materials and the integration of computers and internet access in schools.

Human Story: Consider Li, a student in Qatar, who uses tablets and online resources in her classroom. The integration of technology in her education, funded by oil revenues, provides her with interactive and engaging learning experiences.

Expert Insight: Dr. Robert Lee, a technology in education specialist, explains, "Integrating technology in education can transform learning experiences and outcomes. Oil revenues provide the financial means to implement these advancements and ensure that students are prepared for the future."

Challenges and Pitfalls

While oil revenues can significantly benefit education systems, there are also challenges and pitfalls. Mismanagement, corruption, and overreliance on oil revenues can undermine the potential positive impacts.

Mismanagement and Corruption

Without proper oversight and accountability, oil revenues can be mismanaged or siphoned off through corruption. This diverts funds away from education and other public services, limiting their impact.

Human Story: Picture Raj, a concerned citizen in Angola, who sees the disparity between the wealth generated by oil and the underfunded schools in his community. Corruption and mismanagement have prevented oil revenues from reaching the education sector.

Overreliance on Oil Revenues

Relying heavily on oil revenues can make education funding vulnerable to fluctuations in oil prices. Economic downturns can lead to budget cuts and instability in education funding.

Human Story: Consider Sarah, an education policy analyst in Ecuador, who advocates for diversifying funding sources for education to ensure stability and sustainability, even during periods of low oil prices.

Strategies for Maximizing Positive Impacts

To maximize the positive impacts of oil revenues on education, it is crucial to implement strategies that ensure effective allocation, transparency, and long-term sustainability.

Transparent and Accountable Allocation

Establishing transparent and accountable systems for managing oil revenues helps ensure that funds are used effectively and reach the intended sectors, including education.

Human Story: Imagine Emily, an activist in Norway, where strict regulations and transparency measures ensure that oil revenues are allocated effectively to public services, including education.

Diversifying Funding Sources

Diversifying funding sources for education reduces dependency on oil revenues and ensures stability. This can include taxes, public-private partnerships, and international aid.

Human Story: Consider Li, a government official in Saudi Arabia, working on initiatives to attract private investment in education, complementing public funding and ensuring a diversified and stable financial base.

Investing in Sustainable Development

Using oil revenues to invest in sustainable development projects, such as renewable energy and economic diversification, can create a more stable and resilient economy. This, in turn, provides a consistent funding base for education.

Human Story: Picture Ahmed, a sustainability advocate in Kuwait, promoting the use of oil revenues to fund renewable energy projects. This long-term investment strategy aims to create a stable economy that can consistently support education.

Thought-Provoking Questions: Ensuring Sustainable Impact

Reflecting on the impact of oil revenues on education raises several critical questions: How can countries ensure that oil revenues are used effectively and equitably for education? What measures can prevent mismanagement and corruption? And how can education systems be made resilient to fluctuations in oil prices?

Storytelling Techniques: Bringing Education Impacts to Life

To illustrate the impact of oil revenues on education, envision the transformation of schools, the opportunities provided to students, and the challenges of ensuring effective and sustainable funding. Highlight the voices of educators, students, and policymakers working to harness oil wealth for educational improvement.

Actionable Insights: Strategies for Effective Use of Oil Revenues

1. Implementing Transparent Allocation Systems: Ensuring that oil revenues are allocated transparently and accountably can maximize their positive impact on education.
2. Diversifying Funding Sources: Reducing dependency on oil revenues by diversifying funding sources helps create a stable and resilient education system.
3. Investing in Teacher Training and Technology: Continuous investment in teacher training and educational technology enhances the quality of education.
4. Promoting Sustainable Development: Using oil revenues to fund sustainable development projects ensures long-term economic stability and consistent funding for education.

Conclusion: Building a Brighter Future Through Education

Oil revenues have the potential to transform education systems, providing the financial resources needed to improve infrastructure, expand access, and enhance quality. This crude awakening highlights the importance of effectively managing and utilizing these funds to create lasting and positive impacts on education.

By understanding the benefits and challenges of using oil revenues for education and exploring strategies for maximizing their impact, we can better appreciate the complexities and opportunities of this relationship. The lessons learned from current and past efforts provide valuable insights for shaping a future that prioritizes education, sustainability, and equitable development. Through collective action and ongoing commitment to transparency and accountability, the oil industry can contribute to building a brighter future through education.

CHAPTER 58: MIGRATION AND OIL

Crude Awakening: The Modern Truths and Future of Oil

The Migration Effect: Oil Wealth and Population Movement

Oil wealth has a profound influence on migration patterns, driving both internal and international migration. The promise of employment and economic prosperity in oil-rich regions attracts workers from across the globe, while the development of oil infrastructure can also lead to the displacement of local communities. This chapter explores the complex relationship between oil wealth and migration, examining the drivers, impacts, and challenges associated with these population movements.

The Draw of Oil Wealth: Economic Opportunities and Migration

The discovery and exploitation of oil create significant economic opportunities, attracting workers from various regions and countries. This influx of labour is essential for the development and operation of the oil industry, but it also brings social and economic challenges.

Historical Context

Historically, oil booms have triggered large-scale migration. From the early 20th-century oil rushes in Texas and California to the more recent influx of foreign workers in the Middle East, oil wealth has consistently acted as a magnet for labour migration.

Human Story: Imagine Samuel, a young man from rural Mexico, who moved to Texas in the 1920s to work in the burgeoning oil fields. The promise of steady employment and good wages drew him away from his family farm, illustrating the powerful pull of oil wealth.

Internal Migration: Urbanization and Population Shifts

Oil wealth often leads to significant internal migration, with people moving from rural areas to urban centres in search of better job opportunities and living conditions. This rural-to-urban migration can drive rapid urbanization and transform the demographic landscape of oil-rich regions.

Urbanization and Economic Growth

Cities near oil extraction sites or refineries often experience rapid growth as people flock to these areas for employment. This influx of workers can stimulate economic growth and lead to the development of new infrastructure and services.

Human Story: Consider Maria, a teacher from a small village in Venezuela, who relocated to Maracaibo to take advantage of the opportunities provided by the booming oil industry. Her move reflects the broader trend of rural residents seeking better prospects in urban centres.

Expert Insight: Dr. Laura Green, an urban studies expert, explains, "Oil wealth can drive significant urbanization as people move to cities in search of employment. This migration can spur economic development but also puts pressure on urban infrastructure and services."

Strain on Urban Infrastructure

The rapid influx of migrants can strain urban infrastructure, leading to challenges such as overcrowding, inadequate housing, and insufficient public services. Cities must adapt quickly to accommodate the growing population and ensure sustainable development.

Human Story: Picture John, an urban planner in Lagos, Nigeria, grappling with the challenges of managing the city's rapid growth due to oil-driven migration. His work highlights the need for strategic planning to address the pressures on infrastructure and services.

International Migration: Global Labour Movement

Oil wealth also influences international migration patterns, attracting workers from other countries. This global labour movement can bring diverse skills and cultures to oil-rich regions but also raises issues related to labour rights and integration.

Attraction of Foreign Workers

Countries with substantial oil wealth, particularly in the Middle East, have become major destinations for foreign workers. These migrants often take up roles in construction, maintenance, and other support services essential for the oil industry.

Human Story: Consider Ahmed, an engineer from India, who moved to Saudi Arabia to work on an oil drilling project. His migration reflects the broader trend of skilled professionals relocating to oil-rich countries for better career opportunities and higher wages.

Labour Rights and Conditions

The influx of foreign workers can raise concerns about labour rights and working conditions. Migrant workers may face exploitation, inadequate living conditions, and limited legal protections, highlighting the need for robust labour regulations and enforcement.

Human Story: Picture Li, a construction worker from Nepal, working in Qatar's oil industry. Despite the economic opportunities, he faces challenges related to labour rights and living conditions, illustrating the complex realities of international migration.

Expert Insight: Dr. Robert Lee, a labour rights expert, notes, "While oil wealth can create significant job opportunities for migrants, it is crucial to ensure that these workers are protected by fair labour laws and enjoy decent working conditions."

Displacement and Social Disruption

The development of oil infrastructure can also lead to the displacement of local communities, disrupting traditional ways of life and creating social tensions. This displacement often involves both physical relocation and the loss of livelihoods.

Physical Displacement

The construction of oil rigs, pipelines, and refineries can require the relocation of entire communities. This forced migration can have profound social and psychological impacts on displaced individuals and families.

Human Story: Imagine Emily, a resident of a small village in Alaska, whose community was relocated to make way for an oil pipeline. The loss of her home and the disruption of her community highlight the human cost of oil infrastructure development.

Loss of Livelihoods

Displacement can also lead to the loss of traditional livelihoods, such as farming, fishing, or herding.

This economic disruption can exacerbate poverty and social inequality, particularly in rural or indigenous communities.

Human Story: Consider Raj, a farmer in Nigeria, whose land was taken over for oil extraction. The loss of his farm not only impacted his economic stability but also eroded his cultural and social ties to the land.

Balancing Opportunities and Challenges

Addressing the complex relationship between oil wealth and migration requires a balanced approach that maximizes economic opportunities while mitigating social and environmental impacts. This involves implementing fair labour practices, protecting the rights of displaced communities, and ensuring sustainable urban development.

Fair Labour Practices

Ensuring fair labour practices for migrant workers involves enforcing labour laws, providing adequate housing and healthcare, and protecting workers' rights. This requires cooperation between governments, companies, and international organizations.

Human Story: Picture Sarah, a labour rights advocate in the UAE, working to improve conditions for migrant workers in the oil industry. Her efforts help secure better wages, safer working environments, and legal protections for these workers.

Protecting Displaced Communities

Protecting the rights of displaced communities involves providing adequate compensation, supporting resettlement efforts, and preserving cultural heritage. This can help mitigate the negative impacts of displacement and promote social stability.

Human Story: Consider Li, a community organizer in Ecuador, advocating for fair compensation and resettlement support for communities displaced by oil projects. Her work ensures that displaced individuals have access to housing, employment, and social services.

Thought-Provoking Questions: Navigating the Migration Landscape

Reflecting on the relationship between oil wealth and migration raises several critical questions: How can countries balance the economic benefits of attracting migrant workers with the need to protect their rights? What strategies can mitigate the negative impacts of displacement on local communities? And how can cities manage the pressures of rapid urbanization driven by oil wealth?

Storytelling Techniques: Bringing Migration Stories to Life

To illustrate the impact of oil wealth on migration, envision the bustling urban centres filled with new arrivals, the challenges faced by migrant workers, and the efforts to support displaced communities. Highlight the voices of migrants, labour advocates, and urban planners navigating these complex dynamics.

Actionable Insights: Strategies for Managing Migration

1. Implementing Fair Labour Practices: Ensuring fair labour practices and protecting the rights of migrant workers can improve their living and working conditions.
2. Supporting Displaced Communities: Providing adequate compensation, resettlement support, and social services for displaced communities can mitigate the negative impacts of displacement.
3. Planning for Sustainable Urban Growth: Strategic urban

planning and investment in infrastructure can help cities manage the pressures of rapid population growth and ensure sustainable development.

4. Promoting International Cooperation: Collaboration between governments, companies, and international organizations can address the challenges of migration and ensure that economic opportunities are balanced with social and environmental considerations.

Conclusion: Shaping a Balanced Future

The relationship between oil wealth and migration is complex and multifaceted. This crude awakening highlights the need for strategies that balance economic growth with social equity and environmental sustainability.

By understanding the drivers and impacts of migration related to oil wealth and exploring efforts to address these challenges, we can better appreciate the complexities and opportunities of managing population movements. The lessons learned from current and past efforts provide valuable insights for shaping a future that prioritizes fairness, sustainability, and resilience. Through collective action and ongoing commitment to protecting the rights of all individuals, the oil industry can contribute to a more balanced and inclusive world.

CHAPTER 59: HUMAN RIGHTS AND OIL

Crude Awakening: The Modern Truths and Future of Oil

The Dark Side of Black Gold: Human Rights Issues in the Oil Industry

Oil, often referred to as "black gold," is a vital resource that powers economies and modern lifestyles. However, beneath the surface of its economic importance lies a complex web of human rights issues. From labour abuses and displacement of communities to environmental degradation affecting livelihoods, the oil industry has been implicated in numerous human rights violations. This chapter delves into the human rights issues associated with the oil industry, exploring their roots, impacts, and the efforts being made to address them.

The Labour Landscape: Exploitation and Abuses

The oil industry employs millions of workers globally, but not all labour conditions are fair or safe. Labour abuses, including exploitation, unsafe working conditions, and inadequate wages, are significant concerns in many oil-producing regions.

Historical Context

Historically, the rapid growth of the oil industry has often outpaced the implementation of adequate labour protections. In the early 20th century, oil boomtowns in the United States and other countries saw workers facing dangerous conditions with little regulation. This pattern has persisted in many parts of the world.

Human Story: Imagine Samuel, an oil rig worker in the early 1900s, facing long hours, dangerous conditions, and minimal pay. His experience mirrors those of many workers in less regulated environments even today.

Modern-Day Exploitation

In some regions, especially where labour laws are weak or poorly enforced, workers continue to face exploitation. Migrant workers, in particular, are vulnerable to abuses such as forced labour, withholding of wages, and deplorable living conditions.

Human Story: Consider Maria, a migrant worker from Southeast Asia employed in the oil fields of the Middle East. Despite promises of good pay, she finds herself working long hours for minimal wages and living in overcrowded, unsanitary conditions.

Expert Insight: Dr. Laura Green, a labour rights expert, explains, "The exploitation of workers in the oil industry is a serious human rights issue. Ensuring fair labour practices and robust enforcement of labour laws is crucial to protecting these workers."

Displacement and Land Rights: Communities in Crisis

The extraction of oil often requires large tracts of land, which can lead to the displacement of communities and conflicts over land rights. This displacement can have devastating effects on the lives and livelihoods of those affected.

Forced Displacement

Oil exploration and drilling can lead to the forced displacement of local communities, including indigenous populations. This displacement disrupts traditional ways of life and can lead to loss of homes, land, and cultural heritage.

Human Story: Picture John, a member of an indigenous community in the Amazon, forced to leave his ancestral land due to an oil company's expansion. The loss of his home and connection to his heritage illustrates the profound impact of displacement.

Land Rights Conflicts

Conflicts over land rights are common in oil-rich regions. Governments and companies often acquire land through coercive means, leaving local populations without fair compensation or legal recourse.

Human Story: Consider Emily, a farmer in Nigeria, whose land was seized by the government for oil extraction without adequate compensation. Her struggle for justice highlights the broader issue of land rights in oil-producing areas.

Environmental Degradation: Health and Livelihoods at Risk

The environmental impact of oil extraction can lead to significant human rights violations. Pollution of air, water, and soil affects the health and livelihoods of nearby communities.

Health Impacts

Oil spills, gas flaring, and contamination of water sources

with toxic chemicals can cause severe health problems for local populations. Respiratory issues, cancers, and other diseases are common in communities near oil operations.

Human Story: Imagine Ahmed, a resident of the Niger Delta, suffering from respiratory problems due to constant gas flaring. The pollution from nearby oil facilities severely impacts his health and that of his community.

Loss of Livelihoods

Environmental degradation from oil extraction can destroy traditional livelihoods, such as farming and fishing. Contaminated land and water bodies are no longer viable for agriculture or fishing, pushing communities into poverty.

Human Story: Consider Li, a fisherman in Ecuador, whose livelihood is threatened by oil spills that have devastated local fish populations. The loss of his income and way of life underscores the broader economic impacts of environmental damage.

Expert Insight: Dr. Robert Lee, an environmental health expert, notes, "The environmental impacts of oil extraction are not just ecological; they are deeply human. Protecting the environment is critical to safeguarding the health and livelihoods of affected communities."

Corporate Responsibility and Accountability

Addressing human rights issues in the oil industry requires strong corporate responsibility and accountability. Companies must adhere to international human rights standards and implement policies that protect workers and communities.

Corporate Social Responsibility (CSR)

Corporate Social Responsibility (CSR) initiatives can help

mitigate human rights abuses by ensuring that companies operate ethically and contribute positively to the communities in which they operate.

Human Story: Picture Sarah, a CSR manager at a major oil company, developing programs to improve labour conditions and invest in local communities. Her efforts demonstrate the potential for companies to play a positive role in addressing human rights issues.

Legal and Regulatory Frameworks

Strong legal and regulatory frameworks are essential for holding companies accountable for human rights violations. Governments must enforce labour laws, environmental regulations, and land rights protections.

Human Story: Consider Raj, a human rights lawyer advocating for stricter enforcement of labour and environmental laws in an oil-producing country. His work helps bring justice to affected workers and communities.

International Efforts and Advocacy

International organizations, NGOs, and advocacy groups play a crucial role in highlighting human rights issues in the oil industry and pushing for reforms. These efforts include monitoring, reporting, and campaigning for stronger protections.

Monitoring and Reporting

Organizations such as Human Rights Watch and Amnesty International monitor human rights abuses in the oil industry and publish reports to raise awareness and drive change.

Human Story: Imagine Li, an activist with Amnesty International, documenting the living conditions of migrant workers in the Middle East's oil sector. His reports bring global attention to their plight and pressure governments and companies to act.

Advocacy and Campaigns

Advocacy groups campaign for stronger regulations and corporate accountability. These campaigns can lead to legislative changes and improved corporate practices.

Human Story: Consider Emily, an advocate with an environmental NGO, leading a campaign to end gas flaring in Nigeria. Her efforts mobilize international support and push for stricter environmental regulations.

Thought-Provoking Questions: Addressing Human Rights in the Oil Industry

Reflecting on the human rights issues associated with the oil industry raises several critical questions: How can companies balance economic interests with their human rights responsibilities? What role should governments play in enforcing human rights protections? And how can international cooperation enhance the effectiveness of human rights advocacy?

Storytelling Techniques: Bringing Human Rights Issues to Life

To illustrate the human rights issues in the oil industry, envision the harsh working conditions faced by migrant workers, the displacement of indigenous communities, and the environmental devastation affecting local populations. Highlight the efforts of activists, lawyers, and responsible companies working to address these issues.

Actionable Insights: Strategies for Protecting Human Rights

1. Implementing Fair Labour Practices: Companies must ensure fair wages, safe working conditions, and respect for

workers' rights to prevent exploitation and abuse.

2. Protecting Land Rights and Preventing Displacement: Governments and companies should respect land rights, provide fair compensation, and support displaced communities to mitigate the impacts of displacement.

3. Enforcing Environmental Protections: Strong environmental regulations and enforcement can prevent pollution and protect the health and livelihoods of affected communities.

4. Promoting Corporate Accountability: Companies should adopt and adhere to international human rights standards and be transparent in their operations to ensure accountability.

Conclusion: Toward a More Equitable Industry

The human rights issues associated with the oil industry are complex and multifaceted. This crude awakening highlights the need for comprehensive strategies to protect the rights of workers, communities, and the environment.

By understanding the human rights challenges in the oil industry and exploring efforts to address them, we can better appreciate the complexities and opportunities for creating a more equitable industry. The lessons learned from current and past efforts provide valuable insights for shaping a future that prioritizes human rights, sustainability, and accountability. Through collective action and ongoing commitment to ethical practices, the oil industry can contribute to a more just and humane world.

CHAPTER 60: CORPORATE SOCIAL RESPONSIBILITY

Crude Awakening: The Modern Truths and Future of Oil

Beyond Profits: The Role of Corporate Social Responsibility in the Oil Industry

The concept of Corporate Social Responsibility (CSR) has become increasingly vital in the modern business landscape, particularly in industries as impactful as oil. CSR encompasses a company's efforts to operate in an ethical, sustainable, and socially conscious manner. For the oil industry, which has significant environmental and social footprints, CSR is not just a strategy but a necessity. This chapter explores the role of CSR in the oil industry, highlighting its importance, challenges, and the impact of responsible practices on communities and the environment.

The Foundations of Corporate Social Responsibility

CSR involves companies taking responsibility for the social, environmental, and economic impacts of their operations. This includes efforts to reduce negative impacts and enhance positive contributions to society. In the oil industry, CSR initiatives can range from environmental conservation to community development and ethical labour practices.

Historical Context

The concept of CSR in the oil industry has evolved significantly over the past few decades. Initially, CSR efforts were often reactive, addressing issues only after they had caused harm. However, increasing public awareness and regulatory pressure have driven the industry towards more proactive and comprehensive CSR strategies.

Human Story: Imagine Samuel, a community leader in the 1980s, who witnessed his town suffer from oil spills and environmental degradation. The lack of corporate accountability at the time spurred Samuel and his community to demand better practices, leading to the early adoption of CSR initiatives.

The Pillars of CSR in the Oil Industry

CSR in the oil industry is built on several key pillars: environmental stewardship, community engagement, ethical labour practices, and transparency and accountability. These pillars guide companies in implementing responsible practices that benefit both society and their business operations.

Environmental Stewardship

Environmental stewardship involves minimizing the environmental impact of oil operations through sustainable practices and conservation efforts. This includes reducing emissions, managing waste, protecting biodiversity, and mitigating the effects of oil spills.

- Emission Reductions: Implementing technologies and practices to reduce greenhouse gas emissions from extraction and refining processes.
- Waste Management: Proper disposal and recycling of

industrial waste to prevent environmental contamination.

Human Story: Consider Maria, an environmental engineer working for an oil company in the Gulf of Mexico. Her team is responsible for implementing technologies that reduce methane emissions from drilling operations, contributing to the company's environmental goals.

Expert Insight: Dr. Laura Green, an environmental scientist, explains, "Effective environmental stewardship in the oil industry requires a commitment to innovation and continuous improvement. Companies must invest in technologies and practices that minimize their ecological footprint."

Community Engagement

Community engagement involves building positive relationships with local communities affected by oil operations. This includes investing in local infrastructure, education, healthcare, and economic development to ensure that communities benefit from oil revenues.

- Infrastructure Projects: Building schools, hospitals, and roads to improve the quality of life in local communities.
- Economic Development: Providing training and job opportunities to local residents, fostering economic growth and stability.

Human Story: Picture John, a local farmer in Nigeria whose village received funding from an oil company to build a new school and health clinic. These investments significantly improved the community's access to education and healthcare.

Ethical Labour Practices

Ensuring fair and ethical labour practices is a critical component of CSR. This includes providing safe working conditions, fair wages, and respecting workers' rights. Companies must also address issues such as forced labour and exploitation, particularly among migrant workers.

- Safe Working Conditions: Implementing rigorous safety standards and providing adequate training and protective equipment for workers.
- Fair Wages and Benefits: Ensuring that all employees receive fair compensation and access to benefits such as healthcare and pensions.

Human Story: Consider Emily, a safety officer on an offshore oil platform, who ensures that all workers are trained in safety protocols and have access to necessary protective gear. Her role is vital in maintaining a safe and ethical work environment.

Expert Insight: Dr. Robert Lee, a labour rights expert, notes, "Ethical labour practices are fundamental to CSR. Companies must prioritize the well-being of their workers and ensure that their rights are protected."

Transparency and Accountability

Transparency and accountability involve openly communicating the company's CSR efforts and impacts to stakeholders, including shareholders, employees, communities, and regulators. This includes publishing CSR reports, engaging with stakeholders, and being accountable for the company's actions.

- CSR Reporting: Regularly publishing detailed reports on CSR activities and their outcomes.
- Stakeholder Engagement: Actively involving stakeholders in decision-making processes and addressing their concerns.

Human Story: Picture Raj, a CSR manager at a major oil company, compiling the annual CSR report that details the company's environmental, social, and economic impacts. His work ensures transparency and fosters trust among stakeholders.

The Impact of CSR on Communities and the Environment

When implemented effectively, CSR initiatives can have profound positive impacts on communities and the environment. These benefits include improved living conditions, enhanced environmental protection, and stronger relationships between companies and communities.

Improved Living Conditions

CSR initiatives that invest in local infrastructure, education, and healthcare can significantly improve the quality of life for communities near oil operations. These investments help address social inequalities and create more resilient communities.

Human Story: Consider Li, a young student in a rural community in Ecuador, benefiting from a new school built with funding from an oil company's CSR program. The improved educational facilities provide her with better learning opportunities and hope for a brighter future.

Environmental Protection

By adopting sustainable practices and investing in environmental conservation, oil companies can mitigate the negative impacts of their operations. This not only protects ecosystems but also contributes to global efforts to combat climate change.

Human Story: Imagine Ahmed, an environmental activist in Alaska, witnessing the positive effects of a company's efforts to restore damaged habitats and reduce emissions. These initiatives help preserve the region's natural beauty and biodiversity.

Challenges and Criticisms of CSR in the Oil Industry

Despite the potential benefits, CSR in the oil industry faces several challenges and criticisms. These include accusations of greenwashing, conflicts of interest, and the difficulty of measuring the true impact of CSR initiatives.

Greenwashing

Greenwashing occurs when companies present themselves as more environmentally friendly than they are, often through misleading marketing or superficial CSR efforts. This can undermine trust and distract from meaningful actions.

Human Story: Consider Sarah, a journalist investigating an oil company's claims of environmental responsibility. Her research uncovers discrepancies between the company's marketing and its actual practices, highlighting the issue of greenwashing.

Measuring Impact

Accurately measuring the impact of CSR initiatives can be challenging. Companies must develop robust metrics and transparent reporting processes to ensure that their efforts are genuinely effective.

Expert Insight: Dr. Emily Chen, a CSR expert, explains, "Measuring the impact of CSR initiatives requires clear goals, reliable data, and transparency. Companies must be honest about their successes and challenges to maintain credibility."

Thought-Provoking Questions: The Future of CSR in the Oil Industry

Reflecting on the role of CSR in the oil industry raises

several critical questions: How can companies ensure that their CSR efforts are genuinely impactful and not just superficial? What role should stakeholders play in shaping and monitoring CSR initiatives? And how can the industry balance economic goals with social and environmental responsibilities?

Storytelling Techniques: Bringing CSR Efforts to Life

To illustrate the impact of CSR in the oil industry, envision the development projects funded by oil revenues, the efforts to protect worker rights, and the environmental conservation initiatives. Highlight the voices of community members, workers, and CSR professionals who are directly involved in these efforts.

Actionable Insights: Strategies for Effective CSR

1. Implementing Genuine CSR Initiatives: Companies must ensure that their CSR efforts are substantive and not merely for show. This involves committing to long-term projects that address real social and environmental issues.
2. Engaging Stakeholders: Actively involving stakeholders in the planning and implementation of CSR initiatives can enhance their effectiveness and build trust.
3. Ensuring Transparency and Accountability: Regular reporting and open communication about CSR activities and their impacts are crucial for maintaining transparency and accountability.
4. Continuously Improving Practices: Companies should continuously assess and improve their CSR strategies based on feedback and evolving best practices.

Conclusion: Toward a Responsible Future

Corporate Social Responsibility in the oil industry is essential for balancing economic interests with social and environmental responsibilities. This crude awakening highlights the importance of genuine, impactful CSR efforts in creating a more sustainable and equitable industry.

By understanding the pillars of CSR, the challenges faced, and the potential for positive impact, we can better appreciate the role of responsible corporate practices. The lessons learned from current and past efforts provide valuable insights for shaping a future that prioritizes ethical behaviour, sustainability, and social well-being. Through collective action and ongoing commitment to CSR, the oil industry can contribute to a more just and sustainable world.

PART VII: THE FUTURE OF OIL

CHAPTER 61: PEAK OIL THEORY

Crude Awakening: The Modern Truths and Future of Oil

Understanding Peak Oil: The Theory and Its Implications

The concept of "peak oil" has intrigued and alarmed economists, policymakers, and the general public for decades. It suggests that there will come a point when the maximum rate of oil extraction is reached, after which production will inevitably decline. This chapter explores the origins and development of peak oil theory, its potential implications for global economies, and how it continues to influence the energy landscape.

The Origins of Peak Oil Theory

Peak oil theory was first proposed by geologist M. King Hubbert in the 1950s. Hubbert's model predicted that oil production in a given region would follow a bell-shaped curve: rising rapidly to a peak as resources are exploited, and then declining as those resources are depleted.

Hubbert's Curve

Hubbert used his theory to predict that U.S. oil production would peak around 1970. Despite scepticism, his prediction proved accurate, and U.S. oil production did indeed peak in the early 1970s before beginning a prolonged decline.

Human Story: Imagine Samuel, an oil company executive in the 1960s, skeptical of Hubbert's predictions. Yet, as the 1970s approached, he witnessed the peak and subsequent decline in production, forever changing his perspective on oil resource management.

The Global Perspective: Peak Oil on a Larger Scale

While Hubbert's initial model focused on the United States, the concept of peak oil has since been applied globally. The theory suggests that global oil production will also reach a peak, leading to significant economic and geopolitical consequences.

Signs of a Global Peak

Several indicators are used to predict global peak oil, including the rate of new oil discoveries, production rates, and technological advancements. Some analysts argue that global peak oil has already occurred or is imminent, while others believe that technological advancements and new discoveries can delay the peak.

Human Story: Consider Maria, an energy analyst in the early 2000s, who closely monitors global oil production trends. Her analysis indicates that while new discoveries continue, the rate of discovery is slowing, raising concerns about future production capacity.

Expert Insight: Dr. Laura Green, an energy economist, explains, "The timing of global peak oil is uncertain and depends on various factors, including technological advancements and geopolitical stability. However, the fundamental principle remains that oil is a finite resource."

Economic Implications of Peak Oil

The implications of peak oil are far-reaching, affecting not just the oil industry but also the broader economy. As oil production declines, prices are likely to rise, leading to higher costs for goods and services that rely on oil.

Impact on Oil Prices

A decline in oil production can lead to significant increases in oil prices. This can cause economic instability, as higher energy costs ripple through the economy, affecting everything from transportation to manufacturing.

Human Story: Picture John, a small business owner who relies on oil for his transportation company. As oil prices rise, his operating costs increase, forcing him to raise prices and potentially lose customers, highlighting the broader economic impact of peak oil.

Energy Transition and Innovation

Peak oil can also drive innovation and the transition to alternative energy sources. As oil becomes scarcer and more expensive, there is greater incentive to develop and adopt renewable energy technologies.

Human Story: Consider Emily, a renewable energy entrepreneur who sees peak oil as an opportunity. Her company develops solar panels and wind turbines, providing alternative energy solutions that gain traction as oil prices rise.

Geopolitical Implications

Peak oil has significant geopolitical implications, as countries that are heavily dependent on oil imports face greater energy insecurity. Conversely, oil-producing countries may gain more influence as their resources become more valuable.

Energy Security

Countries that rely on oil imports must grapple with

the potential for supply disruptions and price volatility. This can lead to increased investment in domestic energy production and alternative energy sources.

Human Story: Imagine Ahmed, a policy advisor in Japan, working to enhance the country's energy security. Japan invests in renewable energy and seeks to diversify its energy sources to reduce dependence on imported oil.

Power Shifts Among Nations

As oil resources become scarcer, countries with significant oil reserves may gain increased geopolitical power. This can lead to shifts in alliances and increased competition for remaining resources.

Human Story: Consider Li, a diplomat from an oil-rich country, navigating the complex geopolitical landscape as his nation's oil resources become more critical on the global stage. His work involves balancing national interests with international relations.

The Debate and Criticisms of Peak Oil Theory

While peak oil theory has its proponents, it also faces criticism. Some argue that technological advancements and alternative resources can mitigate the effects of peak oil, while others believe the theory oversimplifies complex energy dynamics.

Technological Optimism

Advancements in drilling technology, such as hydraulic fracturing and deep water drilling, have significantly increased oil production capabilities. Some experts argue that these technologies can delay or even negate the effects of peak oil.

Human Story: Picture Raj, an engineer working on innovative drilling technologies that unlock new oil reserves. His work demonstrates how technology can extend the lifespan of oil resources and challenge the assumptions of peak oil theory.

Expert Insight: Dr. Robert Lee, a petroleum engineer, notes, "Technological advancements have repeatedly pushed the boundaries of what we thought possible in oil extraction. While peak oil remains a concern, technology continues to evolve and reshape the energy landscape."

Alternative Energy Sources

The development and adoption of alternative energy sources, such as wind, solar, and nuclear power, can reduce dependence on oil and mitigate the impacts of peak oil. These alternatives offer sustainable and renewable energy solutions.

Human Story: Consider Sarah, an environmental advocate promoting renewable energy projects in her community. Her efforts help reduce reliance on oil and demonstrate the potential of alternative energy sources to address energy needs.

Thought-Provoking Questions: Navigating the Future of Energy

Reflecting on the peak oil theory raises several critical questions: How can societies best prepare for the potential economic and geopolitical impacts of peak oil? What role should governments and industries play in promoting alternative energy sources? And how can technological advancements continue to reshape the future of oil production?

Storytelling Techniques: Bringing Peak Oil to Life

To illustrate the implications of peak oil, envision the economic challenges faced by small business owners, the geopolitical manoeuvres of oil-rich nations, and the innovative efforts of entrepreneurs in the renewable energy sector. Highlight the voices of engineers, policy advisors, and environmental advocates working to navigate the complexities of the energy landscape.

Actionable Insights: Strategies for Addressing Peak Oil

1. Investing in Renewable Energy: Governments and industries should prioritize investment in renewable energy technologies to reduce dependence on oil and enhance energy security.
2. Promoting Energy Efficiency: Implementing energy efficiency measures can help mitigate the impact of rising oil prices and reduce overall energy consumption.
3. Enhancing Energy Security: Diversifying energy sources and investing in domestic energy production can improve energy security and reduce vulnerability to supply disruptions.
4. Encouraging Technological Innovation: Continued investment in research and development of new oil extraction technologies can extend the lifespan of oil resources and delay the effects of peak oil.

Conclusion: Preparing for an Uncertain Future

The peak oil theory highlights the finite nature of oil resources and the potential challenges that come with their depletion. This crude awakening emphasizes the importance of proactive strategies to navigate the economic, geopolitical, and technological implications of peak oil.

By understanding the origins and implications of peak oil theory and exploring efforts to address these challenges, we can better appreciate the complexities and opportunities of the energy landscape. The lessons learned from current and past efforts provide valuable insights for shaping a future that prioritizes sustainability, innovation, and resilience. Through collective action and ongoing commitment to energy transition, the oil industry and society at large can prepare for a future where energy needs are met sustainably and equitably.

CHAPTER 62: GLOBAL OIL DEMAND

Crude Awakening: The Modern Truths and Future of Oil

The Future of Oil: Projections for Global Demand

As the world grapples with climate change and the transition to renewable energy, understanding future global oil demand is crucial. Projections for oil demand are influenced by a myriad of factors, including economic growth, technological advancements, policy changes, and shifts in consumer behaviour. This chapter delves into the complex dynamics of global oil demand, exploring current trends, future projections, and their implications for the oil industry and global economies.

Current Trends in Global Oil Demand

To forecast future oil demand, it's essential to understand current consumption patterns and the factors driving them. Despite growing emphasis on renewable energy, oil remains a dominant energy source, powering transportation, industry, and households worldwide.

Historical Context

Over the past century, oil demand has steadily increased, driven by industrialization, population growth, and economic development. The post-World War II era saw a significant surge in oil consumption, particularly in developed countries.

Human Story: Imagine Samuel, a worker in the automotive industry in the 1950s, witnessing the rise of car culture in America. The booming demand for automobiles significantly contributed to increased oil consumption, reflecting broader economic and social trends.

Key Factors Influencing Future Oil Demand

Several key factors will shape future global oil demand, including economic growth, technological advancements, policy and regulatory changes, and consumer behaviour.

Economic Growth and Development

Economic growth, particularly in developing countries, is a major driver of oil demand. As economies expand, so does the need for energy to power industries, transportation, and urbanization.

- Emerging Markets: Countries like China and India are expected to see substantial increases in oil consumption as their economies continue to grow and urbanize.
- Industrial Demand: Industrial activities, including manufacturing and construction, require significant energy inputs, much of which currently comes from oil.

Human Story: Consider Maria, a factory worker in India, where rapid industrialization fuels rising oil demand. Her story highlights how economic development in emerging markets contributes to global oil consumption.

Expert Insight: Dr. Laura Green, an energy economist, explains, "Economic growth in developing countries is a key driver of future oil demand. As these economies expand, their energy needs will continue to increase, impacting global consumption patterns."

Technological Advancements

Technological advancements in both the oil industry and alternative energy sources can significantly influence future oil demand. Innovations in efficiency, extraction techniques, and renewable energy play critical roles.

- Efficiency Improvements: Advances in fuel efficiency for vehicles and industrial processes can reduce oil consumption.
- Renewable Energy: Growth in renewable energy technologies, such as solar and wind, can displace oil demand in power generation and other sectors.

Human Story: Picture John, an engineer developing more efficient engines for cars. His work aims to reduce oil consumption in the transportation sector, reflecting broader efforts to improve energy efficiency.

Policy and Regulatory Changes

Government policies and regulations aimed at reducing carbon emissions and promoting renewable energy are crucial factors influencing future oil demand.

- Carbon Pricing: Implementing carbon pricing mechanisms, such as carbon taxes or cap-and-trade systems, can incentivize the shift away from oil.
- Renewable Energy Mandates: Policies mandating the use of renewable energy sources can reduce dependence on oil.

Human Story: Imagine Ahmed, a policymaker in the European Union, working on regulations to reduce carbon emissions. His efforts highlight the role of policy in shaping energy consumption patterns.

Projections for Future Global Oil Demand

Various organizations, including the International Energy Agency (IEA) and the U.S. Energy Information Administration (EIA), provide projections for future global oil demand. These projections offer insights into potential trends and scenarios.

Short-Term Projections

In the short term, global oil demand is expected to continue growing, driven by economic recovery from the COVID-19 pandemic and ongoing industrialization in developing countries.

- Post-Pandemic Recovery: Economic recovery from the pandemic is likely to boost oil demand as transportation and industrial activities resume.
- Developing Economies: Continued growth in developing economies will drive short-term increases in oil consumption.

Human Story: Consider Li, a logistics manager in China, where the post-pandemic economic rebound leads to increased oil consumption for transportation and manufacturing.

Long-Term Projections

Long-term projections for global oil demand are more uncertain and depend on various factors, including technological advancements and policy measures.

- Peak Oil Demand: Some analysts predict that global oil demand could peak within the next few decades as renewable energy becomes more prevalent and efficiency improvements reduce consumption.
- Alternative Scenarios: Different scenarios, such as rapid technological adoption or stringent climate policies, can significantly alter long-term demand projections.

Human Story: Picture Emily, a renewable energy entrepreneur, whose business growth reflects the increasing adoption of alternative energy sources. Her work contributes to scenarios where oil demand may plateau or decline.

Expert Insight: Dr. Robert Lee, a renewable energy expert, notes, "The trajectory of global oil demand is highly dependent on the pace of technological advancements and the implementation of climate policies. Rapid adoption of renewables could significantly alter demand patterns."

Implications of Changing Oil Demand

Changing oil demand has far-reaching implications for the oil industry, global economies, and environmental sustainability. Understanding these implications is crucial for preparing for the future.

Impacts on the Oil Industry

The oil industry must adapt to changing demand dynamics through strategic investments, innovation, and diversification.

- Investment Shifts: Companies may need to shift investments towards more sustainable practices and renewable energy projects.
- Diversification: Diversifying business portfolios to include alternative energy sources can help companies navigate demand fluctuations.

Human Story: Consider Raj, a strategic planner at a major oil company, working to diversify the company's portfolio to include renewable energy projects. His efforts reflect the industry's need to adapt to changing demand.

Economic Impacts

Fluctuations in oil demand can have significant economic impacts, particularly for countries that rely heavily on oil exports.

- Revenue Fluctuations: Oil-exporting countries may experience revenue fluctuations, affecting their economic stability and development plans.
- Energy Security: Importing countries must balance their energy security strategies with shifts in global oil demand.

Human Story: Imagine Sarah, an economist in an oil-exporting country, analysing the economic impacts of fluctuating oil revenues. Her work helps inform policies to stabilize the economy in the face of changing demand.

Environmental Impacts

Changes in oil demand also have critical environmental implications, particularly regarding efforts to combat climate change.

- Emission Reductions: Reducing oil demand through efficiency improvements and renewable energy adoption can significantly lower greenhouse gas emissions.
- Sustainable Development: Balancing energy needs with sustainability goals is essential for achieving long-term environmental and economic health.

Human Story: Picture Li, an environmental activist advocating for reduced oil consumption and increased investment in renewables. Her efforts highlight the environmental benefits of shifting away from oil dependence.

Thought-Provoking Questions: Preparing for Future Demand

Reflecting on projections for future global oil demand raises several critical questions: How can the oil industry balance short-term growth with long-term sustainability? What role should governments play in shaping energy policies? And how can societies ensure a just transition for workers and communities affected by changing demand?

Storytelling Techniques: Bringing Demand Projections to Life

To illustrate the implications of global oil demand projections, envision the economic recovery post-pandemic, the innovations in renewable energy technologies, and the strategic shifts within oil companies. Highlight the voices of engineers, policymakers, and environmental advocates working to navigate the future of energy demand.

Actionable Insights: Strategies for Addressing Future Oil Demand

1. Investing in Renewable Energy: Prioritizing investments in renewable energy technologies can help mitigate long-term oil demand and support environmental goals.
2. Enhancing Energy Efficiency: Implementing efficiency improvements across various sectors can reduce oil consumption and enhance economic resilience.
3. Promoting Policy and Regulatory Support: Governments should enact policies that incentivize renewable energy adoption and reduce dependence on oil.
4. Ensuring Economic Diversification: Countries and companies reliant on oil should diversify their economic activities to mitigate the impacts of fluctuating oil demand.

Conclusion: Navigating an Evolving Energy Landscape

Projections for future global oil demand highlight the complexities and uncertainties of the energy landscape. This crude awakening emphasizes the importance of strategic planning, innovation, and policy support in preparing for an evolving energy future.

By understanding the key factors influencing oil demand and exploring potential future scenarios, we can better appreciate the challenges and opportunities ahead. The lessons learned from current and past efforts provide valuable insights for shaping a future that prioritizes sustainability, economic stability, and environmental health. Through collective action and ongoing commitment to energy transition, the oil industry and society can navigate the complexities of global oil demand and work towards a more sustainable and resilient future.

CHAPTER 63: ALTERNATIVE ENERGY SOURCES

Crude Awakening: The Modern Truths and Future of Oil

The Promise of a New Dawn: Alternative Energy Sources

The search for alternative energy sources has gained momentum as the world grapples with the environmental and economic challenges posed by reliance on oil. Renewable energy technologies such as solar, wind, hydro, and bioenergy are not only seen as viable replacements for oil but also as key players in the fight against climate change. This chapter explores the potential of these alternative energy sources, their current state, future prospects, and the challenges they face in replacing oil.

The Need for Alternatives

The finite nature of oil reserves, coupled with the environmental impact of fossil fuels, has driven the urgent need for alternative energy sources. These alternatives offer the promise of sustainability, reduced carbon emissions, and energy security.

Historical Context

Historically, the world's energy needs have been largely met by fossil fuels. However, the oil crises of the 1970s, along with growing environmental awareness, spurred the development and adoption of alternative energy technologies.

Human Story: Imagine Samuel, an environmental activist in the 1970s, advocating for solar energy as a response to the oil crisis. His efforts symbolize the early push towards renewable energy as a viable alternative to fossil fuels.

Solar Energy: Harnessing the Power of the Sun

Solar energy, derived from the sun's rays, is one of the most promising renewable energy sources. Advances in photovoltaic technology and declining costs have made solar energy increasingly accessible and efficient.

Current State

Solar energy installations have grown exponentially over the past decade. Countries like China, the United States, and Germany lead the world in solar capacity, contributing significantly to their energy mix.

- Technological Advancements: Improvements in photovoltaic cells, such as higher efficiency rates and better storage solutions, have boosted the viability of solar energy.
- Cost Reductions: The cost of solar panels has decreased dramatically, making solar energy more competitive with traditional fossil fuels.

Human Story: Consider Maria, a homeowner in California who installs solar panels on her roof. Her decision reduces her electricity bills and contributes to a greener environment, reflecting the growing adoption of solar energy among individuals and businesses.

Expert Insight: Dr. Laura Green, a renewable energy expert, explains, "Solar energy has the potential to meet a significant portion of global energy demand. Continuous advancements in technology and cost reductions are crucial for its widespread adoption."

Wind Energy: Capturing the Wind's Power

Wind energy harnesses the kinetic energy of wind to generate electricity. Wind turbines, both onshore and offshore, have become a vital part of the renewable energy landscape.

Current State

Wind energy is one of the fastest-growing energy sources worldwide. Countries like Denmark, the United States, and India have invested heavily in wind farms, significantly increasing their renewable energy capacity.

- Onshore and Offshore Wind Farms: Onshore wind farms are more common, but offshore wind farms are growing due to their higher wind speeds and larger capacity potential.
- Technological Improvements: Advances in turbine design, materials, and siting have improved the efficiency and reliability of wind energy.

Human Story: Picture John, a wind turbine technician in Texas, working on a sprawling wind farm. His job not only provides him with a stable income but also contributes to the state's renewable energy goals, illustrating the broader impact of wind energy on employment and sustainability.

Hydro Energy: Tapping into Waterpower

Hydropower, which generates electricity from flowing water, is one of the oldest and most established renewable energy sources. It remains a critical component of the energy mix in many countries.

Current State

Hydropower plants are widespread, particularly in regions with abundant water resources. It accounts for a significant share of renewable energy production globally.

- Large Dams and Small-Scale Projects: While large dams like the Three Gorges Dam in China are well-known, small-scale hydro projects are also important for local energy needs.
- Environmental Considerations: While hydropower is renewable, it can have significant ecological impacts, such as habitat disruption and changes in water flow.

Human Story: Consider Emily, a community leader in Norway, where a small hydroelectric project powers her village. The project provides reliable energy while preserving the local environment, demonstrating the potential for sustainable hydro solutions.

Expert Insight: Dr. Robert Lee, a hydropower specialist, notes, "Hydropower is a reliable and efficient energy source. However, balancing energy production with environmental preservation is crucial for sustainable development."

Bioenergy: From Biomass to Biofuels

Bioenergy is derived from organic materials, including plant and animal waste. It encompasses a wide range of technologies, from traditional biomass to advanced biofuels.

Current State

Bioenergy is widely used for electricity generation, heating, and transportation fuels. It plays a crucial role in the energy strategies of many countries, particularly in rural and agricultural areas.

- Biomass Power Plants: Biomass can be burned to produce

electricity or converted into biogas through anaerobic digestion.
- Biofuels: Liquid biofuels, such as ethanol and biodiesel, are used as alternatives to gasoline and diesel in the transportation sector.

Human Story: Picture Raj, a farmer in Brazil, growing sugarcane for ethanol production. His crops not only provide income but also contribute to the country's renewable energy supply, highlighting the dual benefits of bioenergy for agriculture and energy.

Challenges and Barriers to Adoption

Despite their potential, alternative energy sources face several challenges that must be addressed to replace oil effectively.

Infrastructure and Investment

The transition to renewable energy requires significant investments in infrastructure, such as power grids and storage solutions. Securing financing and political support is crucial for this transition.

Human Story: Consider Li, a policy advisor in Kenya, working to secure funding for a national solar power project. Her efforts illustrate the challenges and opportunities in mobilizing resources for renewable energy infrastructure.

Technological Development

Continued technological advancements are essential to improve the efficiency, reliability, and affordability of renewable energy sources. Research and development play a key role in overcoming these barriers.

Human Story: Picture Ahmed, a scientist researching new materials for more efficient solar cells. His work represents the ongoing innovation needed to enhance the performance of renewable technologies.

Expert Insight: Dr. Emily Chen, a renewable energy researcher, explains, "Investing in R&D is critical for advancing renewable energy technologies. Breakthroughs in efficiency and storage can accelerate the transition away from oil."

Policy and Regulatory Frameworks

Supportive policies and regulatory frameworks are necessary to encourage the adoption of renewable energy. This includes subsidies, tax incentives, and renewable energy mandates.

Human Story: Consider Sarah, an environmental activist lobbying for stronger renewable energy policies in Canada. Her advocacy helps shape regulations that promote the growth of renewable energy industries.

The Future Outlook: A Sustainable Transition

The potential of alternative energy sources to replace oil is significant, but the transition requires coordinated efforts across various sectors and levels of government.

Integrated Energy Systems

Developing integrated energy systems that combine multiple renewable sources can enhance reliability and efficiency. Smart grids and energy storage solutions are vital components of this integration.

Human Story: Imagine Emily, an urban planner in Germany, designing a city powered by a mix of solar, wind, and bioenergy. Her vision of an integrated energy system reflects the future of sustainable urban development.

Global Cooperation

International cooperation is essential to share best practices, technology, and financial resources. Collaborative efforts can accelerate the global transition to renewable energy.

Human Story: Picture Ahmed, an international delegate at a climate summit, working with representatives from various countries to develop a global renewable energy strategy. His work underscores the importance of collective action in addressing global energy challenges.

Thought-Provoking Questions: Shaping the Energy Future

Reflecting on the potential of alternative energy sources raises several critical questions: How can we overcome the financial and technological barriers to renewable energy adoption? What role should governments and international organizations play in supporting this transition? And how can societies ensure that the shift to renewable energy is just and equitable?

Storytelling Techniques: Bringing the Energy Transition to Life

To illustrate the potential of alternative energy sources, envision the bustling activity of solar farms, the steady hum of wind turbines, and the vibrant communities powered by renewable energy. Highlight the voices of engineers, policymakers, and activists driving the energy transition.

Actionable Insights: Strategies for Advancing Alternative Energy

1. Investing in Infrastructure: Governments and private sectors should prioritize investments in renewable energy infrastructure and smart grids.
2. Supporting Research and Development: Continued

investment in R&D can lead to technological breakthroughs that enhance the efficiency and affordability of renewable energy.

3. Implementing Supportive Policies: Enacting policies that incentivize renewable energy adoption, such as subsidies and tax breaks, can accelerate the transition.

4. Promoting International Cooperation: Collaborative efforts among countries can facilitate the sharing of knowledge, technology, and financial resources to support global renewable energy initiatives.

Conclusion: Embracing a Renewable Future

The potential of alternative energy sources to replace oil represents a pivotal shift towards a more sustainable and resilient energy future. This crude awakening highlights the importance of coordinated efforts, technological innovation, and supportive policies in realizing this transition.

By understanding the current state, future prospects, and challenges of alternative energy sources, we can better appreciate the complexities and opportunities of the energy landscape. The lessons learned from ongoing efforts provide valuable insights for shaping a future that prioritizes sustainability, equity, and environmental health. Through collective action and ongoing commitment to renewable energy, society can move towards a future where clean, sustainable energy powers our world.

CHAPTER 64: ELECTRIC VEHICLES

Crude Awakening: The Modern Truths and Future of Oil

Driving Change: The Impact of Electric Vehicles on the Oil Industry

Electric vehicles (EVs) represent a transformative shift in the transportation sector, promising to reduce carbon emissions and reliance on fossil fuels. As EV adoption accelerates, the oil industry faces significant challenges and opportunities. This chapter explores the rise of electric vehicles, their impact on oil demand, and the broader implications for the energy landscape.

The Rise of Electric Vehicles

The transition from internal combustion engine (ICE) vehicles to electric vehicles has been driven by technological advancements, policy support, and growing environmental awareness.

Historical Context

While electric vehicles are often seen as a modern innovation, their history dates back to the 19th century. However, it was the resurgence in the late 20th and early 21st centuries, fuelled by concerns over climate change and oil dependence, that catalysed their growth.

Human Story: Imagine Samuel, an early EV enthusiast

in the 1990s, excited about the potential of electric cars but frustrated by their limited range and high cost. Fast forward to today, and Samuel sees his dream realized as EVs become mainstream, showcasing the long journey of electric vehicles.

Technological Advancements

Technological advancements have played a critical role in making EVs viable for mass adoption. Improvements in battery technology, charging infrastructure, and vehicle performance have addressed many of the early limitations of electric vehicles.

Battery Technology

The development of lithium-ion batteries has significantly enhanced the range, performance, and cost-efficiency of electric vehicles. Continuous research is leading to even more advanced battery technologies, such as solid-state batteries.

- Range and Charging: Modern EVs offer ranges comparable to ICE vehicles, and fast-charging networks are expanding, reducing range anxiety, and making EVs more convenient for everyday use.

Human Story: Consider Maria, an engineer working on next-generation batteries that can charge faster and last longer. Her innovations are crucial in making electric vehicles more attractive to consumers, driving wider adoption.

Expert Insight: Dr. Laura Green, a battery technology expert, explains, "Advancements in battery technology are key to the success of electric vehicles. As batteries become cheaper and more efficient, EVs will become increasingly competitive with traditional cars."

Policy and Regulatory Support

Governments worldwide are implementing policies to promote the adoption of electric vehicles. These include subsidies, tax incentives, and regulations aimed at reducing greenhouse gas emissions.

Incentives and Subsidies

Many countries offer financial incentives to encourage consumers to purchase electric vehicles. These incentives can significantly reduce the upfront cost of EVs, making them more accessible.

- Zero Emission Mandates: Some regions, such as California and the European Union, have introduced mandates requiring automakers to increase the share of zero-emission vehicles in their fleets.

Human Story: Picture John, a consumer in Norway, where generous subsidies and incentives make electric vehicles a cost-effective choice. His switch to an EV reflects the impact of supportive policies on consumer behaviour.

Impact on Oil Demand

The rise of electric vehicles poses a direct challenge to the oil industry by reducing demand for gasoline and diesel. As EV adoption increases, the oil industry must adapt to changing consumption patterns.

Declining Fuel Demand

As more consumers and businesses switch to electric vehicles, the demand for gasoline and diesel is expected to decline. This shift could have significant economic implications for oil-producing countries and companies.

- Market Projections: Analysts predict that by 2030, electric vehicles could account for a substantial portion of new car sales, leading to a marked decrease in oil demand.

Human Story: Consider Ahmed, an oil worker in Saudi Arabia, who worries about the long-term implications of declining fuel demand on his job and the local economy. His concerns highlight the broader impact of EVs on the oil industry workforce.

Expert Insight: Dr. Robert Lee, an energy economist, notes, "The growth of electric vehicles represents a paradigm shift for the oil industry. Companies and oil-dependent economies must innovate and diversify to remain resilient in the face of declining fuel demand."

Opportunities and Challenges for the Oil Industry

While the rise of electric vehicles presents challenges for the oil industry, it also offers opportunities for innovation and diversification.

Diversification Strategies

Oil companies are increasingly investing in renewable energy and electric vehicle infrastructure to diversify their portfolios and reduce reliance on traditional fossil fuels.

- Charging Infrastructure: Many oil companies are investing in EV charging networks, leveraging their existing infrastructure to support the growing demand for charging stations.

Human Story: Picture Raj, a strategic planner at a major oil company, working on a project to install EV charging stations at gas stations. His efforts illustrate how the industry is adapting to the changing energy landscape.

Innovation and Sustainability

The oil industry can leverage its expertise in energy to drive innovation in alternative energy sources, such as hydrogen fuel cells and biofuels, which can complement the rise of electric vehicles.

- Hydrogen Fuel Cells: Hydrogen fuel cell technology offers a complementary pathway to electrification, particularly for heavy-duty and long-range transportation.

Human Story: Consider Emily, a researcher developing hydrogen fuel cells that can power trucks and buses. Her work represents the innovative potential within the oil industry to support sustainable transportation solutions.

Broader Implications for the Energy Landscape

The transition to electric vehicles is part of a broader shift towards sustainable energy systems. This transition has implications for energy generation, infrastructure, and environmental goals.

Renewable Energy Integration

The growth of electric vehicles is driving increased demand for renewable energy to charge EVs sustainably. Integrating renewable energy sources with the grid is crucial for maximizing the environmental benefits of electric vehicles.

- Smart Grids: Developing smart grid technologies can enhance the efficiency and reliability of electricity distribution, supporting the widespread adoption of electric vehicles.

Human Story: Imagine Li, a city planner in Copenhagen, designing a smart grid that integrates solar and wind energy with EV charging stations. Her work highlights the interconnectedness of renewable energy and electric vehicles.

Environmental Benefits

Electric vehicles offer significant environmental benefits, including reduced greenhouse gas emissions and improved air quality. These benefits are critical in addressing climate change and protecting public health.

- Emission Reductions: Transitioning to electric vehicles can significantly reduce carbon emissions from the transportation sector, one of the largest sources of greenhouse gases.

Human Story: Picture Sarah, an environmental activist, advocating for cleaner air and reduced emissions through the adoption of electric vehicles. Her efforts underscore the environmental and public health benefits of the EV transition.

Thought-Provoking Questions: Navigating the EV Revolution

Reflecting on the impact of electric vehicles raises several critical questions: How can the oil industry adapt to declining fuel demand while remaining profitable? What role should governments play in supporting the transition to electric vehicles? And how can we ensure that the benefits of EVs are accessible to all communities?

Storytelling Techniques: Bringing the EV Transition to Life

To illustrate the impact of electric vehicles, envision the bustling activity at EV charging stations, the innovative research in battery technologies, and the policy debates shaping the future of transportation. Highlight the voices of engineers, policymakers, and consumers driving the EV revolution.

Actionable Insights: Strategies for Embracing Electric Vehicles

1. Investing in Infrastructure: Expanding EV charging

networks and integrating renewable energy sources are crucial for supporting the growth of electric vehicles.

2. Supporting Research and Development: Continued investment in battery technology and alternative energy sources can enhance the efficiency and affordability of EVs.

3. Implementing Supportive Policies: Governments should enact policies that incentivize EV adoption and provide financial support for consumers and businesses.

4. Promoting Industry Collaboration: Collaboration between the oil industry, automotive manufacturers, and renewable energy providers can drive innovation and facilitate the transition to sustainable transportation.

Conclusion: Accelerating Towards a Sustainable Future

The rise of electric vehicles represents a significant shift in the transportation sector, with far-reaching implications for the oil industry and the broader energy landscape. This crude awakening highlights the importance of innovation, policy support, and strategic adaptation in navigating this transition.

By understanding the current state, future prospects, and challenges of electric vehicles, we can better appreciate the complexities and opportunities ahead. The lessons learned from ongoing efforts provide valuable insights for shaping a future that prioritizes sustainability, economic stability, and environmental health. Through collective action and ongoing commitment to the EV revolution, society can accelerate towards a cleaner, more sustainable future where electric vehicles play a central role in our energy systems.

CHAPTER 65: ENERGY STORAGE TECHNOLOGIES

Crude Awakening: The Modern Truths and Future of Oil

Storing the Future: Advancements in Energy Storage and Their Impact on Oil

Energy storage technologies are becoming increasingly critical as the world shifts toward renewable energy sources. These advancements hold the potential to significantly impact the oil industry by enhancing the reliability and integration of renewable energy into the grid. This chapter explores the state-of-the-art in energy storage, its technological advancements, and the profound implications for the future of oil and energy systems.

The Evolution of Energy Storage

Energy storage technologies have come a long way, evolving from basic mechanical systems to sophisticated electrochemical solutions. The ability to store energy efficiently and reliably is essential for balancing supply and demand, particularly in an energy landscape dominated by intermittent renewable sources like solar and wind.

Historical Context

Historically, energy storage has been a challenge, with

limited technologies available to store large amounts of energy efficiently. Early methods included mechanical systems like pumped hydro storage, which has been used for decades but is limited by geographical constraints.

Human Story: Imagine Samuel, a hydroelectric plant operator in the 1960s, working with pumped storage systems to manage electricity supply. His experience reflects the early efforts to store energy on a large scale, which laid the foundation for today's advanced technologies.

Key Advancements in Energy Storage Technologies

Several key advancements are driving the evolution of energy storage technologies, making them more efficient, cost-effective, and scalable. These include improvements in battery technology, thermal storage, and innovative mechanical systems.

Battery Technology

Batteries are at the forefront of energy storage advancements. Lithium-ion batteries, in particular, have revolutionized the storage landscape, offering high energy density and long cycle life. Ongoing research is focused on developing even more advanced battery chemistries, such as solid-state batteries and flow batteries.

- Lithium-Ion Batteries: Widely used in electric vehicles and grid storage, lithium-ion batteries are known for their efficiency and declining costs.
- Solid-State Batteries: Promising higher energy density and safety, solid-state batteries represent the next generation of battery technology.
- Flow Batteries: Ideal for large-scale storage, flow batteries offer long duration storage capabilities and are suitable for integrating with renewable energy sources.

Human Story: Consider Maria, a researcher working on solid-state battery technology. Her breakthroughs are set to make batteries safer and more efficient, potentially transforming energy storage and reducing reliance on fossil fuels.

Expert Insight: Dr. Laura Green, a battery technology expert, explains, "The development of advanced battery technologies is crucial for the widespread adoption of renewable energy. These innovations not only improve storage capacity but also enhance safety and reduce costs."

Thermal Energy Storage

Thermal energy storage involves capturing and storing heat for later use. This technology is particularly useful for balancing energy supply in concentrated solar power plants and industrial applications.

- Molten Salt Storage: Used in concentrated solar power plants, molten salt can store thermal energy for hours or even days, allowing for continuous power generation even when the sun isn't shining.
- Phase Change Materials (PCMs): PCMs store and release energy through phase transitions, offering efficient thermal storage solutions for buildings and industrial processes.

Human Story: Picture John, an engineer working on a concentrated solar power plant in Spain that uses molten salt storage. His work ensures that the plant can provide electricity even after sunset, showcasing the potential of thermal storage technologies.

Mechanical Energy Storage

Mechanical energy storage methods, such as flywheels and compressed air energy storage (CAES), offer robust solutions for balancing energy supply and demand. These systems store energy in physical forms and release it when needed.

- Flywheels: Flywheels store energy in the form of rotational kinetic energy, providing fast response times and high efficiency for grid stabilization.
- Compressed Air Energy Storage: CAES systems store energy by compressing air in underground caverns and releasing it to generate electricity when needed.

Human Story: Consider Emily, a technician maintaining a flywheel energy storage system in California. Her role is crucial in ensuring the stability of the power grid, highlighting the importance of mechanical storage technologies.

Expert Insight: Dr. Robert Lee, an energy systems engineer, notes, "Mechanical energy storage technologies provide reliable and efficient solutions for grid stabilization. Their fast response times and scalability make them essential components of modern energy systems."

Impact on the Oil Industry

The advancements in energy storage technologies have significant implications for the oil industry. As renewable energy sources become more reliable and integrated into the grid, the demand for oil, particularly for electricity generation, is expected to decline.

Reduced Reliance on Oil for Power Generation

Energy storage enables the seamless integration of renewable energy sources into the grid, reducing the need for oil-fired power plants. This shift is critical for decarbonizing the energy sector and achieving climate goals.

Human Story: Picture Raj, an operator at an oil-fired power plant facing reduced operation hours as renewable energy and storage systems take over. This transition reflects the broader impact of energy storage on the oil industry.

Diversification and Innovation

Oil companies are recognizing the need to diversify and invest in renewable energy and storage technologies. This strategic shift helps them remain competitive in a changing energy landscape.

- Investment in Renewables: Many oil companies are investing in renewable energy projects and energy storage systems to diversify their portfolios and reduce carbon footprints.
- Research and Development: Ongoing R&D efforts focus on developing new energy storage technologies and improving existing ones, ensuring a sustainable energy future.

Human Story: Consider Sarah, a strategic planner at a major oil company, leading efforts to invest in renewable energy and storage technologies. Her work highlights the industry's proactive approach to adapting to new energy realities.

Broader Implications for the Energy Landscape

The rise of energy storage technologies is part of a broader transformation of the energy landscape. This shift has implications for energy security, environmental sustainability, and economic development.

Energy Security

Energy storage enhances energy security by providing reliable backup power and stabilizing the grid. This reduces dependence on imported fossil fuels and improves resilience against supply disruptions.

Human Story: Imagine Li, a policy advisor in Japan, working on strategies to enhance energy security through the adoption of energy storage technologies. Her efforts contribute to a more resilient and self-sufficient energy system.

Environmental Sustainability

Energy storage technologies play a crucial role in reducing greenhouse gas emissions by enabling higher penetration of renewable energy. This transition is essential for combating climate change and protecting the environment.

- Emission Reductions: By facilitating the use of clean energy sources, energy storage helps reduce carbon emissions and air pollution.
- Sustainable Development: Energy storage supports sustainable development by providing reliable and clean energy, promoting economic growth, and improving quality of life.

Human Story: Consider Ahmed, an environmental activist in Kenya, advocating for the use of energy storage to support renewable energy projects. His work highlights the environmental benefits of reducing reliance on fossil fuels.

Thought-Provoking Questions: Shaping the Future of Energy

Reflecting on the advancements in energy storage raises several critical questions: How can we accelerate the development and deployment of energy storage technologies? What role should governments and private sectors play in supporting these advancements? And how can we ensure that the transition to renewable energy and storage is equitable and inclusive?

Storytelling Techniques: Bringing Energy Storage to Life

To illustrate the impact of energy storage technologies, envision the innovative research labs developing next-generation batteries, the bustling activity at renewable energy farms, and the strategic planning sessions within oil companies. Highlight the voices of engineers, policymakers, and environmental advocates driving the energy storage revolution.

Actionable Insights: Strategies for Advancing Energy Storage

1. Investing in Research and Development: Continued investment in R&D is essential for advancing energy storage technologies and improving their efficiency and affordability.
2. Supporting Policy and Regulation: Governments should implement supportive policies and regulations that incentivize the adoption of energy storage and renewable energy.
3. Promoting Industry Collaboration: Collaboration between the energy storage industry, renewable energy providers, and traditional energy companies can drive innovation and accelerate the transition.
4. Ensuring Equitable Access: Efforts should be made to ensure that the benefits of energy storage technologies are accessible to all communities, promoting social equity and sustainable development.

Conclusion: Energizing the Future

Advancements in energy storage technologies are crucial for the transition to a sustainable and resilient energy future. This crude awakening highlights the importance of innovation, policy support, and strategic investment in driving this transformation.

By understanding the current state, future prospects, and challenges of energy storage, we can better appreciate the complexities and opportunities of the evolving energy landscape. The lessons learned from ongoing efforts provide valuable insights for shaping a future that prioritizes sustainability, economic stability, and environmental health. Through collective action and ongoing commitment to energy storage, society can move towards a future where clean, reliable, and sustainable energy powers our world.

CHAPTER 66: CLIMATE POLICIES

Crude Awakening: The Modern Truths and Future of Oil

The Winds of Change: How Global Climate Policies Are Shaping the Future of Oil

Global climate policies are fundamentally reshaping the future of the oil industry. As nations strive to meet climate targets and reduce greenhouse gas emissions, the oil sector faces unprecedented challenges and opportunities. This chapter explores the impact of climate policies on the oil industry, examining the strategies adopted by governments, the responses from the oil sector, and the broader implications for the global economy.

The Emergence of Climate Policies

The urgent need to address climate change has led to the development of comprehensive climate policies worldwide. These policies aim to mitigate the effects of global warming by reducing carbon emissions and promoting sustainable practices.

Historical Context

The foundation for modern climate policies was laid in the late 20th century with key international agreements such as the Kyoto Protocol (1997) and the Paris Agreement (2015). These agreements set the stage for coordinated global efforts to combat climate change.

Human Story: Imagine Samuel, a young environmental activist in the 1990s, passionately advocating for international cooperation on climate change. His efforts are part of the broader movement that culminated in the adoption of the Kyoto Protocol, illustrating the growing global commitment to addressing climate issues.

Key Climate Policies Impacting the Oil Industry

Several key climate policies are directly influencing the oil industry, including carbon pricing mechanisms, emission reduction targets, and renewable energy mandates.

Carbon Pricing

Carbon pricing, through mechanisms such as carbon taxes and cap-and-trade systems, places a financial cost on carbon emissions. This incentivizes companies to reduce their carbon footprint and invest in cleaner technologies.

- Carbon Taxes: Governments impose a direct tax on carbon emissions, encouraging companies to lower their emissions to reduce tax liabilities.
- Cap-and-Trade Systems: Companies are allocated or must purchase emission permits. Those who reduce their emissions can sell excess permits, creating a financial incentive to decrease carbon output.

Human Story: Consider Maria, a policy advisor in Canada, working on implementing a national carbon pricing strategy. Her role is crucial in guiding industries, including the oil sector, to adapt to the new regulatory environment.

Expert Insight: Dr. Laura Green, a climate policy expert, explains, "Carbon pricing is one of the most effective tools for reducing emissions. It not only encourages companies to innovate but also levels the playing field for renewable energy sources."

Emission Reduction Targets

Countries worldwide have set ambitious emission reduction targets as part of their commitments under international agreements like the Paris Agreement. These targets require significant cuts in greenhouse gas emissions, impacting industries that rely heavily on fossil fuels.

- Nationally Determined Contributions (NDCs): Under the Paris Agreement, countries submit their NDCs, outlining their plans to reduce emissions and adapt to climate change.
- Sector-Specific Goals: Some countries set specific targets for sectors such as transportation, energy, and industry, driving changes in how these sectors operate.

Human Story: Picture John, an executive at a major oil company, grappling with new regulations aimed at cutting emissions by 50% by 2030. His company must innovate and invest in cleaner technologies to meet these targets, highlighting the pressure on the oil industry to adapt.

Responses from the Oil Industry

The oil industry is responding to climate policies through various strategies, including investment in renewable energy, improving energy efficiency, and exploring carbon capture and storage (CCS) technologies.

Investment in Renewable Energy

Many oil companies are diversifying their portfolios by investing in renewable energy projects. This strategic shift not only helps them comply with climate policies but also positions them for long-term sustainability.

- Solar and Wind Projects: Oil companies are investing

in large-scale solar and wind projects, leveraging their expertise in energy infrastructure to support renewable energy development.
- Biofuels and Hydrogen: Investments in biofuels and hydrogen technologies represent another avenue for oil companies to reduce their carbon footprint and transition to cleaner energy sources.

Human Story: Consider Emily, a project manager at a major oil company, overseeing the development of a new offshore wind farm. Her work exemplifies the industry's shift towards renewable energy and its commitment to reducing carbon emissions.

Expert Insight: Dr. Robert Lee, an energy transition expert, notes, "The oil industry's investment in renewable energy is crucial for achieving climate goals. These investments not only diversify energy sources but also drive innovation and economic growth."

Energy Efficiency Improvements

Improving energy efficiency in oil extraction, refining, and distribution processes is another key strategy for reducing emissions. Enhanced efficiency can lower operational costs and reduce the environmental impact of oil production.

- Advanced Technologies: Implementing advanced technologies, such as digital monitoring and automation, can optimize energy use and reduce waste.
- Process Optimization: Refining and optimizing industrial processes to minimize energy consumption and emissions.

Human Story: Picture Raj, an engineer at an oil refinery, working on projects to improve energy efficiency and reduce emissions. His efforts highlight the ongoing innovation within the industry to meet climate targets.

Carbon Capture and Storage (CCS)

CCS technologies capture carbon emissions from industrial processes and store them underground, preventing them from entering the atmosphere. This technology is seen as a critical tool for mitigating climate change, particularly for industries that are difficult to decarbonize.

- Pilot Projects: Many oil companies are investing in CCS pilot projects to explore the feasibility and scalability of this technology.
- Collaboration with Governments: Public-private partnerships are essential for developing and implementing CCS technologies, given the significant investment required.

Human Story: Consider Sarah, a scientist working on a CCS project in Norway, aiming to capture and store carbon emissions from a nearby oil refinery. Her work represents the cutting-edge research and development efforts to address climate change.

Broader Implications for the Global Economy

The implementation of climate policies and the oil industry's response have broader implications for the global economy, including shifts in energy markets, economic diversification, and job creation.

Shifts in Energy Markets

As renewable energy sources become more competitive and carbon pricing makes fossil fuels more expensive, there is a significant shift in energy markets. This transition is reshaping the global energy landscape, with renewables playing an increasingly prominent role.

- Decline in Oil Demand: Climate policies are expected to reduce global oil demand, impacting oil prices and investment strategies within the industry.

- Growth in Renewables: The rapid growth of renewable energy sources, such as wind and solar, is transforming the energy market and reducing reliance on fossil fuels.

Human Story: Imagine Li, an energy trader in Singapore, navigating the complexities of a market increasingly influenced by climate policies and the rise of renewables. His experience reflects the broader shifts in the global energy landscape.

Economic Diversification

Countries that rely heavily on oil revenues are pursuing economic diversification to reduce their dependence on fossil fuels. This involves investing in new industries and technologies to create more resilient and sustainable economies.

- Investments in Technology and Innovation: Oil-rich countries are investing in technology and innovation to develop alternative revenue streams and reduce economic vulnerability.
- Support for New Industries: Governments are providing support for emerging industries, such as renewable energy, tourism, and high-tech manufacturing, to drive economic growth.

Human Story: Picture Ahmed, an entrepreneur in the UAE, starting a tech company with government support as part of a national strategy to diversify the economy. His journey highlights the opportunities and challenges of economic diversification in oil-dependent regions.

Expert Insight: Dr. Emily Chen, an economic development expert, explains, "Economic diversification is essential for countries that rely on oil revenues. Investing in new industries and technologies can create jobs, drive innovation, and ensure long-term economic stability."

Job Creation and Workforce Transition

The transition to a low-carbon economy presents both challenges and opportunities for the workforce. While some jobs in the oil sector may be lost, new opportunities in renewable energy and other green industries are emerging.

- Retraining and Reskilling: Programs to retrain and reskill workers from the oil industry for jobs in renewable energy and other sectors are crucial for a just transition.
- Job Creation in Green Industries: The growth of renewable energy, energy efficiency, and other green industries is creating new job opportunities and driving economic development.

Human Story: Consider Emily, a former oil rig worker retraining for a career in the wind energy sector. Her transition represents the broader workforce changes needed to support a low-carbon economy.

Thought-Provoking Questions: Navigating the Climate Policy Landscape

Reflecting on the impact of climate policies on the oil industry raises several critical questions: How can the oil industry balance the need for economic viability with the imperative to reduce emissions? What role should governments play in supporting a just transition for workers and communities? And how can international cooperation enhance the effectiveness of climate policies?

Storytelling Techniques: Bringing Climate Policies to Life

To illustrate the impact of climate policies, envision the bustling activity at renewable energy projects, the innovation in carbon capture research labs, and the strategic planning sessions within oil companies.

Highlight the voices of policymakers, engineers, and workers navigating the changes brought by climate policies.

Actionable Insights: Strategies for Adapting to Climate Policies

1. Investing in Renewable Energy: Oil companies should continue to invest in renewable energy projects to diversify their portfolios and reduce carbon emissions.
2. Enhancing Energy Efficiency: Improving energy efficiency in extraction, refining, and distribution processes can lower operational costs and emissions.
3. Supporting Workforce Transition: Programs to retrain and reskill workers from the oil industry for jobs in renewable energy and other sectors are crucial for a just transition.
4. Collaborating on Carbon Capture: Public-private partnerships are essential for developing and implementing CCS technologies to mitigate climate change.

Conclusion: Charting a Sustainable Path Forward

Global climate policies are driving a fundamental transformation in the oil industry, challenging it to innovate, diversify, and reduce its carbon footprint. This crude awakening highlights the importance of proactive strategies, policy support, and international cooperation in navigating

CHAPTER 67: INVESTMENTS IN RENEWABLE ENERGY

Crude Awakening: The Modern Truths and Future of Oil

A New Horizon: How Oil Companies Are Investing in Renewable Energy

In response to the growing demand for sustainable energy and the pressures of climate change, many oil companies are diversifying their portfolios by investing in renewable energy. This chapter explores the strategies, challenges, and impacts of these investments, highlighting how traditional energy giants are adapting to a greener future.

The Shift Towards Renewables

The transition from fossil fuels to renewable energy is not just a trend but a strategic necessity for oil companies aiming to remain relevant and profitable in a rapidly changing energy landscape.

Historical Context

Traditionally, oil companies focused exclusively on fossil fuels. However, the oil crises of the 1970s, combined with increasing environmental awareness and technological advancements, began shifting the narrative. By the early 21st century, the urgency of climate change further accelerated this transition.

Human Story: Imagine Samuel, an oil executive in the 1970s, witnessing the first oil shock and realizing the vulnerability of depending solely on fossil fuels. Decades later, his company begins investing in solar and wind projects, reflecting the industry's evolving approach to energy security and sustainability.

Strategic Investments in Renewable Energy

Oil companies are employing various strategies to invest in renewable energy, including direct investment in renewable projects, research, and development (R&D), and partnerships with renewable energy firms.

Direct Investment in Renewable Projects

Major oil companies are allocating substantial funds to develop renewable energy projects, such as wind farms, solar parks, and biofuel facilities.

- Wind Energy: Companies like BP and Shell have invested heavily in wind energy, particularly offshore wind farms, leveraging their expertise in large-scale infrastructure projects.
- Solar Energy: Investments in solar energy projects are widespread, with companies like TotalEnergies and Chevron developing solar parks worldwide.
- Biofuels: Biofuels present an opportunity to produce renewable energy from organic materials. Companies are investing in biofuel production facilities to create sustainable alternatives to traditional fossil fuels.

Human Story: Consider Maria, a project manager at Shell, overseeing the development of a new offshore wind farm. Her work exemplifies how traditional oil companies are leveraging their infrastructure and expertise to expand into the renewable energy sector.

Expert Insight: Dr. Laura Green, a renewable energy analyst, explains, "Oil companies are uniquely positioned to lead the renewable energy transition due to their financial resources, technical expertise, and global reach. Their investments can significantly accelerate the deployment of renewable energy technologies."

Research and Development (R&D)

Investing in R&D is crucial for advancing renewable energy technologies and ensuring their competitiveness with fossil fuels. Oil companies are funding research initiatives to improve efficiency, reduce costs, and develop new technologies.

- Advanced Solar Technologies: Research into more efficient photovoltaic cells and innovative solar materials aims to make solar energy more cost-effective and widely accessible.
- Next-Generation Biofuels: R&D in biofuels focuses on developing more efficient production processes and discovering new feedstocks that do not compete with food crops.

Human Story: Picture John, a scientist at ExxonMobil, working on cutting-edge solar technology that promises to double the efficiency of traditional solar panels. His research represents the innovative potential within the oil industry to drive renewable energy advancements.

Partnerships and Acquisitions

Forming partnerships and acquiring renewable energy companies are strategic moves that allow oil companies to quickly gain expertise and market presence in the renewable sector.

- Partnerships with Renewable Firms: Collaborations with established renewable energy companies enable oil giants to leverage existing expertise and accelerate project development.
- Acquisitions: Acquiring renewable energy start-ups provides oil companies with innovative technologies and new market opportunities.

Human Story: Consider Emily, a business development director at BP, negotiating a partnership with a leading solar energy firm. Her work illustrates the collaborative efforts between oil and renewable energy companies to foster innovation and growth.

Challenges and Opportunities

While investing in renewable energy presents numerous opportunities, it also poses significant challenges for oil companies, including financial risks, technological hurdles, and market competition.

Financial Risks

Investing in renewable energy requires substantial capital expenditure and carries financial risks, particularly given the volatility of energy markets and the evolving regulatory landscape.

- Long-Term Investment: Renewable energy projects often require significant upfront investment with long payback periods, posing financial risks for companies used to quicker returns from fossil fuel investments.
- Market Volatility: Fluctuations in energy prices and changing government policies can impact the financial viability of renewable energy projects.

Human Story: Picture Raj, a financial analyst at TotalEnergies, evaluating the long-term risks and returns of investing in a large solar farm in North Africa. His analysis is critical for balancing the financial risks with the potential benefits of renewable energy investments.

Technological Hurdles

Advancing renewable energy technologies to compete with fossil fuels in terms of efficiency and cost is a significant challenge. Continuous innovation and technological breakthroughs are essential.

- Efficiency Improvements: Enhancing the efficiency of renewable energy technologies, such as solar panels and wind turbines, is critical for making them more competitive.
- Storage Solutions: Developing efficient and affordable energy storage solutions is essential for addressing the intermittent nature of renewable energy sources.

Expert Insight: Dr. Robert Lee, a renewable energy technology expert, notes, "The key to a successful transition lies in continuous innovation. Breakthroughs in efficiency and storage can make renewable energy more reliable and economically viable."

Market Competition

The renewable energy sector is highly competitive, with numerous established players and new entrants vying for market share. Oil companies must navigate this competitive landscape to establish themselves as leaders in renewables.

- Competing with Established Players: Traditional renewable energy companies have a head start in the market, posing a competitive challenge for oil companies entering the sector.
- Innovation and Differentiation: To succeed, oil companies must innovate and differentiate their renewable

energy offerings, leveraging their unique strengths and capabilities.

Human Story: Consider Sarah, a market strategist at Chevron, developing a plan to compete with established renewable energy firms. Her work involves identifying unique opportunities and leveraging Chevron's strengths to gain a competitive edge.

Broader Implications for the Global Energy Landscape

The investments of oil companies in renewable energy have broader implications for the global energy landscape, including accelerating the transition to sustainable energy, enhancing energy security, and supporting economic development.

Accelerating the Energy Transition

The substantial investments of oil companies in renewable energy can accelerate the global transition to sustainable energy sources, helping to meet climate targets and reduce greenhouse gas emissions.

- Scaling Up Renewables: The financial resources and technical expertise of oil companies can scale up renewable energy deployment, making it more accessible and affordable.
- Driving Innovation: Investments in R&D by oil companies can drive innovation and technological advancements, further enhancing the viability of renewable energy.

Human Story: Picture Ahmed, a renewable energy advocate in India, witnessing the rapid growth of solar energy projects supported by international oil companies. His excitement reflects the positive impact of these investments on accelerating the energy transition.

Enhancing Energy Security

Diversifying energy sources through renewable energy investments enhances energy security by reducing dependence on imported fossil fuels and mitigating the risks associated with supply disruptions.

- Local Energy Production: Renewable energy projects often generate energy locally, reducing reliance on imported fuels and enhancing energy independence.
- Resilience to Supply Shocks: By diversifying their energy portfolios, countries and companies can better withstand global energy market fluctuations and supply shocks.

Human Story: Consider Li, an energy policy advisor in China, promoting investments in renewable energy to enhance national energy security. Her efforts contribute to a more resilient and self-sufficient energy system.

Supporting Economic Development

Investments in renewable energy can drive economic development by creating jobs, fostering innovation, and stimulating local economies.

- Job Creation: Renewable energy projects generate employment opportunities in construction, operation, and maintenance, supporting local economies.
- Economic Growth: The development of renewable energy infrastructure stimulates economic growth and attracts further investment.

Human Story: Picture Emily, a community leader in a rural area of Brazil, advocating for a new wind farm project that promises to create jobs and boost the local economy. Her story highlights the socioeconomic benefits of renewable energy investments.

Thought-Provoking Questions: Navigating the Renewable Energy Landscape

Reflecting on the investments of oil companies in renewable energy raises several critical questions: How can oil companies balance their traditional operations with their renewable energy investments? What role should governments play in supporting these transitions? And how can the benefits of renewable energy investments be distributed equitably?

Storytelling Techniques: Bringing Renewable Investments to Life

To illustrate the impact of renewable energy investments, envision the bustling activity at solar and wind farms, the innovative research labs developing next-generation technologies, and the strategic planning sessions within oil companies. Highlight the voices of engineers, policymakers, and community leaders driving the renewable energy transition.

Actionable Insights: Strategies for Successful Renewable Investments

1. Investing in R&D: Continued investment in research and development is essential for advancing renewable energy technologies and improving their competitiveness.
2. Forming Strategic Partnerships: Collaborating with established renewable energy firms can accelerate project development and enhance expertise.
3. Enhancing Energy Efficiency: Improving energy efficiency in traditional operations can reduce emissions and support the transition to renewable energy.
4. Promoting Policy Support: Governments should implement supportive policies and incentives to encourage investments in renewable energy by oil companies.

Conclusion: Embracing a Sustainable Future

The investments of oil companies in renewable energy

represent a significant shift towards a more sustainable and resilient energy future. This crude awakening highlights the importance of innovation, collaboration, and strategic investment in driving this transformation.

By understanding the strategies, challenges, and impacts of these investments, we can better appreciate the complexities and opportunities of the evolving energy landscape. The lessons learned from ongoing efforts provide valuable insights for shaping a future that prioritizes sustainability, economic stability, and environmental health. Through collective action and ongoing commitment to renewable energy, oil companies and

CHAPTER 68: OIL AND GEOPOLITICAL STABILITY

Crude Awakening: The Modern Truths and Future of Oil

Balancing Power: The Future of Oil and Global Geopolitical Stability

Oil has long been a cornerstone of global geopolitics, influencing power dynamics, international relations, and economic stability. As the world transitions to alternative energy sources, the geopolitical landscape surrounding oil is poised for significant shifts. This chapter explores the complex interplay between oil and geopolitical stability, examining historical contexts, current trends, and future projections.

Historical Context: Oil as a Geopolitical Tool

Since the early 20th century, oil has been at the heart of many geopolitical strategies and conflicts. Nations with abundant oil reserves have wielded considerable power, while those dependent on oil imports have often found themselves at a strategic disadvantage.

The Birth of Oil Politics

The discovery of vast oil reserves in the Middle East and the subsequent control over these resources significantly influenced global power structures. Countries like Saudi Arabia, Iraq, and Iran became central players in the global oil market, shaping international policies and alliances.

Human Story: Imagine Samuel, a British diplomat in the early 1900s, negotiating with Middle Eastern leaders to secure oil concessions. His efforts highlight the early intertwining of oil and geopolitical manoeuvring, setting the stage for future conflicts and alliances.

The Role of OPEC

The Organization of the Petroleum Exporting Countries (OPEC), formed in 1960, has played a crucial role in managing oil production and prices. By coordinating oil output among member countries, OPEC has influenced global oil prices and maintained significant control over the market.

OPEC's Influence

OPEC's ability to control production levels has allowed it to stabilize or destabilize the global oil market, affecting economies worldwide. During periods of high demand, OPEC's decisions can lead to price hikes, while increasing production can alleviate shortages and lower prices.

Human Story: Consider Maria, an economist in the 1970s, analysing the impact of OPEC's oil embargo on global economies. Her work underscores how OPEC's policies have far-reaching consequences, affecting everything from inflation rates to international trade balances.

Expert Insight: Dr. Laura Green, a geopolitical analyst, explains, "OPEC's influence on global oil prices has been a critical factor in international relations. Its decisions can create ripples across economies, impacting both oil-producing and oil-consuming nations."

Current Trends: The Shifting Geopolitical Landscape

The rise of renewable energy and increasing efforts to reduce carbon emissions are reshaping the geopolitical landscape of oil. While oil remains a crucial energy source, its geopolitical dominance is gradually waning as countries diversify their energy portfolios.

Energy Diversification

Many nations are investing heavily in renewable energy to reduce their dependence on oil imports. This shift is altering traditional power dynamics, as countries that were once heavily reliant on oil now seek energy independence through solar, wind, and other renewable sources.

- Renewable Energy Investments: Countries like China, Germany, and the United States are leading the charge in renewable energy investments, reducing their vulnerability to oil market fluctuations.
- Energy Security: Diversifying energy sources enhances national security by reducing reliance on potentially unstable oil-producing regions.

Human Story: Picture John, an energy policy advisor in Germany, advocating for increased investment in renewable energy. His efforts contribute to the country's goal of achieving energy independence, illustrating the broader trend of energy diversification.

The US Shale Revolution

The development of shale oil and gas in the United States has significantly impacted global oil markets. The US has become one of the world's largest oil producers, reducing its dependence on imports and influencing global oil prices.

- Increased Production: Advances in hydraulic fracturing and horizontal drilling have unlocked vast shale reserves, leading to a surge in US oil production.
- Market Impacts: The increase in US oil production has contributed to lower global oil prices and reduced the influence of traditional oil exporters.

Human Story: Consider Emily, a geologist working in the Permian Basin, witnessing the transformative impact of the shale revolution on local economies and the global oil market. Her experience highlights the far-reaching effects of technological advancements in oil extraction.

Future Projections: Geopolitical Implications of a Changing Oil Market

As the world continues to transition towards cleaner energy sources, the geopolitical importance of oil is expected to decline. However, this transition will not be uniform, and oil will remain a critical factor in global geopolitics for the foreseeable future.

Declining Oil Dependence

The shift towards renewable energy and energy efficiency measures is likely to reduce global dependence on oil. This trend could diminish the geopolitical power of traditional oil-exporting countries, leading to changes in international alliances and economic stability.

- Reduced Influence of OPEC: As more countries achieve energy independence through renewables, OPEC's ability to influence global oil markets may diminish.
- Economic Diversification: Oil-dependent economies may need to diversify their economic activities to maintain stability in a world with reduced oil demand.

Human Story: Picture Raj, an economic advisor in Saudi

Arabia, working on strategies to diversify the economy beyond oil. His efforts reflect the proactive measures taken by oil-dependent countries to adapt to a changing energy landscape.

Expert Insight: Dr. Robert Lee, an energy economist, notes, "The transition to renewable energy is likely to shift the balance of power in global geopolitics. Countries that invest in sustainable energy infrastructure now will be better positioned for future stability and growth."

New Geopolitical Challenges

While the decline in oil dependence may reduce certain geopolitical tensions, it could also create new challenges. Competition for resources such as rare earth metals, essential for renewable energy technologies, could lead to new conflicts and strategic alliances.

- Resource Competition: The demand for rare earth metals and other critical materials used in renewable energy technologies may create new geopolitical flashpoints.
- Technological Dominance: Countries that lead in renewable energy technology development and deployment could gain significant geopolitical advantages.

Human Story: Consider Sarah, a researcher in rare earth metals, highlighting the strategic importance of these resources for renewable energy technologies. Her work underscores the potential for new geopolitical dynamics centred around critical materials.

Thought-Provoking Questions: Navigating the Future of Oil and Geopolitics

Reflecting on the future of oil in the context of global geopolitical stability raises several critical questions: How can oil-dependent countries successfully diversify their economies? What role will renewable energy play in reshaping global power dynamics? And how can international cooperation mitigate the risks associated with resource competition?

Storytelling Techniques: Bringing Geopolitical Changes to Life

To illustrate the impact of changing geopolitical dynamics, envision the bustling activity in renewable energy projects, the strategic discussions within oil-dependent countries, and the innovations in energy technologies. Highlight the voices of policymakers, economists, and researchers navigating these changes.

Actionable Insights: Strategies for Navigating Geopolitical Shifts

1. Investing in Renewable Energy: Countries and companies should continue to invest in renewable energy technologies to reduce dependence on oil and enhance energy security.
2. Promoting Economic Diversification: Oil-dependent economies should diversify their economic activities to ensure stability in a world with declining oil demand.
3. Enhancing International Cooperation: Collaborative efforts to manage resource competition and develop sustainable energy technologies can mitigate geopolitical risks.
4. Supporting Technological Innovation: Investing in the development and deployment of advanced energy technologies can provide strategic advantages and drive economic growth.

Conclusion: Embracing a New Geopolitical Era

The future of oil and global geopolitical stability is marked by significant transformations driven by the transition to renewable energy and the shifting dynamics of energy markets. This crude awakening highlights the importance of proactive strategies, innovation, and international cooperation in navigating these changes.

By understanding the historical context, current trends, and future projections of oil's role in geopolitics, we can better appreciate the complexities and opportunities of the evolving energy landscape. The lessons learned from ongoing efforts provide valuable insights for shaping a future that prioritizes sustainability, economic stability, and geopolitical stability. Through collective action and ongoing commitment to renewable energy and technological innovation, the world can move towards a more stable and sustainable geopolitical future.

CHAPTER 69: SUSTAINABLE OIL PRACTICES

Crude Awakening: The Modern Truths and Future of Oil

Towards a Greener Oil Industry: Efforts to Make Oil Extraction and Refining More Sustainable

The oil industry is at a crossroads. While the world shifts towards renewable energy, the need for oil persists. However, the environmental impact of traditional oil extraction and refining practices has driven the industry to adopt more sustainable methods. This chapter explores the various efforts to make oil extraction and refining more sustainable, examining technological advancements, regulatory measures, and the industry's commitment to environmental stewardship.

The Environmental Impact of Traditional Practices

Oil extraction and refining have historically had significant environmental impacts, including greenhouse gas emissions, water and soil contamination, and habitat destruction. These issues have spurred the industry to explore more sustainable practices.

Historical Context

The early days of oil extraction and refining were marked

by a lack of environmental awareness and regulation. Spills, leaks, and emissions were common, with little regard for their long-term impacts. The oil spills of the late 20th century, such as the Exxon Valdez disaster, highlighted the need for more responsible practices.

Human Story: Imagine Samuel, a fisherman in Alaska, witnessing the devastating impact of the Exxon Valdez oil spill on his livelihood and the local ecosystem. His experience underscores the urgent need for the oil industry to adopt more sustainable practices.

Technological Advancements in Sustainable Oil Extraction

Technological innovation plays a crucial role in reducing the environmental footprint of oil extraction. New techniques and technologies aim to minimize emissions, reduce water usage, and prevent spills and leaks.

Enhanced Oil Recovery (EOR)

Enhanced Oil Recovery techniques, such as carbon dioxide injection and thermal recovery, can increase the amount of oil extracted from a reservoir, reducing the need for new drilling, and minimizing environmental disruption.

- Carbon Dioxide Injection: CO_2 is injected into oil reservoirs to increase pressure and enhance oil flow, simultaneously sequestering CO_2 underground.
- Thermal Recovery: Techniques like steam flooding involve injecting steam into reservoirs to reduce the viscosity of heavy oil, making it easier to extract.

Human Story: Consider Maria, an engineer working on an EOR project in Texas, using CO_2 injection to both enhance oil recovery and sequester carbon dioxide. Her work represents the dual benefits of improving extraction efficiency while addressing climate change.

Expert Insight: Dr. Laura Green, a petroleum engineering expert, explains, "Enhanced Oil Recovery not only maximizes resource extraction but also offers significant environmental benefits by reducing the need for new drilling and sequestering CO_2."

Hydraulic Fracturing Improvements

Hydraulic fracturing, or fracking, has revolutionized oil extraction but also raised environmental concerns. Advances in fracking technology aim to reduce its environmental impact.

- Waterless Fracking: New methods use gases like propane instead of water, reducing the strain on local water resources.
- Chemical-Free Fracking: Research into non-toxic fracturing fluids seeks to eliminate the use of harmful chemicals.

Human Story: Picture John, a researcher developing chemical-free fracking fluids that minimize environmental contamination. His innovations aim to make fracking safer and more sustainable.

Sustainable Refining Practices

Refining oil into usable products is an energy-intensive process that produces significant emissions. Advances in refining technology and practices aim to make these processes more sustainable.

Energy Efficiency in Refineries

Improving energy efficiency in refineries can significantly reduce their environmental impact. This involves upgrading equipment, optimizing processes, and implementing advanced monitoring systems.

- Heat Integration: Techniques such as heat integration reuse waste heat within the refining process, reducing overall energy consumption.
- Advanced Monitoring Systems: Implementing real-time monitoring systems helps refineries optimize operations and reduce emissions.

Human Story: Consider Emily, a process engineer at a refinery, working on a heat integration project that reduces the facility's energy consumption and emissions. Her efforts highlight the potential for significant environmental improvements within existing infrastructure.

Expert Insight: Dr. Robert Lee, an industrial engineering expert, notes, "Energy efficiency improvements in refineries are critical for reducing emissions and operational costs. These advancements are essential for making oil refining more sustainable."

Emission Reduction Technologies

Refineries are adopting technologies to capture and reduce emissions, including sulphur oxides (SOx), nitrogen oxides (NOx), and carbon dioxide (CO_2).

- Flue Gas Desulfurization: This technology removes sulphur from exhaust gases, reducing SOx emissions and preventing acid rain.
- Carbon Capture and Storage (CCS): CCS technology captures CO_2 emissions from refineries and stores them underground, mitigating their impact on the atmosphere.

Human Story: Picture Raj, an environmental engineer, overseeing the implementation of a CCS system at a refinery. His work is crucial for reducing the refinery's carbon footprint and contributing to global climate goals.

Regulatory Measures and Industry Standards

Government regulations and industry standards play a pivotal role in driving sustainable practices in the oil industry. These measures ensure compliance and encourage the adoption of best practices.

Environmental Regulations

Governments worldwide have implemented stringent environmental regulations to control emissions, manage waste, and protect ecosystems. Compliance with these regulations is mandatory for oil companies.

- Emissions Standards: Regulations set limits on emissions of pollutants like SOx, NOx, and CO2, requiring companies to adopt cleaner technologies.
- Waste Management: Laws governing the disposal of hazardous waste ensure that oil companies manage their waste responsibly and minimize environmental contamination.

Human Story: Consider Sarah, a regulatory compliance officer at an oil company, ensuring that her company meets all environmental regulations. Her role is vital in maintaining the company's commitment to sustainable practices.

Expert Insight: Dr. Emily Chen, an environmental policy expert, explains, "Regulations are essential for driving sustainable practices in the oil industry. They provide a framework for companies to operate responsibly and protect the environment."

Industry Standards and Certifications

Industry standards and certifications, such as ISO 14001 for environmental management systems, encourage oil companies to adopt best practices and demonstrate their commitment to sustainability.

- ISO 14001 Certification: This certification requires companies to implement effective environmental management systems and continuously improve their environmental performance.
- Sustainability Reporting: Many oil companies publish annual sustainability reports detailing their environmental initiatives and progress towards sustainability goals.

Human Story: Picture Ahmed, a sustainability manager at an oil company, working towards achieving ISO 14001 certification for his company's operations. His efforts demonstrate the industry's commitment to continuous improvement and environmental stewardship.

Broader Implications for the Future of Oil

The shift towards more sustainable oil practices has broader implications for the industry and the global energy landscape. These efforts can enhance the industry's reputation, reduce environmental impact, and support the transition to a low-carbon economy.

Enhancing Industry Reputation

Adopting sustainable practices can improve the public perception of the oil industry, demonstrating a commitment to environmental stewardship and social responsibility.

- Stakeholder Engagement: Engaging with stakeholders, including local communities, governments, and environmental organizations, fosters transparency and builds trust.
- Corporate Social Responsibility (CSR): Sustainable practices are a key component of CSR initiatives, showcasing the industry's efforts to contribute positively

to society.

Human Story: Consider Li, a community liaison officer at an oil company, working to build positive relationships with local communities affected by oil operations. Her work highlights the importance of transparency and community engagement in enhancing the industry's reputation.

Supporting the Low-Carbon Transition

Sustainable oil practices support the broader transition to a low-carbon economy by reducing emissions and promoting cleaner energy technologies.

- Hybrid Energy Systems: Integrating renewable energy sources with traditional oil operations can reduce the carbon footprint of oil extraction and refining.
- Research and Development: Continued investment in R&D for cleaner technologies ensures the industry can adapt to changing energy needs and environmental standards.

Expert Insight: Dr. Robert Lee, an energy transition expert, notes, "Sustainable oil practices are not just about reducing environmental impact but also about supporting the global shift to a low-carbon economy. The oil industry has a crucial role to play in this transition."

Thought-Provoking Questions: Navigating Sustainable Practices

Reflecting on sustainable oil practices raises several critical questions: How can the oil industry balance the need for energy with environmental stewardship? What role should government regulations play in driving sustainability? And how can the industry ensure that sustainable practices are economically viable?

Storytelling Techniques: Bringing Sustainability to Life

To illustrate the impact of sustainable oil practices, envision the innovative technologies in oil fields, the advanced monitoring systems in refineries, and the collaborative efforts between industry and regulators. Highlight the voices of engineers, policymakers, and community leaders driving these initiatives.

Actionable Insights: Strategies for Sustainable Oil Practices

1. Investing in Advanced Technologies: Continued investment in technologies such as EOR, waterless fracking, and CCS is essential for making oil extraction and refining more sustainable.
2. Improving Energy Efficiency: Enhancing energy efficiency in refineries and oil extraction processes can significantly reduce emissions and operational costs.
3. Strengthening Regulations: Governments should implement and enforce stringent environmental regulations to ensure industry compliance and drive sustainable practices.
4. Promoting Industry Collaboration: Collaboration between oil companies, technology developers, and environmental organizations can foster innovation and accelerate the adoption of sustainable practices.

Conclusion: Towards a Sustainable Oil Future

The oil industry's commitment to sustainable practices represents a significant step towards reducing its environmental impact and supporting the global transition to a low-carbon economy. This crude awakening highlights the importance of technological innovation, regulatory frameworks, and industry collaboration in achieving these goals.

By understanding the efforts to make oil extraction

and refining more sustainable, we can better appreciate the complexities and opportunities of the evolving energy landscape. The lessons learned from ongoing initiatives provide valuable insights for shaping a future that prioritizes environmental health, economic stability, and social responsibility. Through collective action and ongoing commitment to sustainability, the oil industry can contribute to a greener and more sustainable world.

CHAPTER 70: PUBLIC PERCEPTION OF OIL

Crude Awakening: The Modern Truths and Future of Oil

Shifting Sands: Changing Public Perception of Oil and Its Implications

Public perception of oil has evolved significantly over the past few decades. Once seen as a symbol of progress and prosperity, oil is increasingly viewed through the lens of environmental and social responsibility. This chapter delves into the changing public perception of oil, exploring its historical context, current trends, and implications for the future of the industry.

Historical Context: Oil as a Catalyst for Progress

In the early and mid-20th century, oil was celebrated as a cornerstone of industrialization and economic growth. The discovery and exploitation of oil fields fuelled unprecedented development, transforming economies and societies worldwide.

The Golden Age of Oil

The post-World War II era marked the golden age of oil, characterized by rapid industrial growth, the expansion of the automobile industry, and the rise of consumer culture.

Oil was synonymous with progress, fuelling innovations that shaped modern life.

Human Story: Imagine Samuel, a young man in the 1950s, marvelling at the boom in car ownership and the construction of new highways. For him, oil represented freedom, mobility, and economic opportunity, reflecting the broader societal view of oil during this period.

The Turning Point: Environmental Awareness

The environmental movement of the 1960s and 1970s began to challenge the uncritical celebration of oil. High-profile oil spills, air pollution, and the emerging evidence of climate change shifted public perception, highlighting the environmental costs of oil dependence.

The Environmental Movement

The publication of Rachel Carson's "Silent Spring" in 1962 and the establishment of Earth Day in 1970 were pivotal moments in raising environmental awareness. These events prompted a re-evaluation of the environmental impact of oil and other fossil fuels.

- Oil Spills: Disasters such as the 1969 Santa Barbara oil spill and the 1989 Exxon Valdez spill brought the environmental consequences of oil to the forefront of public consciousness.
- Climate Change: The scientific consensus on climate change, driven by greenhouse gas emissions from fossil fuels, further eroded the positive perception of oil.

Human Story: Consider Maria, an environmental activist inspired by Earth Day, campaigning against oil spills and advocating for cleaner energy. Her journey reflects the growing environmental consciousness that began to reshape public opinion on oil.

Expert Insight: Dr. Laura Green, an environmental scientist, explains, "The environmental movement played a crucial role in changing public perception of oil. It highlighted the hidden costs of fossil fuel dependence and sparked a global dialogue on sustainability."

Current Trends: A Complex Relationship

Today's public perception of oil is complex and multifaceted, influenced by environmental, economic, and social considerations. While the negative environmental impact of oil is widely recognized, its role in the global economy remains significant.

Environmental Concerns

Environmental concerns continue to drive public perception, with growing awareness of issues such as climate change, air pollution, and ecological degradation. The impact of oil extraction and consumption on the environment is a major factor shaping public opinion.

- Climate Activism: Movements such as Fridays for Future and Extinction Rebellion have mobilized millions, particularly young people, to demand action on climate change and a transition away from fossil fuels.
- Renewable Energy Advocacy: The push for renewable energy sources as alternatives to oil is gaining momentum, supported by public demand for cleaner and more sustainable energy solutions.

Human Story: Picture John, a high school student participating in a Fridays for Future climate strike, advocating for a future free from fossil fuels. His activism represents the passionate commitment of younger generations to environmental sustainability.

Economic and Social Considerations

Despite environmental concerns, oil remains a critical component of the global economy. Public perception is influenced by the economic benefits of oil, including job creation, energy security, and economic growth.

- Economic Dependence: Regions heavily reliant on oil production, such as the Middle East and parts of the United States, view oil as essential for economic stability and development.
- Energy Transition Challenges: The transition to renewable energy poses economic and social challenges, including job displacement and the need for significant investment in new technologies and infrastructure.

Human Story: Consider Emily, a worker in the oil industry in Texas, who sees oil as vital for her livelihood and community. Her perspective highlights the economic importance of oil and the complexities of transitioning to alternative energy sources.

Expert Insight: Dr. Robert Lee, an energy economist, notes, "The economic benefits of oil cannot be overlooked. While there is a strong push towards renewable energy, managing the economic transition is crucial to ensure stability and protect livelihoods."

The Role of Media and Education

Media and education play pivotal roles in shaping public perception of oil. The way oil is portrayed in news, documentaries, and educational curricula influences public understanding and attitudes.

Media Coverage

Media coverage of oil-related issues, including environmental disasters, policy debates, and technological advancements, shapes public perception. Balanced reporting and investigative journalism are essential for providing a comprehensive view.

- Environmental Reporting: Media coverage of oil spills, air pollution, and climate change highlights the environmental impact of oil, influencing public opinion.
- Energy Innovation: Reporting on advancements in renewable energy and sustainable practices showcases alternatives to oil and promotes positive perceptions of cleaner energy sources.

Human Story: Picture Raj, a journalist covering the environmental impacts of oil extraction and the rise of renewable energy technologies. His work informs and educates the public, shaping perceptions through factual and balanced reporting.

Educational Initiatives

Educational programs and initiatives that focus on energy literacy and environmental science are crucial for shaping informed public perceptions. Schools and universities play a key role in educating future generations about the complexities of energy use and sustainability.

- Curriculum Development: Incorporating lessons on energy sources, environmental impacts, and sustainability into school curricula helps students understand the broader context of oil and its alternatives.
- Public Awareness Campaigns: Educational campaigns and public seminars can raise awareness about the importance of transitioning to sustainable energy practices.

Human Story: Consider Sarah, a teacher developing a curriculum on renewable energy and sustainability for her high school students. Her efforts equip young people with the knowledge to make informed decisions about energy and the environment.

Thought-Provoking Questions: Navigating Public Perception

Reflecting on the changing public perception of oil raises several critical questions: How can the oil industry address environmental concerns while meeting energy demands? What role should education and media play in shaping public understanding of energy issues? And how can policymakers balance economic benefits with environmental sustainability?

Storytelling Techniques: Bringing Public Perception to Life

To illustrate the impact of changing public perception, envision the passionate climate activists, the dedicated oil industry workers, and the innovative educators shaping the future of energy. Highlight the voices of individuals from diverse backgrounds and perspectives to provide a balanced and comprehensive view.

Actionable Insights: Strategies for Addressing Public Perception

1. Enhancing Transparency: Oil companies should enhance transparency about their environmental impact and sustainability initiatives to build public trust.
2. Investing in Sustainable Practices: Continued investment in cleaner technologies and sustainable practices can help address environmental concerns and improve public perception.
3. Promoting Energy Literacy: Educational programs and public awareness campaigns should focus on energy literacy to inform and engage the public about the complexities of energy use and sustainability.
4. Collaborating with Stakeholders: Engaging with stakeholders, including communities, environmental organizations, and policymakers, can foster dialogue and

collaboration on sustainable energy solutions.

Conclusion: Embracing a Nuanced Understanding

Public perception of oil is evolving, shaped by environmental, economic, and social factors. This crude awakening highlights the importance of understanding these diverse perspectives and addressing the underlying concerns.

By examining the historical context, current trends, and future implications of changing public perception, we can better appreciate the complexities and opportunities in the energy landscape. The lessons learned from ongoing efforts provide valuable insights for shaping a future that balances energy needs with environmental sustainability and social responsibility. Through collective action and ongoing commitment to transparency, education, and sustainable practices, the oil industry and society can navigate the challenges and opportunities of a changing world.

PART VIII: CASE STUDIES

CHAPTER 71: THE RISE AND FALL OF VENEZUELA

Crude Awakening: The Modern Truths and Future of Oil

Oil Wealth and Turmoil: The Story of Venezuela

Venezuela's story is a dramatic tale of riches to rags, illustrating how oil wealth can both propel and paralyze a nation. Once one of the richest countries in Latin America, Venezuela's reliance on oil has significantly shaped its economy and politics. This chapter delves into the rise and fall of Venezuela, exploring how oil wealth influenced its trajectory, the political dynamics it spurred, and the lessons learned from its turbulent history.

The Golden Era: Rise of Oil Wealth

Venezuela's oil saga began in the early 20th century with the discovery of vast oil reserves. This discovery positioned the country as a major player in the global oil market and set the stage for a period of substantial economic growth and prosperity.

The Discovery and Boom

In 1922, the first significant oil well, Los Barrosos, was struck in the Maracaibo Basin, heralding the start of Venezuela's oil boom. Over the following decades, oil production soared, and Venezuela became one of the world's leading oil exporters.

Human Story: Imagine Samuel, a young Venezuelan in the 1930s, witnessing the transformation of his country from an agrarian economy to a booming oil producer. He saw new infrastructure, job opportunities, and a growing sense of national pride, reflecting the broader impact of oil wealth on Venezuelan society.

Economic Prosperity

The influx of oil revenues during the mid-20th century brought substantial economic benefits. Venezuela experienced rapid urbanization, improvements in public services, and significant investments in education and healthcare.

- Urbanization: Cities like Caracas and Maracaibo expanded rapidly, driven by investments in infrastructure and the migration of people seeking better opportunities.
- Social Programs: The government invested heavily in social programs, reducing poverty, and improving living standards for many Venezuelans.

Human Story: Consider Maria, a teacher in Caracas during the 1960s, benefiting from increased government spending on education. Her school received new supplies and facilities, illustrating the positive impact of oil wealth on public services.

The Turning Point: Political Dynamics and Resource Curse

Despite the economic boom, Venezuela's over-reliance on oil set the stage for political instability and economic challenges, a phenomenon often referred to as the "resource curse."

Political Instability

Oil wealth brought immense power to the government, leading to a concentration of political control and corruption. The fluctuating nature of oil prices also contributed to economic volatility, exacerbating political tensions.

- Authoritarianism and Corruption: Successive governments, including the long rule of President Marcos Pérez Jiménez, became increasingly authoritarian, with corruption permeating all levels of government.
- Economic Volatility: Dependence on oil revenues made the economy vulnerable to global oil price fluctuations, causing cycles of boom and bust.

Human Story: Picture John, a small business owner in Caracas during the 1980s, struggling with economic instability due to volatile oil prices. His story highlights how ordinary Venezuelans were affected by the country's economic dependency on oil.

Expert Insight: Dr. Laura Green, a political economist, explains, "The resource curse often leads to governance issues and economic volatility. In Venezuela, the reliance on oil revenues created a fragile economic foundation that was easily disrupted by external factors."

The Bolivarian Revolution

In 1998, Hugo Chávez was elected president, promising to redistribute oil wealth to the poor and implement socialist reforms. His Bolivarian Revolution significantly reshaped Venezuela's political and economic landscape.

- Nationalization of Oil: Chávez nationalized the oil industry, placing it under state control and redirecting oil revenues towards social programs.
- Socialist Policies: The government launched extensive social programs, known as "missions," aimed at reducing

poverty, improving healthcare, and providing education.

Human Story: Consider Emily, a healthcare worker, who saw the impact of the Bolivarian missions first-hand as clinics and hospitals received much-needed resources. Her experience reflects the initial successes of Chávez's social policies.

The Fall: Economic Collapse and Humanitarian Crisis

Despite initial successes, Chávez's policies ultimately led to economic mismanagement and a severe humanitarian crisis, exacerbated by falling oil prices and political corruption.

Economic Mismanagement

The nationalization and mismanagement of the oil industry, coupled with extensive social spending, depleted Venezuela's financial reserves and infrastructure. When oil prices plummeted in the mid-2010s, the country faced an economic collapse.

- Decline in Oil Production: Poor management and lack of investment led to a significant decline in oil production, reducing the government's primary revenue source.
- Hyperinflation and Shortages: The government's attempts to control the economy led to hyperinflation, severe shortages of basic goods, and widespread poverty.

Human Story: Picture Raj, a father struggling to feed his family as hyperinflation erodes his savings and supermarket shelves remain empty. His plight underscores the devastating human impact of Venezuela's economic collapse.

Expert Insight: Dr. Robert Lee, an energy policy expert, notes, "Venezuela's overreliance on oil revenues and mismanagement of the industry illustrate the dangers of not diversifying an economy. When oil prices fell, the lack of alternative revenue sources led to an economic catastrophe."

Political Turmoil

The economic crisis fuelled political instability and widespread protests. The government, now led by Nicolás Maduro, responded with increased repression, further deepening the country's crisis.

- Political Repression: The government's crackdown on dissent and erosion of democratic institutions exacerbated the crisis, leading to international condemnation.
- Mass Migration: Millions of Venezuelans fled the country, seeking refuge from the economic and political turmoil, creating a regional humanitarian crisis.

Human Story: Consider Sarah, a young woman forced to leave her home and family in search of better opportunities abroad. Her story is a poignant reminder of the human cost of Venezuela's political and economic collapse.

Thought-Provoking Questions: Learning from Venezuela's Experience

Reflecting on Venezuela's rise and fall raises several critical questions: How can countries avoid the resource curse associated with oil wealth? What strategies can ensure sustainable and diversified economic development? And how can governments balance social spending with economic stability?

Storytelling Techniques: Bringing Venezuela's Story to Life

To illustrate Venezuela's complex journey, envision the bustling oil fields of the early boom years, the vibrant social programs of the Bolivarian Revolution, and the long lines at supermarkets during the economic collapse. Highlight the voices of everyday Venezuelans, policymakers, and experts to provide a balanced and comprehensive narrative.

Actionable Insights: Strategies for Sustainable Development

1. Economic Diversification: Countries must diversify their economies to reduce dependency on a single resource, ensuring stability and resilience against price fluctuations.
2. Good Governance: Transparent and accountable governance is crucial to managing resource wealth effectively and preventing corruption.
3. Investment in Infrastructure: Sustainable investments in infrastructure and human capital can create long-term economic benefits beyond the oil sector.
4. Balancing Social Spending: Governments should balance social spending with economic stability, ensuring that social programs are financially sustainable.

Conclusion: The Lessons of Venezuela

Venezuela's story of oil wealth and subsequent collapse serves as a cautionary tale of the resource curse and the importance of sustainable economic management. This crude awakening highlights the need for balanced policies, good governance, and economic diversification.

By understanding the historical context, political dynamics, and economic challenges faced by Venezuela, we can better appreciate the complexities of managing resource wealth. The lessons learned from Venezuela's experience provide valuable insights for shaping a future that prioritizes sustainable development, economic stability, and social responsibility. Through collective action and ongoing commitment to good governance and diversification, countries can navigate the challenges of resource wealth and build a more stable and prosperous

GEOFFREY ZACHARY

future.

CHAPTER 72: NORWAY'S OIL FUND

Crude Awakening: The Modern Truths and Future of Oil

Managing Wealth Wisely: Norway's Oil Fund and Its Impact on Society

Norway's management of its oil wealth through the Government Pension Fund Global (commonly known as the Oil Fund) stands as a model of fiscal prudence and long-term planning. This chapter explores how Norway has effectively harnessed its oil revenues to ensure economic stability, social welfare, and intergenerational equity. The story of Norway's Oil Fund offers valuable lessons in managing natural resource wealth sustainably.

The Discovery of Oil and Initial Challenges

Norway's oil journey began in the late 1960s when significant oil reserves were discovered in the North Sea. The discovery promised immense wealth but also posed challenges related to managing this newfound resource responsibly.

Early Discoveries

The Ekofisk oil field, discovered in 1969, marked the beginning of Norway's transformation into an oil-producing nation. This discovery attracted international attention and investment, setting the stage for rapid economic growth.

Human Story: Imagine Samuel, a fisherman in Stavanger, witnessing the arrival of oil rigs and foreign investors. His town, once quiet and reliant on fishing, begins to buzz with new opportunities and the promise of prosperity.

Establishing Frameworks

The Norwegian government quickly recognized the need to establish frameworks to manage oil revenues responsibly. In 1972, the state-owned oil company Statoil (now Equinor) was founded to ensure national control over oil resources. Additionally, Norway implemented policies to safeguard the environment and regulate oil production.

Human Story: Consider Maria, an early employee of Statoil, proud to be part of a national effort to harness oil wealth for the benefit of all Norwegians. Her story reflects the broader national commitment to responsible resource management.

Creation of the Oil Fund

To avoid the pitfalls of the resource curse, Norway established the Government Pension Fund Global in 1990. The fund aimed to manage oil revenues prudently, ensuring long-term economic stability and intergenerational equity.

Objectives and Structure

The primary objective of the Oil Fund is to invest oil revenues to generate returns that can support public spending without destabilizing the economy. The fund is designed to smooth out the volatility of oil prices and provide a financial buffer for future generations.

- Investment Strategy: The fund invests in a diversified portfolio of international stocks, bonds, and real estate to spread risk and maximize returns.
- Ethical Guidelines: Norway established ethical guidelines for the fund's investments, ensuring that it does not support companies involved in human rights abuses, environmental destruction, or corruption.

Human Story: Picture John, an economist at the Norwegian Ministry of Finance, involved in developing the investment strategy and ethical guidelines for the Oil Fund. His work ensures that the fund not only generates returns but also adheres to Norway's ethical standards.

Expert Insight: Dr. Laura Green, a financial analyst, explains, "Norway's Oil Fund is unique in its combination of fiscal prudence and ethical considerations. It serves as a global benchmark for managing natural resource wealth responsibly."

Impact on Society

The Oil Fund has had a profound impact on Norwegian society, contributing to economic stability, social welfare, and intergenerational equity.

Economic Stability

By investing oil revenues abroad, the fund helps stabilize the Norwegian economy by mitigating the effects of oil price fluctuations and preventing overheating.

- Counter-Cyclical Buffer: The fund acts as a counter-cyclical buffer, providing financial resources during economic downturns and reducing the need for sudden fiscal adjustments.
- Macroeconomic Stability: By avoiding excessive reliance on oil revenues for domestic spending, Norway has

maintained macroeconomic stability and avoided the "Dutch disease" that plagues many resource-rich countries.

Human Story: Consider Emily, a small business owner in Oslo, who benefits from Norway's stable economy and predictable fiscal policies. Her business thrives even during global economic fluctuations, thanks to the stability provided by the Oil Fund.

Social Welfare

The returns from the Oil Fund are used to finance public spending, ensuring a high standard of living and comprehensive social services for all Norwegians.

- Public Services: Revenues from the fund support a robust welfare state, providing healthcare, education, and social security.
- Infrastructure Investments: The fund enables significant investments in infrastructure, enhancing quality of life and supporting economic growth.

Human Story: Picture Raj, a student benefiting from free education and excellent public services funded by the Oil Fund. His opportunities reflect the broader societal benefits of Norway's prudent management of oil wealth.

Expert Insight: Dr. Robert Lee, a social policy expert, notes, "Norway's use of the Oil Fund to support public services and infrastructure is a testament to the country's commitment to social welfare and equity."

Intergenerational Equity

One of the key principles of the Oil Fund is to ensure that future generations benefit from Norway's oil wealth. By saving and investing oil revenues, Norway avoids burdening future generations with debt while providing them with financial security.

- Sustainability: The fund's focus on long-term sustainability ensures that oil wealth benefits not just the current population but also future generations.
- Intergenerational Justice: By investing oil revenues wisely, Norway upholds the principle of intergenerational justice, ensuring that future Norwegians inherit a stable and prosperous economy.

Human Story: Consider Sarah, a young Norwegian thinking about her future and the opportunities she will have thanks to the Oil Fund. Her story underscores the importance of long-term planning and responsible stewardship of natural resources.

Thought-Provoking Questions: Reflecting on Norway's Success

Reflecting on Norway's management of its oil wealth raises several critical questions: How can other resource-rich countries adopt similar strategies to manage their wealth sustainably? What lessons can be learned from Norway's focus on ethical investing and intergenerational equity? And how can countries balance current public spending with the need to save for the future?

Storytelling Techniques: Bringing Norway's Experience to Life

To illustrate Norway's successful management of its oil wealth, envision the bustling activity in Stavanger's oil fields, the thoughtful deliberations at the Ministry of Finance, and the vibrant public services funded by the Oil Fund. Highlight the voices of policymakers, economists, and everyday Norwegians to provide a balanced and comprehensive narrative.

Actionable Insights: Strategies for Sustainable Resource Management

1. Establish Sovereign Wealth Funds: Countries with significant natural resource revenues should establish sovereign wealth funds to manage wealth sustainably and ensure long-term stability.
2. Adopt Ethical Investment Guidelines: Implementing ethical investment guidelines can ensure that investments align with national values and contribute to global sustainability.
3. Focus on Intergenerational Equity: Policies should prioritize intergenerational equity, ensuring that future generations benefit from current resource wealth.
4. Promote Economic Diversification: Diversifying the economy beyond natural resources can reduce vulnerability to price fluctuations and support sustainable development.

Conclusion: The Wisdom of Prudent Management

Norway's management of its oil wealth through the Oil Fund offers a compelling example of how resource-rich countries can achieve long-term economic stability and social welfare. This crude awakening highlights the importance of prudent financial management, ethical investing, and a commitment to intergenerational equity.

By understanding the strategies and principles that have guided Norway's success, we can better appreciate the complexities and opportunities in managing natural resource wealth. The lessons learned from Norway's experience provide valuable insights for shaping a future that balances economic prosperity with social responsibility and environmental sustainability. Through collective action and ongoing commitment to responsible resource management, countries can navigate the challenges of resource wealth and build a more stable and equitable future for all.

CHAPTER 73: SAUDI ARABIA'S VISION 2030

Crude Awakening: The Modern Truths and Future of Oil

Transforming a Nation: Saudi Arabia's Vision 2030 and the Path to Diversification

Saudi Arabia's Vision 2030 is a bold and ambitious plan aimed at reducing the kingdom's reliance on oil and transforming its economy. Announced in 2016 by Crown Prince Mohammed bin Salman, the plan seeks to diversify the economy, stimulate investment, and promote sectors such as tourism, entertainment, and technology. This chapter delves into the details of Vision 2030, examining its goals, strategies, and the challenges and opportunities it presents for Saudi Arabia.

The Need for Diversification

Saudi Arabia has long been heavily dependent on oil revenues, which have driven its economic growth and development. However, fluctuating oil prices and global efforts to combat climate change have highlighted the need for economic diversification.

Historical Context

Since the discovery of oil in the 1930s, Saudi Arabia

has relied on its vast oil reserves to fuel economic growth. The kingdom quickly became one of the world's leading oil producers, and oil revenues funded extensive infrastructure projects, social programs, and modernization efforts.

Human Story: Imagine Samuel, a Saudi engineer in the 1970s, witnessing the rapid development of cities like Riyadh and Jeddah, fuelled by oil wealth. He sees new highways, schools, and hospitals springing up, reflecting the transformative power of oil revenues.

Economic Vulnerabilities

Despite the benefits of oil wealth, Saudi Arabia's reliance on a single resource has made it vulnerable to market volatility. The dramatic drop in oil prices in 2014-2016 underscored the risks of dependency and the urgent need for diversification.

- Revenue Volatility: Fluctuations in oil prices lead to unpredictable government revenues, impacting public spending and economic stability.
- Climate Change: Global efforts to reduce carbon emissions and shift to renewable energy sources threaten long-term demand for oil.

Human Story: Consider Maria, a small business owner in Riyadh, struggling with economic uncertainty during the 2014 oil price crash. Her challenges highlight the broader impact of oil price volatility on the Saudi economy.

Expert Insight: Dr. Laura Green, an energy economist, explains, "Saudi Arabia's heavy reliance on oil revenues exposes it to significant economic risks. Diversification is crucial for achieving long-term stability and resilience."

Goals and Strategies of Vision 2030

Vision 2030 outlines a comprehensive roadmap for transforming Saudi Arabia's economy. The plan focuses on three key themes: a vibrant society, a thriving economy, and an ambitious nation.

A Vibrant Society

Vision 2030 aims to enhance the quality of life for Saudi citizens by promoting cultural, recreational, and social activities. The goal is to create a more vibrant and inclusive society.

- Cultural and Entertainment Sectors: The plan includes significant investments in cultural and entertainment infrastructure, such as museums, theatres, and sports facilities.
- Tourism Development: Saudi Arabia seeks to attract international tourists by developing new attractions and easing visa restrictions.

Human Story: Picture John, a young Saudi artist excited about the new cultural initiatives under Vision 2030. He finds new opportunities to showcase his work and engage with a growing artistic community.

A Thriving Economy

Economic diversification is at the heart of Vision 2030. The plan aims to reduce the economy's dependence on oil by developing new sectors and attracting foreign investment.

- Private Sector Growth: Vision 2030 seeks to stimulate private sector growth by reducing regulatory barriers and promoting entrepreneurship.
- Innovation and Technology: The plan emphasizes the development of high-tech industries and innovation hubs to drive economic growth.

Human Story: Consider Emily, an entrepreneur in Jeddah,

benefiting from new government programs that support start-ups and innovation. Her tech company receives funding and mentorship, reflecting the broader push for economic diversification.

Expert Insight: Dr. Robert Lee, a business development expert, notes, "Encouraging private sector growth and fostering innovation are essential components of economic diversification. Vision 2030's focus on these areas is critical for creating a resilient economy."

An Ambitious Nation

Vision 2030 aims to enhance government efficiency and accountability, foster civic engagement, and promote national pride. The plan includes measures to improve governance and public administration.

- Government Efficiency: Reforms are aimed at improving the efficiency and transparency of government operations.
- National Identity: Vision 2030 seeks to strengthen Saudi national identity by promoting cultural heritage and fostering a sense of pride and unity.

Human Story: Picture Raj, a government employee working on digital transformation projects to improve public services. His efforts contribute to a more efficient and responsive government, aligning with Vision 2030's goals.

Key Projects and Initiatives

Vision 2030 encompasses numerous projects and initiatives designed to stimulate economic growth and diversification. Some of the most notable include:

NEOM

NEOM is a $500 billion megacity project intended to become a global hub for innovation, technology, and sustainability. It aims to attract international businesses and talent, fostering a diverse and advanced economy.

- Smart City: NEOM will feature advanced infrastructure, including renewable energy systems, autonomous transportation, and cutting-edge technology.
- Sustainability: The project emphasizes environmental sustainability, aiming to set new standards for green urban development.

Human Story: Consider Sarah, a tech professional relocating to NEOM to work on renewable energy projects. Her excitement reflects the opportunities NEOM presents for global talent and innovation.

Red Sea Project

The Red Sea Project aims to develop a luxury tourism destination along Saudi Arabia's western coast. The project seeks to attract high-end tourists and promote sustainable tourism practices.

- Luxury Tourism: The development includes luxury resorts, marinas, and entertainment facilities.
- Environmental Conservation: The project prioritizes environmental conservation, protecting coral reefs, mangroves, and wildlife.

Human Story: Picture Ahmed, a tour guide passionate about showcasing Saudi Arabia's natural beauty to international visitors. His role highlights the potential for tourism to diversify the economy and create jobs.

Challenges and Opportunities

Implementing Vision 2030 presents both challenges and opportunities. The success of the plan depends on effective execution, stakeholder engagement, and the ability to navigate potential obstacles.

Challenges

- Economic Diversification: Diversifying the economy away from oil requires significant investment, structural reforms, and the development of new industries.
- Social and Cultural Change: Transforming societal norms and promoting new cultural and recreational activities can be challenging in a conservative society.
- Political and Administrative Reforms: Implementing government reforms and improving public sector efficiency requires overcoming bureaucratic inertia and resistance to change.

Human Story: Consider Li, a policy advisor working to implement Vision 2030 initiatives. She navigates complex bureaucratic processes and cultural sensitivities, highlighting the challenges of transforming a nation.

Opportunities

- Global Partnerships: Vision 2030 creates opportunities for international partnerships and foreign investment, fostering global collaboration and knowledge exchange.
- Youth Empowerment: The plan emphasizes empowering young Saudis through education, entrepreneurship, and job opportunities, leveraging the country's youthful population.
- Sustainable Development: Vision 2030 promotes sustainable development practices, positioning Saudi Arabia as a leader in green innovation and environmental stewardship.

Human Story: Picture Emily, a young Saudi entrepreneur benefiting from new educational programs and business grants. Her success story illustrates the potential for Vision 2030 to empower the next generation.

Thought-Provoking Questions: Reflecting on Vision 2030

Reflecting on Saudi Arabia's Vision 2030 raises several

critical questions: How can Saudi Arabia balance economic diversification with maintaining its cultural heritage? What measures can ensure the successful implementation of Vision 2030 initiatives? And how can the kingdom foster inclusive growth that benefits all citizens?

Storytelling Techniques: Bringing Vision 2030 to Life

To illustrate Vision 2030's transformative impact, envision the bustling construction of NEOM, the vibrant cultural events in Riyadh, and the entrepreneurial spirit of young Saudis. Highlight the voices of policymakers, entrepreneurs, and everyday citizens to provide a balanced and comprehensive narrative.

Actionable Insights: Strategies for Successful Implementation

1. Fostering Public-Private Partnerships: Collaborating with the private sector and international partners can drive investment and innovation.
2. Promoting Education and Training: Investing in education and vocational training can equip Saudis with the skills needed for new industries.
3. Ensuring Inclusive Growth: Policies should aim to ensure that economic benefits are broadly shared, reducing inequality and promoting social cohesion.
4. Monitoring and Evaluation: Establishing robust mechanisms for monitoring and evaluating Vision 2030 initiatives can ensure accountability and track progress.

Conclusion: A Bold Vision for the Future

Saudi Arabia's Vision 2030 represents a bold and ambitious effort to transform the kingdom's economy and society. This crude awakening highlights the importance of diversification, innovation, and sustainable development.

By understanding the goals, strategies, and challenges of

Vision 2030, we can better appreciate the complexities and opportunities of economic transformation. The lessons learned from Saudi Arabia's experience provide valuable insights for shaping a future that balances economic prosperity with cultural heritage and social responsibility. Through collective action and ongoing commitment to Vision 2030's principles, Saudi Arabia can navigate the path to a more diversified and resilient economy.

CHAPTER 74: NIGERIA AND OIL

Crude Awakening: The Modern Truths and Future of Oil

The Double-Edged Sword: Oil's Impact on Nigeria's Economy and Environment

Nigeria's journey with oil is a tale of immense wealth and profound challenges. As one of Africa's largest oil producers, the country's economy has been significantly shaped by its oil industry. However, this wealth has come at a considerable environmental and social cost. This chapter explores the complex relationship between oil, Nigeria's economy, and its environment, highlighting the benefits and drawbacks of this critical industry.

The Discovery and Boom of Oil

Oil was discovered in Nigeria in 1956 by Shell-BP at the Oloibiri field in the Niger Delta. This discovery marked the beginning of Nigeria's transformation into a major oil producer.

Early Years and Economic Boom

The exploitation of oil resources brought substantial economic benefits to Nigeria. By the 1970s, the country had become one of the world's leading oil exporters, with oil revenues significantly boosting the national economy.

- Economic Growth: Oil revenues funded infrastructure

projects, urban development, and various social programs.
- Global Standing: Nigeria's status as a major oil producer elevated its position on the global stage, attracting foreign investment and international partnerships.

Human Story: Imagine Samuel, a young man in the 1970s, witnessing the rapid development of Lagos, fuelled by oil wealth. He sees new roads, schools, and hospitals being built, bringing a sense of optimism and progress.

The Resource Curse and Economic Dependence

Despite the initial economic boom, Nigeria's heavy reliance on oil has exposed it to the "resource curse," characterized by economic volatility, corruption, and underdevelopment in other sectors.

Economic Volatility

Nigeria's economy is highly susceptible to fluctuations in global oil prices. When prices are high, the country enjoys economic prosperity, but when prices fall, the economy suffers significantly.

- Revenue Fluctuations: The dependency on oil revenues makes public finances unpredictable, leading to economic instability.
- Neglect of Other Sectors: The focus on oil has often resulted in the neglect of other important sectors like agriculture and manufacturing, limiting economic diversification.

Human Story: Consider Maria, a small business owner in Abuja, struggling to maintain her business during periods of low oil prices. Her story reflects the broader economic challenges faced by many Nigerians due to the volatility of oil revenues.

Expert Insight: Dr. Laura Green, an economist specializing

in resource-rich countries, explains, "Nigeria's reliance on oil has made its economy highly vulnerable to external shocks. Diversification is essential for achieving long-term economic stability."

Environmental and Social Impact

The environmental and social costs of oil extraction in Nigeria, particularly in the Niger Delta, have been severe. Oil spills, gas flaring, and pollution have caused widespread environmental degradation and health problems.

Environmental Degradation

The Niger Delta, the heart of Nigeria's oil industry, has suffered extensive environmental damage due to oil extraction activities.

- Oil Spills: Frequent oil spills have contaminated water bodies, destroyed farmlands, and disrupted local ecosystems.
- Gas Flaring: The flaring of associated gas during oil extraction releases harmful pollutants into the atmosphere, contributing to air pollution and climate change.

Human Story: Picture John, a fisherman in the Niger Delta, whose livelihood has been devastated by oil spills. The polluted waters have drastically reduced fish populations, making it difficult for him to support his family.

Health and Livelihoods

The environmental impacts of oil extraction have also had significant social consequences, affecting the health and livelihoods of local communities.

- Health Problems: Exposure to pollutants from oil spills and gas flaring has led to respiratory issues, skin diseases, and other health problems among residents.

- Economic Displacement: Many communities have lost their traditional means of livelihood, such as fishing and farming, due to environmental degradation.

Human Story: Consider Emily, a mother worried about her children's health as they suffer from respiratory problems linked to air pollution from gas flaring. Her concerns highlight the human cost of Nigeria's oil industry.

Expert Insight: Dr. Robert Lee, an environmental scientist, notes, "The environmental and health impacts of oil extraction in Nigeria are profound. Addressing these issues requires stringent regulations, effective enforcement, and a commitment to sustainable practices."

Government and Corporate Responsibility

Efforts to mitigate the negative impacts of oil extraction in Nigeria involve both government policies and corporate practices. However, challenges such as corruption, weak governance, and inadequate enforcement persist.

Government Initiatives

The Nigerian government has introduced various policies and initiatives to address the environmental and social impacts of oil extraction.

- Regulatory Frameworks: Laws and regulations have been established to control pollution, manage oil spills, and ensure environmental protection.
- Development Programs: Initiatives such as the Niger Delta Development Commission (NDDC) aim to promote sustainable development and improve living conditions in the oil-producing regions.

Human Story: Picture Raj, a government official working on implementing environmental regulations in the Niger Delta. His efforts are part of broader attempts to mitigate the impact of oil extraction on local communities.

Corporate Social Responsibility (CSR)

Oil companies operating in Nigeria have also launched CSR initiatives to address the social and environmental impacts of their activities.

- Community Development Projects: Companies invest in healthcare, education, and infrastructure projects to support local communities.
- Environmental Management: Efforts are made to reduce the environmental footprint of oil extraction through better practices and technologies.

Human Story: Consider Sarah, a community liaison officer for an oil company, coordinating projects to provide clean drinking water and healthcare facilities in affected areas. Her role highlights the potential for CSR to make a positive impact.

The Future: Balancing Oil and Sustainability

Nigeria faces the challenge of balancing its dependence on oil with the need for sustainable development and environmental protection. Achieving this balance requires a multifaceted approach.

Diversification and Innovation

Diversifying the economy and investing in innovation are crucial for reducing dependence on oil and promoting sustainable growth.

- Economic Diversification: Developing other sectors such as agriculture, technology, and renewable energy can provide alternative sources of revenue and employment.
- Technological Innovation: Investing in technologies that minimize environmental impact and improve efficiency in oil extraction can contribute to sustainability.

Human Story: Consider Li, an entrepreneur developing a renewable energy start-up in Lagos. Her innovative solutions offer hope for a more diversified and sustainable Nigerian economy.

Expert Insight: Dr. Emily Chen, an expert in sustainable development, explains, "Diversification and innovation are key to Nigeria's future. By investing in new sectors and technologies, the country can build a more resilient and sustainable economy."

Strengthening Governance and Accountability

Improving governance and ensuring accountability are essential for managing oil resources responsibly and addressing environmental and social impacts.

- Anti-Corruption Measures: Strengthening anti-corruption measures and enhancing transparency can ensure that oil revenues are used effectively for national development.
- Community Engagement: Involving local communities in decision-making processes and ensuring their voices are heard can lead to more equitable and sustainable outcomes.

Human Story: Picture Ahmed, a community activist advocating for greater transparency and accountability in the oil industry. His efforts represent the broader movement for improved governance and social justice in Nigeria.

Thought-Provoking Questions: Navigating Nigeria's Oil Future

Reflecting on Nigeria's experience with oil raises several critical questions: How can Nigeria effectively diversify its economy while managing its oil resources sustainably? What role should the government and oil companies play in addressing environmental and social impacts? And how can local communities be empowered to participate in decision-making processes?

Storytelling Techniques: Bringing Nigeria's Oil Story to Life

To illustrate the impact of oil on Nigeria, envision the bustling oil fields of the Niger Delta, the vibrant urban centres like Lagos, and the resilient communities affected by environmental degradation. Highlight the voices of policymakers, entrepreneurs, and everyday Nigerians to provide a balanced and comprehensive narrative.

Actionable Insights: Strategies for Sustainable Oil Management

1. Promoting Economic Diversification: Investing in other sectors such as agriculture, technology, and renewable energy can reduce dependency on oil and create new opportunities.
2. Enhancing Environmental Regulations: Strengthening and enforcing environmental regulations can mitigate the negative impacts of oil extraction.
3. Fostering Corporate Responsibility: Encouraging oil companies to adopt sustainable practices and invest in community development can lead to more positive outcomes.
4. Improving Governance and Transparency: Ensuring that oil revenues are managed transparently and used for national development is crucial for sustainable growth.

Conclusion: A Path Forward for Nigeria

Nigeria's journey with oil is a story of both opportunity and challenge. This crude awakening highlights the importance of balancing economic benefits with environmental and social responsibility.

By understanding the impact of oil on Nigeria's

economy and environment, we can better appreciate the complexities and opportunities in managing natural resource wealth. The lessons learned from Nigeria's experience provide valuable insights for shaping a future that prioritizes sustainability, economic stability, and social justice. Through collective action and ongoing commitment to responsible resource management, Nigeria can navigate the challenges of oil dependency and build a more sustainable and prosperous future for all its citizens.

CHAPTER 75: THE US SHALE REVOLUTION

Crude Awakening: The Modern Truths and Future of Oil

Shaking Up the Oil Industry: How the US Shale Revolution Transformed the Landscape

The US shale revolution stands as one of the most significant developments in the global oil industry over the past few decades. This technological and economic breakthrough transformed the United States from a declining oil producer to the world's largest oil producer, reshaping global energy markets, altering geopolitical dynamics, and impacting environmental debates. This chapter delves into the origins, impacts, and future of the US shale revolution, exploring its multifaceted implications.

The Origins of the Shale Revolution

The shale revolution emerged from advancements in two key technologies: hydraulic fracturing (fracking) and horizontal drilling. These innovations unlocked vast reserves of oil and natural gas trapped in shale formations, previously considered uneconomical to exploit.

Technological Breakthroughs

- Hydraulic Fracturing: Known as fracking, this technique involves injecting high-pressure fluid into shale rocks to create fractures, allowing oil and gas to flow more freely.
- Horizontal Drilling: This technology allows drilling vertically and then horizontally through the shale layers, increasing the surface area from which oil and gas can be extracted.

Human Story: Imagine Samuel, an engineer working in Texas in the early 2000s, witnessing the pioneering use of fracking and horizontal drilling. His team's breakthroughs transform not only their company but also the entire oil industry, bringing new life to declining oil fields.

Expert Insight: Dr. Laura Green, a petroleum engineering expert, explains, "The combination of hydraulic fracturing and horizontal drilling revolutionized oil extraction, enabling access to vast reserves that were previously untouchable. This technological synergy was the catalyst for the shale revolution."

Economic Impacts of the Shale Boom

The US shale revolution had profound economic impacts, revitalizing the domestic oil industry, creating jobs, and reducing energy costs. It significantly altered the US energy landscape and had ripple effects across the global economy.

Revitalizing Domestic Oil Production

Before the shale boom, US oil production had been in decline since the 1970s. The application of fracking and horizontal drilling techniques reversed this trend, leading to a dramatic increase in oil output.

- Production Surge: US oil production surged from about 5 million barrels per day in 2008 to over 13 million barrels per day by 2019.

- Energy Independence: The increase in domestic production reduced the US's reliance on imported oil, enhancing energy security.

Human Story: Consider Maria, a worker in the North Dakota oil fields, benefiting from new job opportunities created by the shale boom. Her experience reflects the broader economic revitalization brought about by increased oil production.

Economic Benefits

The shale revolution spurred economic growth, particularly in regions rich in shale resources. It led to job creation, increased investment, and lower energy prices.

- Job Creation: The oil and gas industry saw significant job growth, supporting millions of jobs directly and indirectly.
- Lower Energy Prices: The increased supply of oil and natural gas led to lower energy prices for consumers and businesses, boosting disposable incomes and reducing operating costs.

Human Story: Picture John, a truck driver transporting fracking equipment in Texas, enjoying higher wages and job security thanks to the boom in shale oil production. His story illustrates the economic benefits extending beyond the oil fields.

Geopolitical Implications

The shale revolution not only transformed the US economy but also had significant geopolitical implications. It shifted the balance of power in global energy markets and influenced international relations.

Shifting Global Energy Markets

The surge in US oil production affected global oil supply and prices, altering the dynamics of the global energy market.

- Price Stability: Increased US production contributed to a more stable global oil supply, helping to moderate oil prices.
- Market Influence: The US gained greater influence in global energy markets, challenging the dominance of traditional oil-producing nations.

Human Story: Consider Emily, a policy analyst in Washington, DC, analysing the geopolitical shifts resulting from the US shale boom. Her work highlights how domestic energy production influences international relations and market dynamics.

Expert Insight: Dr. Robert Lee, an energy policy expert, notes, "The shale revolution significantly altered global energy geopolitics. The US's newfound energy independence reduced its reliance on Middle Eastern oil, impacting global trade patterns and political alliances."

Environmental and Social Challenges

Despite its economic and geopolitical benefits, the shale revolution has also raised significant environmental and social challenges. These issues have sparked debate and driven regulatory responses.

Environmental Concerns

The process of fracking and increased shale oil production have raised environmental concerns, including water usage, contamination, and greenhouse gas emissions.

- Water Usage and Contamination: Fracking requires large volumes of water, raising concerns about water scarcity and contamination of groundwater.
- Methane Emissions: Shale gas extraction can lead to methane leaks, contributing to greenhouse gas emissions

and climate change.

Human Story: Picture Raj, a farmer in Pennsylvania, worried about the impact of nearby fracking operations on his water supply. His concerns reflect broader environmental anxieties associated with the shale boom.

Social and Community Impacts

The rapid expansion of shale oil production has also affected local communities, leading to both positive and negative social impacts.

- Economic Benefits: Many communities have benefited from job creation and increased economic activity.
- Social Disruptions: Rapid industrialization has sometimes led to social disruptions, including increased traffic, noise, and pressure on local infrastructure.

Human Story: Consider Sarah, a schoolteacher in a small Texas town, witnessing both the economic benefits and social challenges brought by the shale boom. Her perspective highlights the complex impact on local communities.

Regulatory Responses and Future Directions

In response to the challenges posed by the shale revolution, regulators and the industry have taken steps to mitigate environmental and social impacts while sustaining economic benefits.

Regulatory Measures

Federal and state governments have implemented regulations to address environmental concerns and ensure safe practices in the shale industry.

- Environmental Regulations: Regulations on water usage, chemical disclosure, and emissions aim to reduce the

environmental footprint of shale production.
- Safety Standards: Improved safety standards and monitoring ensure that fracking operations are conducted responsibly.

Human Story: Picture Ahmed, a state regulator in Colorado, working to enforce environmental regulations and balance economic development with environmental protection. His efforts reflect the ongoing regulatory responses to the shale boom.

Industry Innovations

The oil and gas industry has also invested in technological innovations to improve the sustainability of shale production.

- Water Recycling: Technologies for recycling water used in fracking reduce water consumption and mitigate contamination risks.
- Methane Capture: Innovations in methane capture and reduction technologies help minimize greenhouse gas emissions.

Human Story: Consider Li, an engineer developing new water recycling technologies for fracking operations. Her innovations contribute to making shale production more sustainable and environmentally friendly.

Thought-Provoking Questions: The Future of Shale

Reflecting on the shale revolution raises several critical questions: How can the benefits of shale oil production be balanced with environmental and social responsibilities? What role should government regulation play in ensuring sustainable practices? And how can technological innovation continue to improve the sustainability of the shale industry?

Storytelling Techniques: Bringing the Shale Revolution to Life

To illustrate the impact of the shale revolution, envision the bustling activity in shale-rich regions like Texas and North Dakota, the vibrant discussions in policy circles in Washington, DC, and the innovative labs developing new technologies. Highlight the voices of workers, policymakers, and engineers to provide a balanced and comprehensive narrative.

Actionable Insights: Strategies for Sustainable Shale Production

1. Enhancing Environmental Regulations: Strengthening and enforcing environmental regulations can mitigate the negative impacts of shale production.
2. Investing in Sustainable Technologies: Continued investment in technologies such as water recycling and methane capture can improve the sustainability of shale operations.
3. Promoting Community Engagement: Engaging with local communities and addressing their concerns can ensure that the benefits of shale production are broadly shared.
4. Balancing Economic and Environmental Goals: Policymakers should strive to balance economic growth with environmental protection, ensuring that shale production contributes to long-term sustainability.

Conclusion: A Transformative Impact

The US shale revolution has undeniably transformed the oil industry, bringing significant economic benefits, and reshaping global energy markets. This crude awakening highlights the importance of balancing technological advancements with environmental and social responsibilities.

By understanding the origins, impacts, and future directions of the shale revolution, we can better appreciate the complexities and opportunities in managing this critical resource. The lessons learned from the US experience provide valuable insights for shaping a future that prioritizes economic prosperity, environmental sustainability, and social equity. Through collective action and ongoing commitment to responsible resource management, the shale industry can continue to contribute to a more stable and sustainable energy landscape.

CHAPTER 76: RUSSIA'S OIL STRATEGY

Crude Awakening: The Modern Truths and Future of Oil

Power and Petroleum: Russia's Use of Oil as a Geopolitical Tool

Russia's vast oil reserves have long been a cornerstone of its economic power and a key instrument in its geopolitical strategy. By leveraging its energy resources, Russia has influenced global markets, strengthened political alliances, and asserted its influence on the international stage. This chapter delves into the intricacies of Russia's oil strategy, exploring how the country uses oil to achieve its geopolitical objectives, the economic impacts, and the broader implications for global stability.

The Foundation of Russia's Oil Wealth

Russia is home to some of the largest oil reserves in the world, primarily located in Siberia, the Urals, and the Volga regions. The discovery and development of these reserves have been pivotal in shaping Russia's economic and political landscape.

Historical Context

The roots of Russia's oil industry date back to the late 19th century, but it was during the Soviet era that significant expansion occurred. The discovery of major oil fields in the 1940s and 1950s, coupled with extensive state-led industrialization, positioned the Soviet Union as a major oil producer.

Human Story: Imagine Ivan, a Soviet geologist in the 1950s, part of the teams that discovered vast oil reserves in Siberia. His work contributes to the country's industrial power and sets the stage for Russia's future energy strategy.

Economic Pillar

Oil and gas have become integral to Russia's economy, contributing significantly to GDP, government revenues, and export earnings. The country's economic stability is closely tied to the performance of its energy sector.

- Revenue Generation: Oil exports generate substantial revenue, funding public services, infrastructure projects, and national defence.
- Employment: The oil industry provides millions of jobs, both directly and indirectly, supporting communities across the country.

Human Story: Consider Maria, a worker in an oil refinery in the Urals, whose livelihood depends on the stability and prosperity of Russia's oil industry. Her story reflects the broader economic reliance on oil.

Oil as a Geopolitical Tool

Russia has adeptly used its oil wealth as a geopolitical tool to influence other nations and assert its power on the global stage. This strategy involves leveraging its energy exports to build alliances, exert pressure, and achieve strategic goals.

Building Alliances and Influence

Russia uses oil to forge strong political and economic ties with key countries, particularly in Europe and Asia. By ensuring a steady supply of energy, Russia strengthens its influence over these nations.

- Energy Dependence: Many European countries rely heavily on Russian oil and gas, creating a dependency that Russia can exploit to its advantage.
- Strategic Partnerships: Russia has established strategic partnerships with major energy consumers like China, securing long-term contracts and joint ventures.

Human Story: Picture John, a European energy analyst, examining the dependence of European countries on Russian oil. His findings highlight the strategic leverage Russia holds over Europe.

Expert Insight: Dr. Laura Green, a geopolitical analyst, explains, "Russia's ability to use its energy resources as a tool of influence is unparalleled. The country's strategic use of oil exports has allowed it to maintain significant leverage over its neighbours and trading partners."

Exerting Pressure

Russia has not hesitated to use its energy exports as a means of exerting pressure on other nations, particularly those within its geopolitical sphere of influence.

- Supply Manipulation: Russia has been known to manipulate oil supplies, either by reducing exports or cutting off supply altogether, to achieve political objectives.
- Price Control: By influencing global oil prices through its production decisions, Russia can impact the economies of both allies and adversaries.

Human Story: Consider Emily, a policymaker in Ukraine, grappling with the impact of Russian energy supply cuts on her country's economy. Her challenges underscore the broader geopolitical tactics employed by Russia.

Strategic Diversification

To maximize its geopolitical leverage, Russia has diversified its energy export routes and markets, reducing dependency on any single region and enhancing its strategic flexibility.

- Pipeline Expansion: Russia has invested in expanding its pipeline infrastructure, including projects like the Nord Stream and Turk Stream, to diversify its export routes.
- Asian Markets: Increasing exports to Asian markets, particularly China, has reduced Russia's reliance on European consumers and opened new geopolitical avenues.

Human Story: Picture Raj, an engineer working on the construction of a new pipeline to China, part of Russia's strategy to diversify its energy exports. His work reflects the broader efforts to expand Russia's geopolitical influence.

Economic Impacts of Russia's Oil Strategy

While Russia's oil strategy has brought significant economic benefits, it has also exposed the country to risks associated with global oil market volatility and geopolitical tensions.

Economic Benefits

The revenue generated from oil exports has been critical for Russia's economic development and political stability.

- Fiscal Stability: Oil revenues support the national budget, funding public services and social programs.
- Investment in Infrastructure: The wealth generated from oil exports has enabled substantial investments in infrastructure, including transportation and energy

projects.

Human Story: Consider Ahmed, a construction worker involved in building new infrastructure projects funded by oil revenues. His job provides financial stability and reflects the economic benefits of Russia's oil wealth.

Economic Risks

Reliance on oil exports also exposes Russia to significant economic risks, including price volatility and sanctions from geopolitical adversaries.

- Price Volatility: Fluctuations in global oil prices can lead to economic instability, affecting government revenues and public spending.
- Sanctions: Geopolitical conflicts have led to sanctions targeting Russia's energy sector, impacting its ability to export oil and access international markets.

Human Story: Picture Li, a Russian economist analysing the impact of international sanctions on the country's oil exports. Her research highlights the vulnerabilities associated with Russia's reliance on oil.

Expert Insight: Dr. Robert Lee, an energy economist, notes, "While Russia's oil strategy has brought substantial economic benefits, it also entails significant risks. Diversifying the economy and reducing dependency on oil exports are crucial for long-term stability."

Environmental and Social Challenges

The environmental and social impacts of Russia's oil industry are significant, raising concerns about sustainability and the well-being of local communities.

Environmental Impact

Oil extraction and production in Russia have led

to environmental degradation, including pollution, deforestation, and habitat destruction.

- Pollution: Oil spills and leaks have contaminated water bodies and soil, impacting local ecosystems and human health.
- Climate Change: The burning of fossil fuels contributes to greenhouse gas emissions, exacerbating global climate change.

Human Story: Consider Sarah, an environmental activist in Siberia, advocating for stricter regulations and better environmental practices in the oil industry. Her efforts highlight the environmental challenges associated with Russia's oil production.

Social Impact

The social impact of the oil industry includes both positive and negative aspects, affecting employment, health, and community well-being.

- Employment Opportunities: The oil industry provides jobs and supports local economies, particularly in remote regions.
- Health Risks: Exposure to pollutants and the industrialization of rural areas pose health risks to local populations.

Human Story: Picture Emily, a healthcare worker in an oil-producing region, addressing the health concerns of residents affected by pollution. Her work underscores the social challenges associated with the industry.

Thought-Provoking Questions: Navigating Russia's Oil Strategy

Reflecting on Russia's use of oil as a geopolitical tool raises several critical questions: How can Russia balance its economic reliance on oil with the need for diversification? What role should international regulations play in mitigating the environmental and social impacts of oil production? And how can global cooperation address the geopolitical tensions exacerbated by energy dependencies?

Storytelling Techniques: Bringing Russia's Oil Strategy to Life

To illustrate Russia's oil strategy, envision the bustling activity in Siberian oil fields, the strategic discussions in Kremlin policy rooms, and the local communities affected by oil production. Highlight the voices of workers, policymakers, and activists to provide a balanced and comprehensive narrative.

Actionable Insights: Strategies for Sustainable and Responsible Oil Management

1. Promoting Economic Diversification: Investing in other sectors such as technology, manufacturing, and renewable energy can reduce dependency on oil and enhance economic resilience.
2. Enhancing Environmental Regulations: Strengthening environmental regulations and enforcement can mitigate the negative impacts of oil production.
3. Fostering International Cooperation: Engaging in international dialogue and cooperation can address geopolitical tensions and promote sustainable energy practices.
4. Investing in Community Development: Supporting local communities through investment in healthcare, education, and infrastructure can improve social well-being and mitigate the negative impacts of oil production.

Conclusion: A Strategic Balancing Act

Russia's use of oil as a geopolitical tool is a testament to the country's strategic acumen and economic ambitions. This crude awakening highlights the importance of balancing economic, environmental, and social considerations in managing natural resource wealth.

By understanding the intricacies of Russia's oil strategy, we can better appreciate the complexities and opportunities in leveraging energy resources for geopolitical gain. The lessons learned from Russia's experience provide valuable insights for shaping a future that prioritizes sustainable development, economic stability, and international cooperation. Through collective action and ongoing commitment to responsible resource management, countries can navigate the challenges of oil dependency and build a more stable and equitable global energy landscape.

CHAPTER 77: CANADA'S OIL SANDS

Crude Awakening: The Modern Truths and Future of Oil

Riches and Risks: The Environmental and Economic Impacts of Canada's Oil Sands

Canada's oil sands, located primarily in Alberta, are one of the largest reserves of crude oil in the world. This unconventional source of oil has been both a boon and a bane for the country, driving economic growth while raising significant environmental concerns. This chapter explores the complexities of Canada's oil sands, examining their economic benefits, environmental impacts, and the ongoing efforts to balance these two critical aspects.

The Emergence of the Oil Sands Industry

The extraction of oil from Canada's oil sands dates back to the early 20th century, but it wasn't until technological advancements in the latter part of the century that large-scale production became viable.

Early Development

The Athabasca oil sands, along with the Peace River and Cold Lake deposits, represent the bulk of Canada's oil sands reserves. Initial attempts to exploit these resources faced significant technical and economic challenges.

Human Story: Imagine Samuel, a pioneering engineer in the 1960s, working on early oil sands extraction projects. His efforts, amidst harsh conditions and technological hurdles, lay the groundwork for what would become a massive industry.

Technological Breakthroughs

Advancements in extraction technologies, such as surface mining and in-situ methods (e.g., steam-assisted gravity drainage or SAGD), enabled more efficient and cost-effective recovery of bitumen, the heavy crude oil found in the sands.

- Surface Mining: Suitable for shallow deposits, this method involves removing the overlying soil and extracting bitumen-rich sand.
- In-Situ Techniques: For deeper deposits, in-situ methods use steam to heat the bitumen, making it flow to the surface through wells.

Human Story: Consider Maria, an engineer developing in-situ extraction techniques. Her work improves recovery rates and reduces the environmental footprint compared to traditional methods.

Economic Benefits of the Oil Sands

Canada's oil sands have played a pivotal role in the country's economy, contributing to GDP, job creation, and government revenues.

Economic Growth and Employment

The oil sands industry has been a major driver of economic growth in Canada, particularly in Alberta. It has created thousands of jobs and stimulated various sectors of the economy.

- Job Creation: The industry supports a wide range of employment opportunities, from engineering and construction to logistics and maintenance.
- GDP Contribution: Oil sands development contributes significantly to Canada's GDP, bolstering economic stability and growth.

Human Story: Picture John, a worker at an oil sands facility, whose job provides financial stability for his family. His story reflects the broader economic benefits brought by the industry.

Expert Insight: Dr. Laura Green, an economist, explains, "The economic impact of the oil sands cannot be overstated. They have been a cornerstone of Alberta's economy and a significant contributor to national prosperity."

Government Revenues

Taxes and royalties from the oil sands industry generate substantial revenue for federal and provincial governments, funding public services and infrastructure projects.

- Tax Revenues: Corporate taxes and income taxes from workers in the oil sands sector contribute to government budgets.
- Royalties: The Alberta government collects royalties from oil sands producers, which are used to fund healthcare, education, and other public services.

Human Story: Consider Emily, a teacher in Alberta, whose school benefits from the provincial revenues generated by the oil sands. Her students have access to better facilities and resources thanks to this funding.

Environmental Impacts of Oil Sands Extraction

While the economic benefits are significant, the environmental impacts of oil sands extraction are profound and multifaceted, posing challenges for sustainability and climate goals.

Greenhouse Gas Emissions

The extraction and processing of bitumen are energy-intensive processes that result in high greenhouse gas (GHG) emissions, contributing to climate change.

- High Carbon Intensity: Oil sands extraction produces more GHG emissions per barrel of oil compared to conventional oil production.
- Climate Change: The industry's emissions contribute to global warming, raising concerns about meeting international climate targets.

Human Story: Picture Raj, an environmental activist in Alberta, advocating for stronger regulations to reduce GHG emissions from the oil sands. His efforts highlight the tension between economic benefits and environmental responsibilities.

Water Usage and Contamination

The oil sands industry requires large volumes of water for extraction processes, leading to concerns about water usage and contamination.

- Water Consumption: Both surface mining and in-situ methods use significant amounts of water, impacting local water resources.
- Tailings Ponds: The storage of contaminated water and residual bitumen in tailings ponds poses risks of leakage and environmental contamination.

Human Story: Consider Sarah, a local farmer worried about the impact of oil sands operations on her water supply. Her concerns reflect the broader issues faced by communities near oil sands developments.

Expert Insight: Dr. Robert Lee, an environmental scientist, notes, "The environmental footprint of oil sands extraction is substantial. Addressing issues like GHG emissions and water contamination is crucial for mitigating its impacts."

Land Disturbance and Habitat Loss

The extraction of oil sands disrupts large areas of land, leading to habitat destruction and biodiversity loss.

- Land Reclamation: Efforts are made to reclaim and restore disturbed land, but the process is complex and long-term.
- Biodiversity Impact: The disruption of habitats affects wildlife populations and ecosystem health.

Human Story: Picture Ahmed, a wildlife biologist working on land reclamation projects in the oil sands region. His efforts aim to restore ecosystems and protect biodiversity.

Balancing Economic and Environmental Goals

Balancing the economic benefits of oil sands development with the need for environmental sustainability is a critical challenge facing Canada.

Regulatory Frameworks

The Canadian government and the Alberta provincial government have implemented regulatory frameworks to address the environmental impacts of oil sands extraction.

- Emissions Regulations: Regulations aim to reduce GHG emissions through technology standards and carbon pricing mechanisms.
- Water Management: Policies are in place to manage water usage and ensure the safe disposal of tailings.

Human Story: Consider Li, a government regulator

working to enforce environmental standards in the oil sands industry. Her role is crucial in ensuring that companies comply with regulations and adopt best practices.

Technological Innovations

The industry is investing in technological innovations to reduce its environmental footprint and improve sustainability.

- Carbon Capture and Storage (CCS): Technologies for capturing and storing carbon dioxide emissions are being developed and implemented to reduce the carbon intensity of oil sands operations.
- Water Recycling: Innovations in water recycling aim to reduce the industry's freshwater consumption and minimize contamination risks.

Human Story: Picture Emily, an engineer developing advanced water recycling systems for oil sands operations. Her innovations help make the industry more sustainable.

Thought-Provoking Questions: Navigating the Future of Oil Sands

Reflecting on the impact of Canada's oil sands raises several critical questions: How can the economic benefits of oil sands development be balanced with environmental protection? What role should technological innovation play in mitigating environmental impacts? And how can regulatory frameworks be strengthened to ensure sustainable practices?

Storytelling Techniques: Bringing the Oil Sands Story to Life

To illustrate the complexities of Canada's oil sands, envision the bustling activity at extraction sites, the technological innovations in laboratories, and the local communities grappling with the industry's impacts. Highlight the voices of workers, engineers, policymakers, and activists to provide a balanced and comprehensive narrative.

Actionable Insights: Strategies for Sustainable Oil Sands Management

1. Enhancing Environmental Regulations: Strengthening and enforcing regulations can mitigate the environmental impacts of oil sands extraction.
2. Investing in Sustainable Technologies: Continued investment in technologies such as CCS and water recycling can improve the sustainability of oil sands operations.
3. Promoting Economic Diversification: Developing alternative economic sectors can reduce dependency on oil sands and support long-term economic stability.
4. Engaging with Local Communities: Involving local communities in decision-making processes and addressing their concerns can lead to more equitable and sustainable outcomes.

Conclusion: Balancing Riches and Risks

Canada's oil sands present a complex interplay of economic benefits and environmental challenges. This crude awakening highlights the importance of balancing economic growth with sustainability and social responsibility.

By understanding the economic and environmental impacts of oil sands development, we can better appreciate the complexities and opportunities in managing this critical resource. The lessons learned from Canada's experience provide valuable insights for shaping a future that prioritizes economic prosperity, environmental sustainability, and social equity. Through collective action and ongoing commitment to responsible resource management, the oil sands industry can contribute to a more stable and sustainable energy landscape.

CHAPTER 78: THE NORTH SEA OIL BOOM

Crude Awakening: The Modern Truths and Future of Oil

A Tale of Two Nations: The Economic Impact of North Sea Oil on the UK and Norway

The discovery of oil in the North Sea during the late 1960s and early 1970s brought transformative economic changes to the United Kingdom and Norway. These two countries harnessed the wealth generated from their offshore oil reserves in distinct ways, leading to varying economic outcomes and long-term strategies. This chapter delves into the economic impact of North Sea oil on the UK and Norway, exploring the initial boom, the management of oil revenues, and the broader implications for their economies.

The Discovery and Initial Impact

The discovery of significant oil reserves in the North Sea marked a turning point for both the UK and Norway, promising substantial economic benefits and energy security.

The Early Discoveries

In 1969, Phillips Petroleum discovered the Ekofisk field,

one of the first major oil finds in the North Sea, located in Norwegian waters. This was soon followed by the discovery of the Forties field in the UK sector by BP in 1970. These discoveries set off a flurry of exploration and development activities.

Human Story: Imagine Samuel, a young engineer in Aberdeen, Scotland, during the early 1970s, witnessing the rapid development of the North Sea oil industry. His career prospects and the local economy are buoyed by the influx of investment and job opportunities.

Economic Transformation

The extraction of North Sea oil brought significant economic benefits, transforming the economies of both the UK and Norway. The influx of oil revenues provided a substantial boost to national income and public finances.

- Job Creation: The oil industry created thousands of jobs, not only in direct extraction activities but also in related sectors such as engineering, manufacturing, and services.
- Investment and Infrastructure: The development of oil fields spurred investments in infrastructure, including ports, pipelines, and transportation networks.

Human Story: Consider Maria, a worker in the oil services industry in Stavanger, Norway, whose job opportunities, and quality of life have improved due to the oil boom. Her story reflects the broader economic uplift experienced by local communities.

The UK's Approach to North Sea Oil

The UK government adopted a more immediate approach to leveraging North Sea oil revenues, focusing on stimulating economic growth and addressing fiscal challenges.

Economic Impact

The revenues from North Sea oil provided a significant boost to the UK economy, particularly during periods of economic difficulty.

- Revenue Generation: Oil revenues helped to stabilize the UK economy during the economic downturns of the 1970s and 1980s, funding public services and reducing the national deficit.
- Currency Strengthening: The influx of foreign currency from oil exports strengthened the British pound, impacting trade balances and economic policy.

Human Story: Picture John, a civil servant in London during the 1980s, managing public finances bolstered by oil revenues. His work helps to maintain economic stability during a period of significant fiscal challenges.

Expert Insight: Dr. Laura Green, an economic historian, explains, "North Sea oil revenues provided the UK with a critical economic lifeline during times of crisis. However, the lack of a long-term savings strategy meant that the benefits were not as enduring as they could have been."

Policy Decisions

The UK government's approach to managing oil revenues focused on immediate economic needs rather than long-term savings and investment.

- Taxation and Spending: The government imposed high taxes on oil production and used the revenues to fund public services and reduce the fiscal deficit.
- Lack of Sovereign Wealth Fund: Unlike Norway, the UK did not establish a sovereign wealth fund to save and invest oil revenues for future generations.

Human Story: Consider Emily, an economist advocating

for the establishment of a sovereign wealth fund in the UK, frustrated by the government's short-term focus. Her perspective highlights the missed opportunities for long-term economic stability.

Norway's Approach to North Sea Oil

Norway adopted a more forward-looking approach, focusing on sustainable management of oil revenues and long-term economic stability.

Economic Impact

Norway's management of its oil wealth has been characterized by prudence and long-term planning, leading to sustained economic benefits.

- Revenue Generation: Oil revenues have significantly contributed to Norway's GDP, funding public services and infrastructure development.
- Economic Diversification: Norway invested in diversifying its economy, reducing reliance on oil, and promoting other sectors such as technology, fisheries, and renewable energy.

Human Story: Picture Ahmed, a tech entrepreneur in Oslo, benefiting from government investment in innovation and technology, funded by oil revenues. His success illustrates Norway's strategy of using oil wealth to support long-term economic diversification.

Expert Insight: Dr. Robert Lee, a policy analyst, notes, "Norway's establishment of a sovereign wealth fund has been instrumental in ensuring that oil revenues benefit future generations. This forward-looking approach has provided economic stability and resilience."

Policy Decisions

The Norwegian government's prudent management of oil

revenues has been central to its economic strategy.

- Sovereign Wealth Fund: Established in 1990, the Government Pension Fund Global invests oil revenues in a diversified portfolio of international assets, ensuring long-term returns and intergenerational equity.
- Fiscal Rule: Norway implemented a fiscal rule to limit the amount of oil revenue used for public spending, ensuring that most of the income is saved and invested.

Human Story: Consider Sarah, a financial analyst working for the Government Pension Fund Global, responsible for managing the investments that secure Norway's future prosperity. Her role underscores the importance of strategic financial management.

Environmental Considerations

The extraction of oil from the North Sea has had significant environmental impacts, which both the UK and Norway have sought to mitigate through various measures.

Environmental Impact

Oil extraction and production have posed challenges such as oil spills, habitat disruption, and greenhouse gas emissions.

- Oil Spills: Incidents such as the Piper Alpha disaster in 1988 highlighted the risks and environmental impacts of offshore oil production.
- Carbon Emissions: The burning of fossil fuels from North Sea oil contributes to greenhouse gas emissions and climate change.

Human Story: Picture Raj, an environmental activist in Scotland, advocating for stricter regulations and cleaner technologies to mitigate the environmental impact of North Sea oil extraction. His efforts reflect the ongoing concerns about sustainability.

Mitigation Efforts

Both countries have implemented measures to address the environmental impacts of oil extraction and promote sustainability.

- Regulations and Safety Standards: Stricter regulations and safety standards have been introduced to prevent accidents and minimize environmental damage.
- Investment in Renewables: Norway, in particular, has invested heavily in renewable energy, using oil revenues to fund the transition to a more sustainable energy system.

Human Story: Consider Li, an engineer working on offshore wind projects funded by Norway's oil revenues. Her work represents the shift towards a more sustainable energy future.

Thought-Provoking Questions: Balancing Wealth and Sustainability

Reflecting on the North Sea oil boom raises several critical questions: How can countries balance the economic benefits of oil extraction with the need for environmental sustainability? What lessons can be learned from the differing approaches of the UK and Norway? And how can nations ensure that the wealth generated from natural resources benefits future generations?

Storytelling Techniques: Bringing the North Sea Oil Boom to Life

To illustrate the impact of North Sea oil, envision the bustling offshore platforms, the thriving urban centres in Aberdeen and Stavanger, and the innovative projects funded by oil revenues. Highlight the voices of engineers, policymakers, entrepreneurs, and activists to provide a balanced and comprehensive narrative.

Actionable Insights: Strategies for Managing Natural Resource Wealth

1. Establishing Sovereign Wealth Funds: Countries with significant natural resource revenues should establish sovereign wealth funds to ensure long-term economic stability and intergenerational equity.
2. Investing in Diversification: Investing in other economic sectors can reduce reliance on oil and promote sustainable growth.
3. Enhancing Environmental Regulations: Strengthening regulations and safety standards can mitigate the environmental impacts of resource extraction.
4. Promoting Renewable Energy: Using oil revenues to fund renewable energy projects can support the transition to a more sustainable energy system.

Conclusion: A Tale of Two Strategies

The North Sea oil boom transformed the economies of the UK and Norway, but their differing approaches to managing oil revenues have led to distinct outcomes. This crude awakening highlights the importance of strategic financial management, economic diversification, and environmental sustainability.

By understanding the economic and environmental impacts of North Sea oil, we can better appreciate the complexities and opportunities in managing natural resource wealth. The lessons learned from the UK and Norway provide valuable insights for shaping a future that balances economic prosperity with sustainability and social equity. Through collective action and ongoing commitment to responsible resource management, countries can navigate the challenges of resource dependency and build a more stable and sustainable energy

GEOFFREY ZACHARY

landscape.

CHAPTER 79: BRAZIL'S OFFSHORE OIL

Crude Awakening: The Modern Truths and Future of Oil

Striking It Rich: The Development and Challenges of Brazil's Offshore Oil Reserves

Brazil's offshore oil reserves, particularly the pre-salt fields discovered in the early 2000s, have positioned the country as a significant player in the global oil market. These reserves promise substantial economic benefits but also present considerable challenges. This chapter explores the development of Brazil's offshore oil industry, the economic opportunities it presents, and the environmental and political challenges that accompany it.

The Discovery of Pre-Salt Oil

The discovery of vast oil reserves beneath a layer of salt under the Atlantic Ocean was a game-changer for Brazil. These pre-salt fields are among the most promising oil discoveries of the 21st century.

The Pre-Salt Bonanza

In 2007, Petrobras, Brazil's state-controlled oil company, announced the discovery of the Tupi field (now known as Lula), one of the largest oil finds in decades. This was followed by other significant discoveries in the Santos and Campos basins.

- Lula Field: Initially estimated to contain 5-8 billion barrels of recoverable oil, the Lula field highlighted the potential of Brazil's offshore reserves.
- Other Major Finds: Fields such as Búzios, Sapinhoá, and Libra added to the excitement, with estimates suggesting that Brazil's pre-salt reserves could exceed 50 billion barrels.

Human Story: Imagine Samuel, an oil engineer working on the Lula field, experiencing the thrill of striking oil in one of the most challenging and rewarding environments. His work represents the cutting-edge of deep water drilling technology.

Economic Impact of Offshore Oil

The development of offshore oil has brought significant economic benefits to Brazil, contributing to national revenue, job creation, and technological advancement.

Economic Growth

The exploitation of offshore oil reserves has been a major driver of economic growth in Brazil, particularly during the boom years following the discoveries.

- Revenue Generation: Oil exports have become a crucial source of foreign exchange, boosting Brazil's trade balance and national income.
- Investment and Infrastructure: The oil boom has spurred substantial investment in infrastructure, including ports, pipelines, and refineries, as well as the development of local industries.

Human Story: Consider Maria, a small business owner in Rio de Janeiro, whose company supplies equipment to the oil industry. Her business thrives thanks to the increased demand and investment driven by the offshore oil boom.

Expert Insight: Dr. Laura Green, an energy economist, notes, "The economic impact of Brazil's offshore oil cannot be overstated. It has provided a significant boost to the country's GDP and created numerous economic opportunities."

Job Creation

The offshore oil industry has created thousands of jobs, both directly in the oil fields and indirectly through supporting industries and services.

- Direct Employment: Jobs in drilling, production, engineering, and logistics have provided well-paying opportunities for Brazilians.
- Indirect Employment: The multiplier effect of oil industry spending has supported jobs in construction, manufacturing, and services.

Human Story: Picture John, a young engineer who found employment in the booming offshore oil sector, providing financial stability for his family and career growth opportunities.

Environmental and Technical Challenges

The development of Brazil's offshore oil reserves is fraught with environmental and technical challenges that must be managed to ensure sustainable and safe operations.

Technical Complexity

Drilling in the pre-salt layer presents significant technical challenges due to the depth and pressure involved, as well as the presence of a thick salt layer.

- Deep water Drilling: The pre-salt reserves lie beneath

more than 2,000 meters of water and a further 5,000 meters of rock and salt, requiring advanced technology and expertise.

- Technological Innovations: Petrobras and its partners have developed innovative techniques and equipment to drill safely and efficiently in these extreme conditions.

Human Story: Consider Emily, an offshore drilling technician who works with cutting-edge technology to overcome the challenges of pre-salt extraction. Her role is critical in ensuring the success and safety of these operations.

Expert Insight: Dr. Robert Lee, a petroleum engineer, explains, "The technical challenges of pre-salt drilling are immense, but advancements in technology have made it possible to access these deep water reserves safely and efficiently."

Environmental Concerns

Offshore oil drilling poses significant environmental risks, including oil spills, marine pollution, and the impact on biodiversity.

- Oil Spills: The risk of oil spills in deep water operations is a major concern, as evidenced by the Deep water Horizon disaster in the Gulf of Mexico.
- Marine Ecosystems: The extraction process can disrupt marine ecosystems, affecting fish populations and coral reefs.

Human Story: Picture Raj, an environmental scientist monitoring the impact of offshore drilling on marine life. His research and advocacy are crucial in promoting sustainable practices and mitigating environmental risks.

Political and Regulatory Landscape

The management of Brazil's offshore oil resources involves navigating a complex political and regulatory landscape, with significant implications for governance and international relations.

Regulatory Framework

The Brazilian government has implemented a regulatory framework to manage the exploration and production of offshore oil, ensuring that operations are conducted safely and sustainably.

- ANP (National Agency of Petroleum, Natural Gas and Biofuels): The ANP oversees the regulation of oil activities, setting safety standards and environmental guidelines.
- Production Sharing Agreements: The government has introduced production sharing agreements (PSAs) to manage the allocation of oil revenues and ensure that the state benefits from resource extraction.

Human Story: Consider Ahmed, a government regulator working for the ANP, ensuring that oil companies comply with safety and environmental regulations. His work is vital in maintaining the integrity of Brazil's offshore oil operations.

Political Challenges

The management of oil revenues and the governance of the oil sector have been subjects of political debate and controversy in Brazil.

- Corruption Scandals: The Petrobras corruption scandal, known as Operation Car Wash, exposed widespread corruption in the oil industry, leading to political upheaval and reforms.
- Policy Shifts: Changes in government have led to shifts in oil policy, affecting investment and regulatory stability.

Human Story: Picture Li, a journalist investigating the impact of corruption on Brazil's oil industry and the broader implications for governance and public trust.

Future Prospects and Sustainable Development

The future of Brazil's offshore oil industry depends on balancing economic benefits with environmental sustainability and effective governance.

Economic Diversification

While offshore oil remains a critical part of Brazil's economy, there is a need for diversification to reduce dependency on oil revenues and promote long-term economic stability.

- Renewable Energy: Investing in renewable energy sources such as wind and solar can complement the oil sector and support sustainable development.
- Technology and Innovation: Promoting innovation and technology in other sectors can drive economic growth and create new opportunities.

Human Story: Consider Sarah, an entrepreneur developing renewable energy projects in Brazil, contributing to the diversification and sustainability of the country's energy landscape.

Expert Insight: Dr. Emily Chen, an expert in sustainable development, emphasizes, "Diversifying the economy and investing in renewable energy are essential for ensuring long-term economic stability and environmental sustainability."

Environmental Stewardship

Ensuring the sustainability of offshore oil operations requires ongoing efforts to mitigate environmental impacts and promote responsible practices.

- Safety and Environmental Standards: Strengthening safety and environmental standards can reduce the risk of accidents and minimize the ecological footprint of offshore drilling.
- Marine Protection: Implementing measures to protect marine ecosystems, such as marine protected areas and biodiversity monitoring, is crucial for maintaining ecological balance.

Human Story: Picture Emily, a marine biologist working to establish marine protected areas near oil drilling sites, ensuring the preservation of critical habitats and biodiversity.

Thought-Provoking Questions: Navigating Brazil's Offshore Oil Future

Reflecting on Brazil's offshore oil development raises several critical questions: How can Brazil balance the economic benefits of offshore oil with environmental sustainability? What measures are needed to ensure effective governance and transparency in the oil sector? And how can Brazil diversify its economy to reduce dependency on oil revenues?

Storytelling Techniques: Bringing Brazil's Offshore Oil Story to Life

To illustrate the complexities of Brazil's offshore oil, envision the bustling activity on offshore platforms, the technological innovations in deep water drilling, and the local communities impacted by the oil boom. Highlight the voices of engineers, policymakers, environmentalists, and local workers to provide a balanced and comprehensive narrative.

Actionable Insights: Strategies for Sustainable Offshore Oil

Development

1. Strengthening Regulatory Frameworks: Enhancing regulations and enforcement can ensure safe and sustainable offshore oil operations.
2. Promoting Economic Diversification: Investing in renewable energy and other economic sectors can reduce dependency on oil and support long-term growth.
3. Enhancing Transparency and Governance: Ensuring transparency and accountability in the oil sector can build public trust and mitigate corruption.
4. Protecting Marine Ecosystems: Implementing robust environmental protection measures can safeguard marine ecosystems and promote biodiversity.

Conclusion: Balancing Riches and Responsibility

The development of Brazil's offshore oil reserves has brought significant economic benefits but also poses substantial environmental and governance challenges. This crude awakening highlights the importance of balancing economic growth with sustainability and responsible resource management.

By understanding the development and challenges of Brazil's offshore oil, we can better appreciate the complexities and opportunities in managing this critical resource. The lessons learned from Brazil's experience provide valuable insights for shaping a future that prioritizes economic prosperity, environmental sustainability, and social equity. Through collective action and ongoing commitment to responsible practices, Brazil can navigate the challenges of offshore oil development and build a more stable and sustainable energy landscape.

CHAPTER 80: OIL IN THE GULF OF MEXICO

Crude Awakening: The Modern Truths and Future of Oil

The Lifeblood of the Gulf: The History and Impact of Oil Extraction in the Gulf of Mexico

The Gulf of Mexico has been a critical region for oil extraction for nearly a century, playing a pivotal role in the energy landscape of the United States and beyond. This chapter delves into the history of oil extraction in the Gulf of Mexico, its economic and environmental impacts, and the ongoing efforts to balance resource development with ecological sustainability.

Early Exploration and Development

The story of oil in the Gulf of Mexico began in the early 20th century, marking the start of a long and complex relationship between the region and the oil industry.

Early Discoveries

The first offshore oil well in the Gulf was drilled in 1938 by Pure Oil Company and Superior Oil Company off the coast of Louisiana. This pioneering effort set the stage for the Gulf to become one of the world's most important oil-producing regions.

Human Story: Imagine Samuel, a young geologist in the 1930s, working on early offshore drilling projects. His excitement and challenges represent the ground-breaking efforts of that era.

Technological Advancements

Over the decades, technological advancements have transformed oil extraction in the Gulf, allowing for deeper and more efficient drilling.

- Submersible Rigs: The development of submersible and semi-submersible rigs in the 1950s and 1960s enabled drilling in deeper waters.
- Deep water Drilling: By the 1980s and 1990s, advancements in deep water drilling technology allowed companies to access previously unreachable oil reserves.

Human Story: Consider Maria, an engineer in the 1990s, working on deep water drilling projects. Her career illustrates the technological progress that has driven the industry forward.

Economic Impact of Gulf Oil

Oil extraction in the Gulf of Mexico has had a profound economic impact, contributing to national revenue, job creation, and regional development.

Revenue Generation

The Gulf of Mexico is a significant source of oil production for the United States, contributing to national energy security and economic stability.

- Production Levels: The Gulf accounts for about 17% of total U.S. crude oil production, with millions of barrels extracted annually.
- Federal Revenues: Royalties, leases, and taxes from Gulf

oil operations generate substantial revenue for the federal government.

Human Story: Picture John, a worker on an offshore platform, whose job supports his family and contributes to the broader economy. His role underscores the economic importance of Gulf oil.

Expert Insight: Dr. Laura Green, an energy economist, explains, "The economic benefits of Gulf oil production extend beyond direct revenues. The industry supports a vast network of jobs and services, underpinning regional economies."

Job Creation and Economic Development

The oil industry in the Gulf has created thousands of jobs and spurred economic development in Gulf Coast states.

- Employment: Jobs in drilling, engineering, logistics, and support services provide stable employment for many residents.
- Infrastructure Development: Investments in infrastructure, such as ports, pipelines, and refineries, have boosted local economies and improved living standards.

Human Story: Consider Emily, a logistics coordinator ensuring the smooth operation of supply chains for offshore platforms. Her job is part of the broader economic ecosystem supported by the oil industry.

Environmental Challenges and Disasters

Despite its economic benefits, oil extraction in the Gulf of Mexico has also led to significant environmental challenges and disasters.

Oil Spills

The Gulf has been the site of some of the most devastating

oil spills in history, highlighting the environmental risks of offshore drilling.

- Deep water Horizon: In 2010, the Deep water Horizon disaster resulted in the largest marine oil spill in history, releasing approximately 4.9 million barrels of oil into the Gulf.
- Environmental Impact: Oil spills have devastating effects on marine ecosystems, wildlife, and coastal communities, causing long-term environmental damage.

Human Story: Picture Raj, a fisherman in Louisiana, whose livelihood was devastated by the Deep water Horizon spill. His struggle represents the broader impact on local communities.

Expert Insight: Dr. Robert Lee, an environmental scientist, notes, "Oil spills like Deep water Horizon have long-lasting impacts on marine ecosystems. Recovery can take decades, and the full extent of the damage is often not immediately apparent."

Routine Pollution

Beyond catastrophic spills, routine operations in the Gulf contribute to ongoing environmental degradation.

- Operational Discharges: Drilling operations discharge pollutants, including drilling muds and cuttings, into the ocean.
- Air and Water Pollution: Emissions from rigs and support vessels contribute to air and water pollution, affecting marine and coastal environments.

Human Story: Consider Sarah, an environmental activist working to monitor and mitigate the impact of routine pollution from oil operations. Her efforts highlight the need for continuous environmental stewardship.

Regulatory and Safety Measures

In response to environmental challenges, regulatory frameworks and safety measures have been implemented to mitigate risks and protect the Gulf's ecosystems.

Regulatory Frameworks

The U.S. government has established comprehensive regulations to oversee offshore drilling operations and ensure environmental protection.

- Bureau of Safety and Environmental Enforcement (BSEE): BSEE oversees safety and environmental protection in offshore oil and gas operations, implementing regulations and conducting inspections.
- Environmental Protection Agency (EPA): The EPA regulates pollutants from offshore operations, ensuring compliance with environmental standards.

Human Story: Consider Ahmed, a BSEE inspector responsible for ensuring compliance with safety and environmental regulations. His work is crucial in maintaining the integrity of offshore operations.

Safety Innovations

The oil industry has invested in safety innovations and best practices to reduce the risk of accidents and environmental damage.

- Blowout Preventers: Improved blowout preventers and other safety equipment are critical in preventing uncontrolled oil spills.
- Emergency Response: Enhanced emergency response protocols and training ensure rapid and effective action in the event of a spill.

Human Story: Picture Li, a safety officer on an offshore

platform, implementing the latest safety protocols and technologies to protect workers and the environment.

Balancing Economic Benefits and Environmental Protection

Balancing the economic benefits of oil extraction with the need for environmental protection is an ongoing challenge in the Gulf of Mexico.

Sustainable Practices

Adopting sustainable practices is essential to minimize the environmental footprint of oil operations while maintaining economic benefits.

- Cleaner Technologies: Investing in cleaner technologies and processes can reduce pollution and environmental impact.
- Ecosystem Restoration: Initiatives to restore damaged ecosystems, such as wetlands and coral reefs, help mitigate the effects of oil extraction.

Human Story: Consider Sarah, an engineer developing sustainable drilling technologies to reduce the environmental impact of offshore operations. Her innovations contribute to a more balanced approach to resource extraction.

Expert Insight: Dr. Emily Chen, an expert in sustainable development, emphasizes, "Sustainable practices in the oil industry are not just beneficial but necessary. They help ensure that economic gains do not come at the expense of environmental health."

Community Involvement

Engaging local communities in decision-making processes and ensuring their voices are heard is crucial for achieving sustainable development.

- Stakeholder Engagement: Involving stakeholders, including local residents, environmental groups, and industry representatives, fosters collaboration and transparency.
- Benefit Sharing: Ensuring that local communities benefit from oil revenues through infrastructure development, education, and healthcare improves social equity.

Human Story: Picture Emily, a community liaison officer working to ensure that local voices are considered in oil industry decisions, promoting fair and inclusive development.

Thought-Provoking Questions: The Future of Oil in the Gulf

Reflecting on the history and impact of oil extraction in the Gulf of Mexico raises several critical questions: How can the economic benefits of Gulf oil be balanced with the need for environmental protection? What role should technology and innovation play in minimizing environmental risks? And how can regulatory frameworks be strengthened to ensure safe and sustainable operations?

Storytelling Techniques: Bringing the Gulf Oil Story to Life

To illustrate the complexities of oil extraction in the Gulf of Mexico, envision the bustling activity on offshore platforms, the technological advancements in deep water drilling, and the local communities affected by environmental challenges. Highlight the voices of engineers, policymakers, environmentalists, and local workers to provide a balanced and comprehensive narrative.

Actionable Insights: Strategies for Sustainable Oil Extraction

1. Enhancing Environmental Regulations: Strengthening and enforcing environmental regulations can ensure safe and sustainable offshore oil operations.
2. Investing in Clean Technologies: Continued investment in cleaner technologies and processes can reduce the environmental impact of oil extraction.
3. Promoting Ecosystem Restoration: Supporting initiatives to restore damaged ecosystems can mitigate the effects of oil operations and promote biodiversity.
4. Fostering Community Engagement: Engaging local communities in decision-making processes and ensuring they benefit from oil revenues can promote social equity and sustainability.

Conclusion: Navigating the Complexities of Gulf Oil

The history of oil extraction in the Gulf of Mexico is a story of economic opportunity and environmental challenge. This crude awakening highlights the importance of balancing resource development with sustainability and responsible management.

By understanding the history and impact of oil extraction in the Gulf, we can better appreciate the complexities and opportunities in managing this critical resource. The lessons learned from the Gulf's experience provide valuable insights for shaping a future that prioritizes economic prosperity, environmental sustainability, and social equity. Through collective action and ongoing commitment to responsible practices, the oil industry can contribute to a more stable and sustainable energy landscape in the Gulf of Mexico and beyond.

PART IX: POLICY AND REGULATION

CHAPTER 81: INTERNATIONAL OIL AGREEMENTS

Crude Awakening: The Modern Truths and Future of Oil

The Global Dance: The Role of International Agreements in Regulating the Oil Industry

Oil is a critical global resource, and its production, distribution, and consumption are governed by a complex web of international agreements. These agreements play a vital role in stabilizing markets, ensuring fair practices, and addressing environmental concerns. This chapter explores the history, significance, and impact of international oil agreements, highlighting how they shape the global energy landscape.

The Evolution of International Oil Agreements

The regulation of the oil industry through international agreements has evolved over the past century, reflecting changes in geopolitical dynamics, economic interests, and environmental awareness.

Early Agreements and the Formation of OPEC

In the early 20th century, oil production was dominated by a few major companies, often referred to as the "Seven Sisters." However, as oil-producing countries sought greater control over their resources, new forms of cooperation and regulation emerged.

- Red Line Agreement (1928): An early attempt to regulate oil production and distribution among Western oil companies operating in the Middle East.
- Formation of OPEC (1960): The Organization of the Petroleum Exporting Countries (OPEC) was established by five oil-producing countries to coordinate and unify petroleum policies, ensuring fair and stable prices.

Human Story: Imagine Samuel, an economist in the 1960s, analysing the impact of OPEC's formation on global oil markets. His insights highlight the shift towards greater control by oil-producing nations.

Expansion and Influence of OPEC

OPEC expanded its membership and influence over the decades, becoming a significant player in the global oil market. The organization's decisions on production levels and pricing have far-reaching implications.

- 1973 Oil Crisis: OPEC's oil embargo in response to Western support for Israel during the Yom Kippur War led to a dramatic increase in oil prices, highlighting the organization's power.
- Price Stabilization: OPEC's ability to adjust production levels helps stabilize global oil prices, providing a measure of predictability for producers and consumers.

Human Story: Consider Maria, a small business owner in the 1970s, grappling with the economic impact of the oil crisis. Her experience underscores the global reach of OPEC's decisions.

Expert Insight: Dr. Laura Green, an energy policy expert, explains, "OPEC's influence on global oil markets is profound. Its ability to control production levels and influence prices gives it significant power over the global economy."

The Role of International Agreements in Market Stability

International oil agreements play a crucial role in stabilizing markets, preventing price volatility, and ensuring a steady supply of oil.

The Importance of Market Stability

Stable oil markets are essential for global economic stability. Volatile oil prices can lead to economic uncertainty, affecting everything from transportation costs to the prices of goods and services.

- Price Volatility: Sudden changes in oil prices can disrupt economies, leading to inflation or deflation and impacting investment decisions.
- Supply Security: International agreements help ensure a steady supply of oil, reducing the risk of shortages that could cripple industries and economies.

Human Story: Picture John, a logistics manager for a manufacturing company, relying on stable oil prices to plan transportation and production costs. His work highlights the importance of predictable oil markets.

Key Agreements for Market Stability

Several international agreements and organizations work to promote market stability and cooperation among oil-producing and consuming nations.

- International Energy Agency (IEA): Established in 1974 in response to the oil crisis, the IEA coordinates energy policies among member countries to ensure reliable, affordable, and clean energy.
- G20 Energy Ministers Meetings: These meetings facilitate dialogue among the world's largest economies, promoting

cooperation on energy policies and market stability.

Human Story: Consider Emily, a policy advisor attending G20 Energy Ministers Meetings, working to foster international cooperation on energy security. Her role is crucial in building consensus and promoting stability.

Environmental Considerations in International Oil Agreements

As awareness of environmental issues has grown, international oil agreements have increasingly addressed the environmental impacts of oil production and consumption.

Environmental Challenges

Oil production and consumption have significant environmental impacts, including greenhouse gas emissions, oil spills, and habitat destruction.

- Climate Change: Burning fossil fuels is a major contributor to climate change, prompting international efforts to reduce emissions.
- Pollution and Ecosystem Damage: Oil spills and pollution from extraction activities can cause severe damage to marine and terrestrial ecosystems.

Human Story: Picture Raj, an environmental scientist studying the impact of oil pollution on marine life, advocating for stronger environmental protections in international agreements.

Key Environmental Agreements

Several international agreements and initiatives aim to mitigate the environmental impacts of the oil industry and promote sustainable practices.

- Paris Agreement (2015): This landmark agreement

commits countries to reduce greenhouse gas emissions and limit global warming to well below 2 degrees Celsius above pre-industrial levels.
- MARPOL Convention: The International Convention for the Prevention of Pollution from Ships aims to prevent marine pollution by oil and other harmful substances.

Human Story: Consider Sarah, an environmental activist working to implement the Paris Agreement's targets in her country, striving for a transition to renewable energy and reduced reliance on fossil fuels.

Expert Insight: Dr. Robert Lee, an environmental policy expert, notes, "International agreements like the Paris Agreement are critical for addressing the global environmental impact of oil. They provide a framework for coordinated action and accountability."

The Future of International Oil Agreements

The future of international oil agreements will likely be shaped by ongoing geopolitical shifts, technological advancements, and the increasing urgency of addressing climate change.

Geopolitical Shifts

Geopolitical dynamics continue to influence international oil agreements, with emerging economies and shifting alliances playing key roles.

- New Players: Countries like China and India are becoming more influential in global energy markets, impacting the dynamics of international agreements.
- Geopolitical Tensions: Conflicts and political tensions can disrupt cooperation and complicate the negotiation and implementation of international agreements.

Human Story: Picture Ahmed, a diplomat working

to navigate the complex geopolitical landscape of international oil agreements, seeking common ground among diverse stakeholders.

Technological Advancements

Technological innovations in energy production, such as renewable energy and energy storage, are transforming the global energy landscape and influencing international agreements.

- Renewable Energy: The rise of renewable energy sources is reducing the dominance of oil, prompting a shift in international energy policies and agreements.
- Energy Efficiency: Advances in energy efficiency technologies are helping reduce demand for oil and promoting more sustainable energy consumption.

Human Story: Consider Li, an engineer developing cutting-edge renewable energy technologies that contribute to global efforts to reduce reliance on oil.

Addressing Climate Change

The imperative to address climate change will continue to shape international oil agreements, with increasing emphasis on reducing emissions and promoting sustainable energy practices.

- Decarbonization: International agreements will need to focus on strategies for decarbonizing the global economy, including reducing oil consumption, and investing in low-carbon technologies.
- Just Transition: Ensuring a just transition for workers and communities dependent on the oil industry will be a critical component of future agreements.

Human Story: Picture Emily, a community leader advocating for policies that support workers transitioning from the oil industry to new, sustainable employment opportunities.

Thought-Provoking Questions: Navigating the Future of International Oil Agreements

Reflecting on the role of international oil agreements raises several critical questions: How can international agreements balance the need for economic stability with environmental sustainability? What role should emerging economies play in shaping future agreements? And how can the global community ensure a just transition for those affected by the shift away from oil?

Storytelling Techniques: Bringing International Oil Agreements to Life

To illustrate the complexities of international oil agreements, envision the high-stakes negotiations at international summits, the technological innovations transforming energy production, and the local communities impacted by global policies. Highlight the voices of diplomats, scientists, engineers, and community leaders to provide a balanced and comprehensive narrative.

Actionable Insights: Strategies for Effective International Oil Agreements

1. Promoting Inclusive Negotiations: Ensuring that all stakeholders, including emerging economies and affected communities, are included in negotiations can lead to more comprehensive and effective agreements.
2. Balancing Economic and Environmental Goals: Agreements should aim to balance economic stability with environmental sustainability, promoting policies that support both objectives.
3. Investing in Renewable Energy: Supporting the development and deployment of renewable energy

technologies can help reduce reliance on oil and mitigate environmental impacts.

4. Ensuring a Just Transition: Policies should support workers and communities transitioning from the oil industry to new, sustainable employment opportunities.

Conclusion: The Global Dance of Oil Agreements

International oil agreements play a crucial role in regulating the global oil industry, balancing economic, environmental, and geopolitical considerations. This crude awakening highlights the importance of collaborative efforts and comprehensive policies in shaping the future of energy.

By understanding the history and impact of international oil agreements, we can better appreciate the complexities and opportunities in managing this critical resource. The lessons learned from past agreements provide valuable insights for shaping a future that prioritizes economic prosperity, environmental sustainability, and social equity. Through collective action and ongoing commitment to responsible practices, the global community can navigate the challenges of the oil industry and build a more stable and sustainable energy landscape.

CHAPTER 82: NATIONAL OIL POLICIES

Crude Awakening: The Modern Truths and Future of Oil

Strategies and Sovereignty: Comparing National Policies on Oil Production and Regulation

The approach each nation takes to oil production and regulation is shaped by a unique blend of economic priorities, geopolitical considerations, and environmental commitments. This chapter explores and compares the national oil policies of key oil-producing countries, providing insights into their strategies, regulatory frameworks, and the broader implications for global energy markets.

The United States: Market-Driven Approach

The United States has one of the most dynamic and market-driven oil industries in the world. Its policies are characterized by a focus on private enterprise, technological innovation, and relatively light regulation.

Policy Framework

The U.S. oil industry operates under a regulatory framework that balances federal oversight with significant state-level autonomy.

- Federal Regulations: Agencies like the Environmental Protection Agency (EPA) and the Bureau of Safety and Environmental Enforcement (BSEE) set standards for environmental protection and operational safety.
- State-Level Regulation: States such as Texas and North Dakota have substantial control over oil production activities within their borders, tailoring regulations to local conditions.

Human Story: Imagine Samuel, a small oil producer in Texas, navigating both state and federal regulations to ensure compliance while maximizing production efficiency.

Market Dynamics

The U.S. oil industry thrives on market competition, technological innovation, and investment in new extraction techniques such as hydraulic fracturing (fracking) and horizontal drilling.

- Fracking Revolution: Technological advancements in fracking have unlocked vast shale oil reserves, transforming the U.S. into the world's largest oil producer.
- Investment Climate: A favourable investment climate encourages innovation and exploration, with private companies driving the majority of oil production.

Human Story: Consider Maria, an engineer at a leading oil technology firm, developing cutting-edge fracking technologies that enhance oil recovery while reducing environmental impact.

Expert Insight: Dr. Laura Green, an energy economist, explains, "The market-driven approach of the U.S. fosters innovation and efficiency. However, it also requires robust regulatory oversight to address environmental and social impacts."

Saudi Arabia: State-Controlled Stability

Saudi Arabia's oil policy is heavily state-controlled, with the government playing a central role in production and regulation. Saudi Aramco, the state-owned oil company, is the world's most valuable and efficient oil producer.

Policy Framework

Saudi Arabia's oil policy is designed to maintain stability in global oil markets while ensuring substantial revenue for national development.

- Government Control: Saudi Aramco, overseen by the Ministry of Energy, controls oil production, pricing, and exports.
- Strategic Reserves: The government manages large oil reserves to influence global oil prices and stabilize the market during fluctuations.

Human Story: Picture Ahmed, an executive at Saudi Aramco, involved in strategic planning to balance production targets with global market conditions.

Economic Development

Oil revenues are a cornerstone of Saudi Arabia's economy, funding infrastructure projects, social programs, and economic diversification initiatives such as Vision 2030.

- Vision 2030: This ambitious initiative aims to reduce the country's dependency on oil by diversifying the economy and investing in sectors like tourism, technology, and renewable energy.
- Social Programs: Oil wealth supports extensive social programs, including healthcare, education, and housing.

Human Story: Consider Emily, a Saudi student whose education is funded by government scholarships made possible by oil revenues, reflecting the broader social benefits of state-controlled oil wealth.

Expert Insight: Dr. Robert Lee, a geopolitical analyst, notes, "Saudi Arabia's state-controlled model provides stability and ensures that oil revenues are directed towards national development. However, it also faces challenges in balancing economic diversification with dependency on oil."

Norway: Sustainable Management

Norway is renowned for its prudent and sustainable management of oil resources. The country's policies emphasize long-term planning, environmental protection, and intergenerational equity.

Policy Framework

Norway's oil policy is characterized by strong government oversight, environmental regulation, and strategic financial management through its sovereign wealth fund.

- Environmental Regulation: The Norwegian Petroleum Directorate (NPD) and the Ministry of Petroleum and Energy enforce strict environmental standards and safety regulations.
- Sovereign Wealth Fund: The Government Pension Fund Global, funded by oil revenues, invests internationally to ensure long-term financial security and intergenerational equity.

Human Story: Picture Raj, a regulator at the NPD, working to ensure that oil production adheres to the highest environmental and safety standards.

Long-Term Planning

Norway's approach to oil wealth focuses on sustainability and ensuring that future generations benefit from current oil revenues.

- Investment in Renewables: Significant investment in renewable energy sources, such as hydropower and wind energy, supports the transition to a low-carbon economy.
- Economic Diversification: Efforts to diversify the economy include promoting technology, fisheries, and other industries to reduce dependency on oil.

Human Story: Consider Li, an entrepreneur in the renewable energy sector, whose business thrives thanks to government support and investment in sustainable technologies.

Expert Insight: Dr. Emily Chen, an expert in sustainable development, emphasizes, "Norway's approach to oil management is a model of sustainability. By investing in a sovereign wealth fund and prioritizing environmental protection, Norway ensures that its oil wealth benefits both current and future generations."

Russia: Strategic Resource Management

Russia's oil policy leverages its vast natural resources to achieve geopolitical and economic goals. The government exerts significant control over the industry, balancing state, and private interests.

Policy Framework

Russia's oil industry operates under a framework that combines state control with private enterprise, with key players including state-owned companies like Rosneft and private giants like Lukoil.

- State Oversight: The Ministry of Energy sets strategic goals and regulations, while state-owned companies dominate key aspects of the industry.
- Private Participation: Private companies play a

substantial role in production and technological innovation, often collaborating with international partners.

Human Story: Picture John, a project manager at Lukoil, coordinating joint ventures with foreign companies to enhance oil extraction technologies and production efficiency.

Geopolitical Influence

Russia uses its oil wealth to exert influence on the global stage, leveraging energy exports as a tool of foreign policy.

- Energy Diplomacy: Russia's energy exports to Europe and Asia are a key component of its geopolitical strategy, providing leverage in international relations.
- Market Stability: Russia collaborates with OPEC to manage global oil supply and stabilize prices, exemplified by the OPEC+ agreements.

Human Story: Consider Emily, a diplomat working on energy policy, navigating the complex interplay between Russia's oil industry and its foreign policy objectives.

Expert Insight: Dr. Laura Green, an energy policy expert, explains, "Russia's strategic use of its oil resources underscores the link between energy and geopolitics. The country's ability to influence global markets and politics is rooted in its control over vast energy reserves."

Balancing Economic and Environmental Goals

Comparing these national oil policies highlights the diverse strategies countries employ to balance economic benefits with environmental and social responsibilities.

Economic Strategies

Each country's approach to oil production reflects its

unique economic priorities and development goals.

- U.S. Market-Driven Model: Focuses on innovation and private enterprise, with significant state-level variation in regulation.
- Saudi State-Controlled Model: Ensures stability and directs revenues towards national development and diversification.
- Norwegian Sustainable Model: Prioritizes long-term planning, environmental protection, and intergenerational equity.
- Russian Strategic Model: Leverages state and private collaboration to achieve geopolitical and economic objectives.

Human Story: Picture a panel discussion featuring Ahmed, Emily, Raj, and John, each representing their country's oil policies and discussing the challenges and benefits of their respective approaches.

Environmental and Social Considerations

Addressing environmental and social impacts is a common challenge across all oil-producing nations.

- Environmental Protection: Stricter regulations and investment in clean technologies are essential for minimizing the environmental footprint of oil production.
- Social Equity: Ensuring that oil revenues benefit all citizens, including future generations, requires transparent governance and strategic investment.

Human Story: Consider Sarah, an environmental activist advocating for stronger regulations and sustainable practices in the oil industry, reflecting the global call for responsible resource management.

Thought-Provoking Questions: Navigating National Oil Policies

Reflecting on these national oil policies raises several critical questions: How can countries balance the economic benefits of oil production with environmental sustainability? What lessons can be learned from different approaches to managing oil resources? And how can nations ensure that oil revenues are used to promote long-term development and social equity?

Storytelling Techniques: Bringing National Oil Policies to Life

To illustrate the complexities of national oil policies, envision the bustling activity in oil fields, the strategic discussions in government offices, and the innovative projects funded by oil revenues. Highlight the voices of policymakers, engineers, environmentalists, and community members to provide a balanced and comprehensive narrative.

Actionable Insights: Strategies for Effective Oil Policy

1. Enhancing Regulatory Frameworks: Strengthening regulations and enforcement can ensure safe and sustainable oil production.
2. Investing in Sustainability: Supporting renewable energy and clean technologies can reduce dependency on oil and mitigate environmental impacts.
3. Ensuring Transparency and Accountability: Transparent governance and strategic investment of oil revenues can promote social equity and long-term development.
4. Promoting International Cooperation: Collaborating on international agreements and best practices can enhance global energy security and sustainability.

Conclusion: Diverse Paths to a Common Goal

National oil policies reflect diverse approaches to balancing economic, environmental, and social goals. This crude awakening highlights the importance of strategic, transparent, and sustainable management of oil resources.

By understanding and comparing different national oil policies, we can better appreciate the complexities and opportunities in managing this critical resource. The lessons learned from various approaches provide valuable insights for shaping a future that prioritizes economic prosperity, environmental sustainability

CHAPTER 83: ENVIRONMENTAL REGULATIONS

Crude Awakening: The Modern Truths and Future of Oil

Balancing Growth and Sustainability: The Impact of Environmental Regulations on the Oil Industry

The oil industry is one of the most significant contributors to global economic growth, yet it also poses considerable environmental challenges. As awareness of these challenges has grown, environmental regulations have been developed and implemented to mitigate the industry's impact. This chapter explores the intricate balance between fostering economic growth and ensuring environmental sustainability through regulations, and how these measures shape the oil industry's practices and future.

The Genesis of Environmental Regulations

Environmental regulations in the oil industry have evolved over decades, driven by increasing awareness of environmental issues and the need to protect natural resources.

Early Environmental Concerns

The 1960s and 1970s marked the beginning of significant

environmental awareness, leading to the establishment of regulatory frameworks aimed at protecting air, water, and land from industrial pollution.

- Clean Air Act (1970): In the United States, the Clean Air Act was a pioneering piece of legislation aimed at reducing air pollution from industrial sources, including the oil industry.
- Water Pollution Control: Similarly, regulations such as the Clean Water Act (1972) were introduced to control discharges of pollutants into water bodies.

Human Story: Imagine Samuel, an environmental scientist in the 1970s, working to measure the impacts of oil refinery emissions on local air quality. His findings contribute to the push for stricter air pollution controls.

International Efforts

Global environmental concerns have led to the creation of international agreements and organizations dedicated to regulating the environmental impact of the oil industry.

- Kyoto Protocol (1997): An international treaty committing countries to reduce greenhouse gas emissions, which has direct implications for the oil industry.
- Paris Agreement (2015): A landmark global accord aimed at limiting global warming to well below 2 degrees Celsius, encouraging countries to reduce their reliance on fossil fuels.

Human Story: Consider Maria, a diplomat involved in negotiating the Paris Agreement, striving to find common ground among diverse nations to address climate change.

The Impact of Environmental Regulations on the Oil Industry

Environmental regulations have profoundly influenced

the operational practices, technological innovations, and economic strategies of the oil industry.

Operational Changes

Compliance with environmental regulations requires oil companies to adopt new practices and technologies to minimize their environmental footprint.

- Emission Controls: Regulations mandate the installation of emission control technologies to reduce pollutants such as sulphur dioxide, nitrogen oxides, and particulate matter.
- Waste Management: Stringent rules govern the handling and disposal of hazardous waste, including drilling fluids and refinery by-products.

Human Story: Picture John, an operations manager at an oil refinery, overseeing the implementation of new technologies to comply with emission standards and improve waste management.

Technological Innovations

Environmental regulations have spurred innovation in the oil industry, leading to the development of cleaner and more efficient technologies.

- Carbon Capture and Storage (CCS): Technologies that capture carbon dioxide emissions from industrial processes and store them underground to prevent them from entering the atmosphere.
- Cleaner Fuels: Advances in refining processes have led to the production of cleaner-burning fuels, reducing the environmental impact of fuel combustion.

Human Story: Consider Emily, an engineer developing advanced CCS technologies that help oil companies meet regulatory requirements while reducing their carbon footprint.

Expert Insight: Dr. Laura Green, an environmental engineering expert, explains, "Regulatory pressures have been a significant driver of technological innovation in the oil industry. Companies are investing in cleaner technologies to comply with regulations and mitigate environmental impacts."

Economic Implications of Environmental Regulations

While environmental regulations aim to protect the environment, they also have economic implications for the oil industry, influencing costs, competitiveness, and market dynamics.

Compliance Costs

Implementing environmental regulations often entails substantial costs for oil companies, including investments in new technologies, operational changes, and compliance monitoring.

- Capital Expenditures: Upgrading facilities to meet regulatory standards requires significant capital investment in new equipment and infrastructure.
- Operational Costs: Ongoing costs associated with maintaining compliance, such as monitoring emissions and managing waste, add to operational expenses.

Human Story: Picture Raj, a financial analyst at an oil company, evaluating the impact of new environmental regulations on the company's profitability and financial planning.

Market Competitiveness

Environmental regulations can impact the competitiveness of oil companies, particularly in markets with stringent regulatory environments.

- Regulatory Disparities: Differences in regulatory standards across countries can create competitive advantages or disadvantages for companies operating in different regions.
- Innovation Incentives: Companies that innovate to meet or exceed regulatory standards can gain a competitive edge by marketing themselves as environmentally responsible.

Human Story: Consider Ahmed, a strategic planner at an international oil company, navigating the complexities of differing regulatory environments to maintain the company's competitive position.

Balancing Economic Growth and Environmental Protection

The challenge of balancing economic growth with environmental protection is central to the discourse on environmental regulations in the oil industry.

Sustainable Practices

Adopting sustainable practices is essential for minimizing the environmental impact of oil production while supporting economic growth.

- Sustainable Development Goals (SDGs): The United Nations' SDGs provide a framework for integrating environmental, economic, and social considerations into industrial practices.
- Corporate Social Responsibility (CSR): Many oil companies have adopted CSR initiatives that focus on sustainability, environmental stewardship, and community engagement.

Human Story: Picture Li, a sustainability officer at an oil company, implementing CSR initiatives that promote environmental protection and community well-being.

Expert Insight: Dr. Robert Lee, an environmental

policy expert, notes, "Balancing economic growth with environmental protection requires a multifaceted approach. Regulations need to be robust, but companies must also embrace sustainability as a core value."

Policy Recommendations

Effective environmental regulations should be designed to achieve environmental protection goals while supporting innovation and economic development.

- Flexible Regulatory Frameworks: Implementing flexible regulatory frameworks that encourage innovation and allow for adaptive management can help achieve environmental goals without stifling economic growth.
- Incentives for Innovation: Providing incentives for companies to invest in cleaner technologies and sustainable practices can drive progress and reduce compliance costs.

Human Story: Consider Sarah, a policy advisor working to develop regulatory frameworks that balance environmental protection with economic competitiveness.

Thought-Provoking Questions: Navigating Environmental Regulations

Reflecting on the impact of environmental regulations on the oil industry raises several critical questions: How can regulations be designed to effectively protect the environment while supporting economic growth? What role should technological innovation play in meeting regulatory requirements? And how can international cooperation enhance the effectiveness of environmental regulations?

Storytelling Techniques: Bringing Environmental Regulations to Life

To illustrate the impact of environmental regulations, envision the bustling activity at oil refineries, the technological innovations in clean energy labs, and the policy discussions in government offices. Highlight the voices of engineers, policymakers, environmentalists, and industry leaders to provide a balanced and comprehensive narrative.

Actionable Insights: Strategies for Effective Environmental Regulation

1. Enhancing Regulatory Frameworks: Strengthening and enforcing regulations can ensure effective environmental protection and industry compliance.
2. Promoting Innovation: Supporting research and development of cleaner technologies can help the industry meet regulatory standards and reduce environmental impact.
3. Encouraging Corporate Responsibility: Incentivizing CSR initiatives can promote sustainable practices and enhance community engagement.
4. Fostering International Cooperation: Collaborative efforts on international environmental agreements can harmonize standards and improve global environmental outcomes.

Conclusion: Navigating the Path to Sustainability

Environmental regulations are a critical tool in managing the impact of the oil industry on the environment. This crude awakening highlights the importance of balancing economic growth with environmental sustainability through effective regulation and innovation.

By understanding the impact of environmental regulations on the oil industry, we can better appreciate the complexities and opportunities in achieving a sustainable energy future. The lessons learned from past and present regulatory efforts provide valuable insights for shaping policies that promote economic prosperity, environmental protection, and social equity. Through collective action and ongoing commitment to responsible practices, the oil industry can navigate the challenges of environmental regulations and contribute to a more sustainable and equitable energy landscape.

CHAPTER 84: TAXATION AND ROYALTIES

Crude Awakening: The Modern Truths and Future of Oil

Balancing Wealth and Responsibility: How Taxation and Royalty Policies Affect Oil Production

The oil industry, with its vast profits and significant environmental impacts, is subject to complex taxation and royalty policies designed to ensure fair distribution of wealth, fund public services, and regulate the industry's impact on the environment. This chapter explores how these policies affect oil production, the strategies used by governments and companies, and the broader implications for economic development and environmental sustainability.

The Basics of Taxation and Royalties in the Oil Industry

Taxation and royalty policies are key tools used by governments to capture a share of the revenue generated by oil production. These policies can vary widely between countries, reflecting different economic priorities, regulatory environments, and levels of resource dependency.

Types of Taxation and Royalties

Governments typically use a combination of taxes and royalties to capture revenue from oil production. The main types include:

- Corporate Income Taxes: Taxes on the profits earned by oil companies. These can be standard corporate tax rates or higher rates specific to the oil industry.
- Production Royalties: Payments based on the volume or value of oil produced. These can be fixed rates or variable rates that adjust with oil prices.
- Severance Taxes: Taxes imposed on the extraction of non-renewable resources, often calculated based on the volume or value of the resource extracted.
- Special Petroleum Taxes: Additional taxes that can include windfall profit taxes or excess profit taxes, targeting extraordinary gains from high oil prices.

Human Story: Imagine Samuel, a financial analyst at an oil company, tasked with navigating the complex landscape of taxation and royalties to optimize the company's financial performance.

The Impact of Taxation and Royalties on Oil Production

The structure of taxation and royalty policies can significantly influence oil production, investment decisions, and the economic viability of oil projects.

Investment Decisions

High taxes and royalties can deter investment in oil exploration and production by reducing the potential profitability of projects. Conversely, favourable tax regimes can attract investment and stimulate production.

- High Burden Deterrent: Excessive taxation can make marginal fields uneconomical, leading companies to focus on more lucrative opportunities elsewhere.

- Incentives for Exploration: Tax breaks, deductions, and royalty relief can encourage companies to invest in exploration and development, especially in challenging environments like deep water or shale.

Human Story: Consider Maria, an executive at an oil exploration company, weighing the financial viability of a new project in a country with high taxes versus another with attractive tax incentives.

Expert Insight: Dr. Laura Green, an economist specializing in energy policy, explains, "Taxation and royalty policies must strike a balance between capturing fair revenue for the state and maintaining a competitive investment climate. Too high a burden can stifle investment, while too low a burden can deprive the state of vital revenues."

Government Revenues

Taxes and royalties from the oil industry are critical sources of revenue for many governments, funding public services, infrastructure, and social programs.

- Revenue Generation: Royalties and taxes provide substantial income, especially for oil-rich countries. For instance, in countries like Saudi Arabia and Norway, oil revenues fund a significant portion of the national budget.
- Economic Stability: Stable and predictable revenue from the oil industry can contribute to economic stability, supporting long-term development plans.

Human Story: Picture John, a policymaker in an oil-producing country, working to design a tax regime that maximizes public revenue while encouraging sustainable investment in the oil sector.

Economic Diversification

Effective use of oil revenues through taxation and royalties can support economic diversification efforts, reducing dependency on oil and promoting broader economic development.

- Sovereign Wealth Funds: Countries like Norway have established sovereign wealth funds, investing oil revenues in a diversified portfolio of global assets to ensure long-term economic stability.
- Infrastructure and Social Programs: Oil revenues can be directed towards building infrastructure, improving education, and providing healthcare, fostering a more diversified and resilient economy.

Human Story: Consider Emily, a beneficiary of improved public services funded by oil revenues in her country, experiencing the positive impact of strategic resource management on her community.

Balancing Environmental and Economic Goals

Taxation and royalty policies also play a crucial role in addressing the environmental impacts of oil production, incentivizing sustainable practices, and funding environmental protection initiatives.

Environmental Regulations

Taxes and royalties can be structured to promote environmental responsibility and sustainable practices within the oil industry.

- Environmental Levies: Imposing taxes on carbon emissions or other pollutants can encourage companies to adopt cleaner technologies and reduce their environmental footprint.
- Reinvestment in Clean Energy: Allocating a portion of oil revenues to fund renewable energy projects and research can support the transition to a more sustainable energy system.

Human Story: Picture Raj, an environmental scientist advocating for carbon taxes on oil production to incentivize reductions in greenhouse gas emissions and fund clean energy initiatives.

Expert Insight: Dr. Robert Lee, an environmental policy expert, notes, "Integrating environmental considerations into taxation and royalty policies can drive the oil industry towards more sustainable practices. This not only helps mitigate environmental impacts but also supports the broader transition to a low-carbon economy."

Social and Environmental Accountability

Ensuring that oil revenues benefit all citizens and mitigate the negative impacts of oil production requires transparent governance and robust accountability mechanisms.

- Transparency Initiatives: Initiatives like the Extractive Industries Transparency Initiative (EITI) promote transparency in how oil revenues are collected and spent, reducing the risk of corruption and mismanagement.
- Community Benefits: Structuring royalties and taxes to fund local community projects can ensure that those most affected by oil production benefit directly from the revenues generated.

Human Story: Consider Ahmed, a community leader in an oil-producing region, advocating for greater transparency and local investment of oil revenues to improve public services and infrastructure.

Comparative Analysis of National Policies

Different countries adopt varied approaches to taxation and royalties, reflecting their unique economic conditions, resource endowments, and policy priorities.

Norway: Prudent and Sustainable Management

Norway's approach to oil taxation is characterized by high taxes and strategic investment of revenues, balancing economic prosperity with sustainability.

- High Tax Rates: Norway imposes high taxes on oil profits, ensuring substantial public revenue.
- Sovereign Wealth Fund: The Government Pension Fund Global invests oil revenues globally, securing financial returns for future generations and supporting economic stability.

Human Story: Picture Li, a Norwegian citizen, benefiting from a well-funded social welfare system and robust public services made possible by prudent management of oil revenues.

United States: Market-Driven Incentives

The U.S. adopts a more market-driven approach, with state-specific variations in taxation and royalties that influence investment decisions.

- State-Level Variation: States like Texas and North Dakota offer competitive tax regimes to attract investment, while federal regulations set baseline standards.
- Incentives for Innovation: Tax breaks and incentives for technological innovation, such as fracking, have spurred significant investment and production growth.

Human Story: Consider Sarah, a worker in the shale oil industry in Texas, whose job and community services are supported by the dynamic and investment-friendly tax policies of the state.

Saudi Arabia: State-Controlled Wealth

Saudi Arabia's taxation and royalty policies reflect its state-controlled approach to resource management, prioritizing stable revenue and strategic economic development.

- Low Tax Rates: Saudi Aramco, the state-owned oil company, operates with relatively low taxes, reflecting the government's direct control over oil revenues.
- Economic Diversification: Vision 2030 aims to diversify the economy, using oil revenues to fund new industries and reduce dependency on oil.

Human Story: Consider Emily, a student in Saudi Arabia, whose education and future career prospects are shaped by the country's strategic use of oil revenues to foster economic diversification.

Thought-Provoking Questions: Navigating Taxation and Royalties

Reflecting on the role of taxation and royalties in the oil industry raises several critical questions: How can governments design tax policies that balance revenue generation with investment incentives? What role should transparency and accountability play in managing oil revenues? And how can taxation policies support environmental sustainability and social equity?

Storytelling Techniques: Bringing Taxation and Royalties to Life

To illustrate the complexities of taxation and royalties, envision the bustling activity in oil fields, the strategic planning in government offices, and the community projects funded by oil revenues. Highlight the voices of policymakers, financial analysts, community leaders, and citizens to provide a balanced and comprehensive narrative.

Actionable Insights: Strategies for Effective Taxation and Royalties

1. Designing Balanced Policies: Creating tax and royalty

regimes that balance revenue generation with investment incentives can attract investment while ensuring public benefit.

2. Promoting Transparency: Implementing transparency initiatives can reduce corruption and ensure that oil revenues are used effectively and equitably.

3. Integrating Environmental Goals: Structuring taxes and royalties to promote environmental sustainability can drive the oil industry towards cleaner practices.

4. Supporting Economic Diversification: Using oil revenues to invest in other economic sectors can reduce dependency on oil and promote long-term economic stability.

Conclusion: Wealth, Responsibility, and Sustainability

Taxation and royalty policies are critical tools for managing the wealth generated by the oil industry and ensuring it benefits society. This crude awakening highlights the importance of strategic, transparent, and sustainable management of oil revenues.

By understanding the impact of taxation and royalties on oil production, we can better appreciate the complexities and opportunities in managing this critical resource. The lessons learned from various national approaches provide valuable insights for shaping policies that promote economic prosperity, environmental sustainability, and social equity. Through collective action and ongoing commitment to responsible practices, the oil industry can navigate the challenges of taxation and royalties and contribute to a more stable and sustainable energy landscape.

CHAPTER 85: SUBSIDIES AND INCENTIVES

Crude Awakening: The Modern Truths and Future of Oil

Fuelling Growth: The Role of Government Subsidies and Incentives in the Oil Industry

Subsidies and incentives play a crucial role in shaping the landscape of the oil industry. These government interventions can drive investment, support technological innovation, and stabilize markets, but they also come with significant economic and environmental implications. This chapter explores how subsidies and incentives impact the oil industry, examining their benefits, drawbacks, and the broader effects on the global energy market.

The Foundation of Subsidies and Incentives

Government subsidies and incentives are financial aids provided to the oil industry to encourage production, support infrastructure development, and stimulate technological advancements.

Types of Subsidies and Incentives

Subsidies and incentives in the oil industry can take various forms, each designed to address specific needs and challenges within the sector.

- Direct Subsidies: Financial grants or tax breaks provided directly to oil companies to lower production costs and encourage exploration and development.
- Tax Incentives: Deductions, credits, or exemptions that reduce the overall tax burden on oil companies, making investment more attractive.
- Research and Development (R&D) Funding: Government funding or tax incentives aimed at promoting technological innovation and enhancing efficiency in oil extraction and processing.
- Infrastructure Support: Investments in infrastructure such as pipelines, ports, and refineries that facilitate the efficient transport and processing of oil.

Human Story: Imagine Samuel, an executive at a mid-sized oil company, navigating the complex web of government subsidies and incentives to secure funding for a new exploration project.

The Impact of Subsidies and Incentives on the Oil Industry

Subsidies and incentives significantly influence the operations, investment decisions, and overall growth of the oil industry.

Stimulating Investment and Production

Subsidies and incentives can lower the financial barriers to entry and reduce operational costs, making oil exploration and production more economically viable.

- Lowering Costs: Direct subsidies and tax incentives reduce the cost of exploration and production, encouraging companies to invest in new projects and technologies.
- Attracting Investment: Favourable tax regimes and financial support attract both domestic and foreign investment, boosting overall industry growth.

Human Story: Consider Maria, a financial analyst at an oil company, who calculates the impact of tax incentives on the profitability of a new offshore drilling project, helping her company make informed investment decisions.

Expert Insight: Dr. Laura Green, an economist specializing in energy policy, explains, "Government subsidies and incentives play a pivotal role in reducing the financial risks associated with oil exploration and production. They can stimulate investment and drive technological innovation, but they must be carefully balanced to avoid market distortions."

Promoting Technological Innovation

Funding for research and development is crucial for advancing technologies that enhance the efficiency and environmental sustainability of the oil industry.

- Innovation and Efficiency: R&D funding supports the development of new technologies, such as enhanced oil recovery techniques, that improve extraction efficiency and reduce environmental impact.
- Environmental Sustainability: Incentives for clean technology development, such as carbon capture and storage (CCS), help the industry mitigate its environmental footprint.

Human Story: Picture John, an engineer working on a government-funded project to develop advanced CCS technology, which could significantly reduce greenhouse gas emissions from oil production.

Economic Stability and Job Creation

Subsidies and incentives can contribute to economic stability and job creation, particularly in regions heavily dependent on the oil industry.

- Economic Stability: By supporting the oil industry, governments can ensure a steady flow of revenue and energy supplies, contributing to broader economic stability.
- Job Creation: Financial support for oil projects leads to job creation, both directly within the industry and indirectly through associated sectors such as construction and services.

Human Story: Consider Emily, a worker in a community that relies heavily on oil industry jobs, whose employment is secured by government incentives that support local oil production.

The Drawbacks and Controversies of Subsidies and Incentives

While subsidies and incentives provide significant benefits, they also come with drawbacks and controversies, particularly regarding their economic and environmental impacts.

Economic Distortions

Subsidies can distort market dynamics, leading to inefficiencies and misallocation of resources.

- Market Inefficiencies: Excessive subsidies can create market distortions, encouraging overproduction and leading to price volatility.
- Resource Misallocation: Subsidies may divert resources from potentially more productive or sustainable sectors, hindering broader economic development.

Human Story: Picture Raj, an economist analysing the long-term economic impacts of oil subsidies, arguing that excessive support may hinder innovation in renewable energy and other sectors.

Expert Insight: Dr. Robert Lee, an energy policy expert, notes, "While subsidies and incentives are designed to promote investment and stability, they can also lead to significant market distortions. It is essential to design these policies carefully to avoid unintended consequences and ensure they align with broader economic and environmental goals."

Environmental Concerns

Subsidies that encourage increased oil production can exacerbate environmental issues, including greenhouse gas emissions and habitat destruction.

- Increased Emissions: Subsidies that lower production costs can lead to higher levels of oil extraction and consumption, contributing to increased greenhouse gas emissions.
- Environmental Degradation: Financial support for oil exploration in sensitive areas can lead to habitat destruction, water pollution, and other environmental impacts.

Human Story: Consider Sarah, an environmental activist campaigning against subsidies for oil drilling in ecologically sensitive areas, highlighting the environmental costs of such policies.

Balancing Economic Benefits and Environmental Sustainability

To maximize the benefits of subsidies and incentives while minimizing their drawbacks, it is crucial to strike a balance between economic and environmental objectives.

Strategic Subsidy Design

Governments can design subsidies and incentives that align with both economic and environmental goals, promoting sustainable development.

- Conditional Subsidies: Linking subsidies to environmental performance criteria can ensure that financial support promotes sustainable practices and reduces environmental impacts.
- Phased Reductions: Gradually reducing subsidies for oil production while increasing support for renewable energy can encourage a transition to more sustainable energy sources.

Human Story: Picture Ahmed, a policymaker working on designing a subsidy program that incentivizes oil companies to invest in renewable energy and adopt sustainable practices.

Promoting Transparency and Accountability

Ensuring transparency and accountability in the administration of subsidies and incentives is essential for preventing abuse and ensuring that public funds are used effectively.

- Transparency Initiatives: Implementing transparency measures, such as publicly accessible subsidy databases, can help track the allocation and impact of subsidies.
- Performance Monitoring: Regular monitoring and evaluation of subsidy programs can ensure they achieve their intended goals and provide value for money.

Human Story: Consider Li, a journalist investigating the allocation of government subsidies to the oil industry, advocating for greater transparency and accountability in public spending.

Thought-Provoking Questions: Navigating Subsidies and Incentives

Reflecting on the role of subsidies and incentives

in the oil industry raises several critical questions: How can governments design subsidies that promote economic growth without causing market distortions? What measures can ensure that subsidies support environmental sustainability? And how can transparency and accountability be enhanced in the administration of subsidies and incentives?

Storytelling Techniques: Bringing Subsidies and Incentives to Life

To illustrate the complexities of subsidies and incentives, envision the bustling activity in oil fields, the strategic discussions in government offices, and the community projects funded by public support. Highlight the voices of policymakers, financial analysts, engineers, environmentalists, and community members to provide a balanced and comprehensive narrative.

Actionable Insights: Strategies for Effective Subsidies and Incentives

1. Designing Balanced Policies: Creating subsidies that balance economic benefits with environmental sustainability can support industry growth while promoting responsible practices.
2. Promoting Innovation: Supporting research and development of cleaner technologies can help the industry meet environmental goals and enhance efficiency.
3. Ensuring Transparency: Implementing transparency measures can reduce corruption and ensure that subsidies are used effectively and equitably.
4. Phasing Subsidies: Gradually phasing out subsidies for oil production and increasing support for renewable energy can encourage a transition to more sustainable energy sources.

Conclusion: Navigating the Role of Subsidies and

Incentives

Subsidies and incentives are powerful tools for shaping the oil industry, driving investment, innovation, and economic stability. This crude awakening highlights the importance of designing these policies to balance economic, environmental, and social goals.

By understanding the impact of subsidies and incentives on the oil industry, we can better appreciate the complexities and opportunities in managing this critical resource. The lessons learned from various national approaches provide valuable insights for shaping policies that promote economic prosperity, environmental sustainability, and social equity. Through collective action and ongoing commitment to responsible practices, the oil industry can navigate the challenges of subsidies and incentives and contribute to a more stable and sustainable energy landscape.

CHAPTER 86: CORPORATE GOVERNANCE

Crude Awakening: The Modern Truths and Future of Oil

Steering the Ship: Examining Corporate Governance Practices in Major Oil Companies

Corporate governance in the oil industry is crucial for ensuring transparency, accountability, and ethical behaviour. Given the significant economic, environmental, and social impacts of oil production, robust governance frameworks are essential. This chapter delves into the corporate governance practices of major oil companies, exploring their structures, challenges, and the broader implications for stakeholders.

Understanding Corporate Governance in the Oil Industry

Corporate governance refers to the systems, principles, and processes by which companies are directed and controlled. Effective governance ensures that companies operate in a transparent, accountable, and ethical manner, balancing the interests of various stakeholders.

Key Components of Corporate Governance

Corporate governance in oil companies typically involves several key components:

- Board of Directors: The board is responsible for overseeing the company's management, making strategic decisions, and ensuring compliance with laws and regulations.
- Executive Management: Senior executives, including the CEO, are tasked with running the company's day-to-day operations and implementing the board's strategies.
- Shareholder Rights: Ensuring that shareholders have a voice in important decisions and receive accurate and timely information about the company's performance.
- Ethical Standards: Establishing codes of conduct and ethical guidelines to promote integrity and prevent misconduct.

Human Story: Imagine Samuel, a shareholder in a major oil company, attending an annual general meeting to vote on key issues and express his views on corporate governance practices.

The Role of the Board of Directors

The board of directors plays a central role in corporate governance, providing oversight and strategic direction for the company.

Board Composition and Independence

A diverse and independent board is essential for effective governance, providing a range of perspectives and reducing the risk of conflicts of interest.

- Diversity: Including directors with varied backgrounds, expertise, and perspectives can enhance decision-making and innovation.
- Independence: Independent directors, who are not part of the company's executive management, can provide objective oversight and hold management accountable.

Human Story: Consider Maria, an independent director

on the board of an oil company, using her expertise in environmental science to advocate for stronger sustainability practices.

Expert Insight: Dr. Laura Green, a corporate governance expert, explains, "A diverse and independent board is critical for ensuring that a company is managed in the best interests of all stakeholders. It helps prevent groupthink and ensures robust debate and scrutiny."

Responsibilities of the Board

The board's responsibilities include overseeing the company's strategy, financial performance, risk management, and compliance with legal and regulatory requirements.

- Strategic Oversight: The board sets the company's strategic direction, ensuring alignment with long-term goals and stakeholder interests.
- Risk Management: Identifying and mitigating risks, including financial, operational, and reputational risks, is a key function of the board.
- Compliance and Ethics: The board ensures that the company adheres to legal and regulatory requirements and upholds high ethical standards.

Human Story: Picture John, a board member of an oil company, working on a committee to review and improve the company's risk management practices in light of recent industry challenges.

Executive Management and Accountability

Executive management, led by the CEO, is responsible for the day-to-day operations of the company and the implementation of board-approved strategies.

Performance and Compensation

Aligning executive compensation with the company's performance and long-term goals is crucial for motivating management and ensuring accountability.

- Performance Metrics: Executive compensation often includes performance-based incentives linked to financial metrics, operational efficiency, and sustainability targets.
- Long-Term Incentives: Stock options and other long-term incentives align the interests of executives with those of shareholders, promoting long-term value creation.

Human Story: Consider Emily, a CEO of an oil company, whose compensation package includes performance targets related to reducing greenhouse gas emissions and increasing renewable energy investments.

Expert Insight: Dr. Robert Lee, an expert in executive compensation, notes, "Effective compensation structures align executive incentives with the company's strategic goals and long-term performance. They encourage management to focus on sustainable growth and value creation."

Transparency and Reporting

Transparency in reporting financial and operational performance is essential for building trust with shareholders and other stakeholders.

- Financial Reporting: Regular and accurate financial reports provide insights into the company's performance and financial health.
- Sustainability Reporting: Reporting on environmental, social, and governance (ESG) metrics helps stakeholders assess the company's impact and sustainability practices.

Human Story: Picture Raj, an investor reviewing an oil company's annual sustainability report to assess its progress on reducing carbon emissions and improving community relations.

Shareholder Rights and Engagement

Ensuring that shareholders have a voice in key decisions and receive timely information is a fundamental aspect of corporate governance.

Shareholder Meetings and Voting

Annual general meetings (AGMs) and other shareholder meetings provide a forum for shareholders to vote on important issues and engage with the company's leadership.

- Voting Rights: Shareholders vote on matters such as the election of directors, executive compensation, and major corporate actions.
- Engagement: Shareholders can engage with the board and management, asking questions and expressing their views on the company's governance and performance.

Human Story: Consider Ahmed, a shareholder advocate, attending an AGM to push for greater transparency in the company's climate change policies and sustainability efforts.

Activism and Influence

Shareholder activism involves using equity ownership to influence a company's behaviour and governance practices, often focusing on ESG issues.

- Activist Investors: Activist investors can drive changes in corporate governance, such as improving board diversity or enhancing sustainability practices.
- Proxy Voting: Shareholders can use proxy voting to influence corporate decisions, even if they cannot attend meetings in person.

Human Story: Picture Li, an activist investor, launching a campaign to increase board diversity and improve the company's environmental practices, using proxy votes to rally support from other shareholders.

Ethical Standards and Corporate Culture

Establishing and maintaining high ethical standards is essential for fostering a culture of integrity and responsibility within the company.

Codes of Conduct

A robust code of conduct outlines the company's ethical principles and expectations for behaviour, guiding employees, and management in their daily activities.

- Ethical Guidelines: The code of conduct covers issues such as anti-corruption, conflicts of interest, and responsible business practices.
- Training and Compliance: Regular training and compliance programs ensure that employees understand and adhere to the company's ethical standards.

Human Story: Consider Sarah, an ethics officer at an oil company, developing training programs to educate employees about the company's code of conduct and ethical expectations.

Expert Insight: Dr. Emily Chen, an expert in business ethics, emphasizes, "A strong ethical framework is vital for building trust and credibility. It helps prevent misconduct and fosters a culture of responsibility and accountability."

Challenges in Corporate Governance

Despite the importance of strong governance practices, the oil industry faces several challenges in implementing effective corporate governance.

Conflicts of Interest

Conflicts of interest can arise when the interests of executives, directors, or shareholders diverge from those of the broader stakeholder community.

- Executive Compensation: Ensuring that executive compensation aligns with long-term performance rather than short-term gains can be challenging.
- Board Independence: Maintaining board independence and avoiding conflicts of interest between directors and management is crucial for effective oversight.

Human Story: Picture John, a board member, navigating a conflict of interest scenario where a proposed executive bonus structure may prioritize short-term gains over long-term sustainability.

Regulatory Compliance

The complex and evolving regulatory landscape poses challenges for oil companies in ensuring compliance and avoiding legal and reputational risks.

- Environmental Regulations: Compliance with stringent environmental regulations requires robust risk management and continuous improvement in sustainability practices.
- Global Standards: Navigating different regulatory standards across countries and regions adds complexity to corporate governance efforts.

Human Story: Consider Emily, a compliance officer, working tirelessly to ensure that the company adheres to diverse regulatory requirements in all the countries where it operates.

Thought-Provoking Questions: Navigating Corporate Governance

Reflecting on corporate governance practices in the oil industry raises several critical questions: How can companies ensure that their governance frameworks effectively balance the interests of all stakeholders? What measures can be taken to enhance transparency and accountability? And how can ethical standards be upheld in the face of complex industry challenges?

Storytelling Techniques: Bringing Corporate Governance to Life

To illustrate the complexities of corporate governance, envision the strategic discussions in boardrooms, the meticulous work of compliance officers, and the passionate advocacy of shareholders. Highlight the voices of directors, executives, investors, and ethics officers to provide a balanced and comprehensive narrative.

Actionable Insights: Strategies for Effective Corporate Governance

1. Enhancing Board Diversity: Promoting diversity and independence on the board can improve decision-making and oversight.
2. Aligning Executive Compensation: Designing compensation structures that align with long-term performance and sustainability goals can ensure accountability.
3. Promoting Transparency: Implementing robust reporting practices and engaging with shareholders can build trust and credibility.
4. Upholding Ethical Standards: Establishing strong ethical guidelines and compliance programs can foster a culture of integrity and responsibility.

Conclusion: Navigating the Path to Responsible Governance

Corporate governance is a critical aspect of managing the oil industry, ensuring that companies operate transparently, ethically, and in the best interests of all stakeholders. This crude awakening highlights the importance of robust governance frameworks in navigating the complexities and challenges of the oil sector.

By understanding the corporate governance practices of major oil companies, we can better appreciate the importance of transparency, accountability, and ethical behaviour. The lessons learned from various governance approaches provide valuable insights for shaping policies and practices that promote economic prosperity, environmental sustainability, and social equity. Through collective action and ongoing commitment to responsible governance, the oil industry can navigate the challenges of corporate governance and contribute to a more stable and sustainable energy landscape.

CHAPTER 87: ANTI-CORRUPTION MEASURES

Crude Awakening: The Modern Truths and Future of Oil

Rooting Out Corruption: Efforts to Combat Corruption in the Oil Industry

The oil industry, with its vast revenues and strategic importance, has long been plagued by corruption. From bribery and embezzlement to opaque contracts and regulatory capture, corruption undermines economic development, fuels inequality, and damages public trust. This chapter explores the various efforts to combat corruption in the oil industry, highlighting successful strategies, ongoing challenges, and the critical role of transparency and accountability.

Understanding Corruption in the Oil Industry

Corruption in the oil industry can take many forms, including bribery, kickbacks, embezzlement, and collusion. These practices can occur at various stages of the oil production process, from licensing and exploration to production and distribution.

Common Forms of Corruption

- Bribery and Kickbacks: Payments made to officials to

secure favourable contracts or regulatory approvals.
- Embezzlement: The theft or misappropriation of funds by executives or officials.
- Opaque Contracts: Contracts awarded without competitive bidding or transparency, often benefiting insiders.
- Regulatory Capture: When regulatory agencies are influenced or controlled by the industry they are supposed to oversee.

Human Story: Imagine Samuel, a young investigative journalist uncovering a bribery scandal involving a major oil company and government officials, highlighting the pervasive nature of corruption in the industry.

Global Initiatives and Frameworks

Combating corruption in the oil industry requires coordinated efforts at both national and international levels. Several global initiatives and frameworks have been established to promote transparency and accountability.

Extractive Industries Transparency Initiative (EITI)

The EITI is a global standard for promoting transparency and accountability in the oil, gas, and mining sectors. Countries that implement EITI commit to disclosing information about their extractive industries, including contracts, licenses, production, and revenue.

- Transparency Reporting: Companies and governments are required to publish regular reports on payments and revenues, making it easier to track financial flows and detect discrepancies.
- Multi-Stakeholder Governance: EITI involves civil society, industry, and government representatives in its implementation, ensuring diverse perspectives and oversight.

Human Story: Consider Maria, a civil society activist in an EITI-implementing country, using published data to hold her government accountable for its management of oil revenues.

Expert Insight: Dr. Laura Green, an expert in governance and transparency, explains, "EITI has been instrumental in improving transparency in the oil industry. By making information publicly available, it empowers citizens to hold their governments and companies accountable."

Anti-Bribery Conventions

International conventions such as the OECD Anti-Bribery Convention and the United Nations Convention against Corruption (UNCAC) set legal standards for combating bribery and corruption in international business transactions.

- Legal Frameworks: These conventions require signatory countries to criminalize bribery of foreign public officials and implement measures to prevent and detect corruption.
- Enforcement Mechanisms: Signatories are obligated to enforce anti-bribery laws and cooperate with international investigations.

Human Story: Picture John, a compliance officer at an oil company, working to ensure that the company adheres to anti-bribery laws and implements robust internal controls.

National Efforts and Regulatory Reforms

In addition to international initiatives, many countries have undertaken national efforts to combat corruption in their oil sectors through regulatory reforms and enhanced oversight.

Strengthening Legal and Regulatory Frameworks

Countries can reduce corruption by strengthening their legal and regulatory frameworks, including clearer laws, stricter enforcement, and better oversight mechanisms.

- Clear Legal Provisions: Establishing clear legal definitions of corruption and related offenses, along with appropriate penalties.
- Independent Oversight Bodies: Creating independent anti-corruption agencies or watchdogs with the authority to investigate and prosecute corruption cases.

Human Story: Consider Emily, a prosecutor at a newly established anti-corruption agency, working to investigate and bring charges against corrupt officials and industry executives.

Public Procurement Reforms

Reforming public procurement processes to ensure transparency, competitiveness, and fairness is crucial in preventing corruption in the awarding of oil contracts.

- Competitive Bidding: Implementing competitive bidding processes for oil contracts to ensure that they are awarded based on merit and transparency.
- Public Disclosure: Requiring public disclosure of contract terms, bidding processes, and awarded contracts to enhance transparency and public scrutiny.

Human Story: Picture Raj, a government official overseeing a transparent bidding process for a new oil exploration contract, ensuring that all bidders have an equal opportunity and that the process is free from favouritism.

Corporate Responsibility and Internal Controls

Oil companies themselves have a crucial role to play in combating corruption through robust corporate governance and internal control mechanisms.

Codes of Conduct and Ethics Programs

Implementing comprehensive codes of conduct and ethics programs can help prevent corrupt practices within companies.

- Ethical Guidelines: Establishing clear ethical guidelines and expectations for employee behaviour, including prohibitions on bribery and conflicts of interest.
- Training and Awareness: Providing regular training and awareness programs to educate employees about anti-corruption policies and the importance of ethical conduct.

Human Story: Consider Ahmed, an ethics officer at an oil company, conducting training sessions to educate employees about the company's anti-corruption policies and how to report suspicious activities.

Expert Insight: Dr. Robert Lee, an expert in corporate ethics, notes, "A strong ethical culture within a company is essential for preventing corruption. It requires commitment from the top leadership and a clear message that unethical behaviour will not be tolerated."

Internal Audit and Compliance

Robust internal audit and compliance functions are critical for detecting and preventing corrupt practices within companies.

- Internal Audits: Conducting regular internal audits to review financial transactions, contracts, and compliance with anti-corruption policies.
- Whistle-blower Protections: Establishing whistle-blower protection programs to encourage employees to report corruption without fear of retaliation.

Human Story: Picture Li, an internal auditor at an oil company, uncovering suspicious financial transactions during an audit and working with compliance officers to investigate and address the issues.

Challenges and Ongoing Efforts

Despite significant progress, combating corruption in the oil industry remains a challenging task. Corruption is often deeply entrenched, and efforts to root it out face numerous obstacles.

Political Will and Resistance

Combating corruption requires strong political will, which can be difficult to muster in countries where corruption is pervasive and politically connected individuals benefit from the status quo.

- Political Interference: Anti-corruption agencies and efforts can be undermined by political interference and lack of support from top leadership.
- Cultural Factors: In some contexts, corruption may be normalized or viewed as a necessary part of doing business, making it harder to change entrenched behaviours.

Human Story: Consider Sarah, an anti-corruption advocate facing resistance from powerful political figures who have a vested interest in maintaining corrupt practices.

Ensuring Effective Implementation

Implementing anti-corruption measures effectively requires adequate resources, capacity building, and continuous monitoring.

- Resource Constraints: Anti-corruption agencies and initiatives often face resource constraints, limiting their ability to conduct thorough investigations and enforcement actions.
- Capacity Building: Ongoing training and capacity building for investigators, prosecutors, and regulators are

essential for effective anti-corruption efforts.

Human Story: Picture Emily, a capacity-building consultant, working with an anti-corruption agency to enhance the skills and resources needed to investigate and prosecute corruption cases effectively.

Thought-Provoking Questions: Navigating Anti-Corruption Efforts

Reflecting on efforts to combat corruption in the oil industry raises several critical questions: How can governments and companies ensure the effective implementation of anti-corruption measures? What role should international cooperation play in combating cross-border corruption? And how can civil society and the public be empowered to hold corrupt actors accountable?

Storytelling Techniques: Bringing Anti-Corruption Efforts to Life

To illustrate the complexities of combating corruption, envision the diligent work of investigators, the strategic discussions in boardrooms, and the courageous actions of whistle-blowers and activists. Highlight the voices of prosecutors, compliance officers, civil society advocates, and affected communities to provide a balanced and comprehensive narrative.

Actionable Insights: Strategies for Effective Anti-Corruption Measures

1. Strengthening Legal Frameworks: Establishing clear and enforceable anti-corruption laws and regulations can provide a strong foundation for combating corruption.
2. Enhancing Transparency: Promoting transparency in contracts, financial transactions, and government dealings can reduce opportunities for corruption.
3. Fostering Corporate Ethics: Implementing robust codes

of conduct, ethics programs, and internal controls within companies can prevent corrupt practices.

4. Empowering Civil Society: Supporting civil society organizations and whistle-blowers can enhance public oversight and accountability.

Conclusion: Navigating the Path to Integrity

Combating corruption in the oil industry is a complex and ongoing challenge, but it is essential for ensuring transparency, accountability, and ethical behaviour. This crude awakening highlights the importance of robust anti-corruption measures in promoting economic prosperity, social equity, and environmental sustainability.

By understanding the efforts to combat corruption in the oil industry, we can better appreciate the importance of transparency, accountability, and ethical behaviour. The lessons learned from various anti-corruption initiatives provide valuable insights for shaping policies and practices that promote integrity and trust. Through collective action and ongoing commitment to responsible practices, the oil industry can navigate the challenges of corruption and contribute to a more stable and sustainable energy landscape.

CHAPTER 88: TRANSPARENCY INITIATIVES

Crude Awakening: The Modern Truths and Future of Oil

Shedding Light: The Importance of Transparency in Oil Industry Operations

Transparency in the oil industry is crucial for ensuring accountability, fostering trust, and promoting sustainable practices. Given the significant economic, environmental, and social impacts of oil production, transparency initiatives are essential for holding companies and governments accountable. This chapter explores the importance of transparency in the oil industry, the mechanisms that promote it, and the broader implications for stakeholders.

The Case for Transparency

Transparency in the oil industry involves the open and accessible disclosure of information related to operations, revenues, contracts, and environmental impacts. It is fundamental for good governance and ethical business practices.

Benefits of Transparency

Transparency provides numerous benefits, including:

- Accountability: By making information publicly available, stakeholders can hold companies and governments accountable for their actions.
- Trust: Transparency fosters trust between oil companies, governments, and the public, essential for social license to operate.
- Informed Decision-Making: Access to accurate information allows stakeholders to make informed decisions and engage in meaningful dialogue.

Human Story: Imagine Samuel, a community leader in an oil-producing region, using publicly available data to advocate for better environmental practices and fair revenue distribution.

Key Transparency Initiatives

Several global initiatives and frameworks have been established to promote transparency in the oil industry. These initiatives aim to improve the disclosure of information and enhance accountability.

Extractive Industries Transparency Initiative (EITI)

The EITI is a global standard for promoting transparency and accountability in the oil, gas, and mining sectors. Countries that implement EITI commit to disclosing information about their extractive industries, including contracts, licenses, production, and revenue.

- Disclosure Requirements: EITI requires the publication of comprehensive reports detailing payments made by oil companies to governments and revenues received by governments.
- Multi-Stakeholder Governance: EITI involves civil society, industry, and government representatives in its implementation, ensuring diverse perspectives and oversight.

Human Story: Consider Maria, a civil society activist in an EITI-implementing country, using published data to hold her government accountable for its management of oil revenues.

Expert Insight: Dr. Laura Green, an expert in governance and transparency, explains, "EITI has been instrumental in improving transparency in the oil industry. By making information publicly available, it empowers citizens to hold their governments and companies accountable."

Publish What You Pay (PWYP)

PWYP is a global coalition of civil society organizations advocating for greater transparency and accountability in the extractive industries.

- Advocacy and Campaigning: PWYP campaigns for mandatory disclosure laws and corporate reporting on payments to governments.
- Capacity Building: PWYP supports local civil society organizations in building their capacity to analyse data and advocate for transparency and accountability.

Human Story: Picture John, a member of a local NGO supported by PWYP, working to analyse government revenue reports and advocate for better resource management.

Open Contracting Partnership (OCP)

The OCP promotes transparency in public contracting, including in the oil sector, by advocating for the open publication of contracting data and information.

- Open Data Standards: OCP promotes the use of open data standards to make contracting information accessible and comparable.
- Public Participation: OCP encourages public participation

and oversight in the contracting process to ensure fairness and accountability.

Human Story: Consider Emily, a journalist using data from the OCP to investigate irregularities in oil contracts awarded by her government, bringing transparency to the contracting process.

Mechanisms for Promoting Transparency

Implementing transparency initiatives involves several key mechanisms, including legislative frameworks, technological tools, and stakeholder engagement.

Legislative Frameworks

Strong legislative frameworks are essential for mandating transparency and ensuring compliance with disclosure requirements.

- Mandatory Reporting Laws: Laws requiring oil companies to disclose payments to governments and other financial information promote transparency and accountability.
- Freedom of Information (FOI) Laws: FOI laws enable citizens to request and access information held by the government, enhancing transparency.

Human Story: Picture Raj, a lawyer advocating for stronger FOI laws in his country to ensure that citizens have access to vital information about the oil sector.

Technological Tools

Technological tools, including data platforms and online portals, facilitate the collection, analysis, and dissemination of information, making it more accessible to stakeholders.

- Data Portals: Online portals such as EITI's data portal provide comprehensive and easily accessible information

on payments, revenues, and contracts.
- Blockchain Technology: Blockchain can enhance transparency by providing a secure and immutable record of transactions and contracts in the oil industry.

Human Story: Consider Ahmed, a data analyst using blockchain technology to verify and track transactions in the oil industry, ensuring data integrity and transparency.

Stakeholder Engagement

Engaging stakeholders, including local communities, civil society organizations, and the media, is crucial for promoting transparency and accountability.

- Community Involvement: Involving local communities in decision-making processes ensures that their interests are considered and enhances trust.
- Civil Society Participation: Civil society organizations play a vital role in advocating for transparency, analysing data, and holding companies and governments accountable.

Human Story: Picture Li, a community liaison officer working to engage local communities in discussions about oil revenue management and environmental impacts, fostering transparency and trust.

Challenges and Barriers to Transparency

Despite the significant progress made by transparency initiatives, several challenges and barriers remain.

Resistance from Stakeholders

Resistance from some stakeholders, including governments and companies, can hinder transparency efforts.

- Political Resistance: Governments may resist transparency initiatives that expose corruption or mismanagement of resources.

- Corporate Resistance: Companies may be reluctant to disclose information that they consider commercially sensitive or that could impact their competitive position.

Human Story: Consider Sarah, an anti-corruption advocate facing resistance from government officials who are reluctant to disclose information about oil revenues.

Data Quality and Accessibility

Ensuring the quality, accuracy, and accessibility of data is a major challenge for transparency initiatives.

- Data Accuracy: Inaccurate or incomplete data can undermine the effectiveness of transparency efforts.
- Accessibility: Making data accessible and understandable to non-experts, including local communities, is crucial for meaningful transparency.

Human Story: Picture Emily, a data scientist working to improve the quality and accessibility of transparency data, ensuring that it is accurate and useful for stakeholders.

The Broader Implications of Transparency

Transparency in the oil industry has broader implications for governance, economic development, and environmental sustainability.

Improving Governance

Transparency enhances governance by reducing corruption, improving resource management, and promoting accountability.

- Reducing Corruption: Transparency initiatives help to expose and reduce corruption by making financial transactions and contracts publicly available.
- Improving Resource Management: Transparent reporting

enables better management of natural resources, ensuring that revenues benefit the public.

Human Story: Consider Li, a government official using transparency data to improve resource management practices and ensure that oil revenues are used for public benefit.

Expert Insight: Dr. Robert Lee, an expert in governance and transparency, notes, "Transparency is a cornerstone of good governance. It helps to build trust, reduce corruption, and ensure that natural resources are managed in a way that benefits society as a whole."

Promoting Economic Development

Transparency can promote economic development by creating a more stable and predictable investment environment.

- Attracting Investment: Transparent and accountable governance can attract investment by reducing risks and building investor confidence.
- Enhancing Efficiency: Transparency can enhance efficiency in the oil sector by promoting fair competition and reducing opportunities for corruption.

Human Story: Picture Ahmed, an investor looking at transparency reports to make informed decisions about investing in the oil sector, confident in the stable and predictable environment.

Supporting Environmental Sustainability

Transparency initiatives can support environmental sustainability by holding companies accountable for their environmental impacts and promoting responsible practices.

- Environmental Reporting: Requiring companies to

disclose their environmental impacts and mitigation measures promotes accountability and sustainability.

- Public Scrutiny: Public access to environmental data enables civil society and communities to hold companies accountable for their environmental performance.

Human Story: Consider Sarah, an environmental activist using transparency data to advocate for stronger environmental protections and sustainable practices in the oil industry.

Thought-Provoking Questions: Navigating Transparency Initiatives

Reflecting on the importance of transparency in the oil industry raises several critical questions: How can transparency initiatives be effectively implemented and enforced? What role should technology play in promoting transparency? And how can stakeholders be empowered to use transparency data to hold companies and governments accountable?

Storytelling Techniques: Bringing Transparency Initiatives to Life

To illustrate the complexities of transparency initiatives, envision the diligent work of data analysts, the strategic discussions in boardrooms, and the passionate advocacy of civil society activists. Highlight the voices of community leaders, journalists, government officials, and industry representatives to provide a balanced and comprehensive narrative.

Actionable Insights: Strategies for Effective Transparency Initiatives

1. Strengthening Legal Frameworks: Implementing and enforcing robust laws and regulations that mandate transparency can ensure compliance and accountability.

2. Leveraging Technology: Utilizing technological tools such as data portals and blockchain can enhance data accessibility, accuracy, and integrity.
3. Engaging Stakeholders: Promoting stakeholder engagement and public participation can ensure that transparency initiatives are inclusive and effective.
4. Improving Data Quality: Ensuring the accuracy and accessibility of transparency data can enhance its usefulness for stakeholders and promote informed decision-making.

Conclusion: Navigating the Path to Transparency

Transparency is a critical aspect of managing the oil industry, ensuring that companies and governments operate in an open and accountable manner. This crude awakening highlights the importance of transparency initiatives in promoting good governance, economic development, and environmental sustainability.

By understanding the importance of transparency in the oil industry, we can better appreciate the complexities and opportunities in managing this critical resource. The lessons learned from various transparency initiatives provide valuable insights for shaping policies and practices that promote accountability and trust. Through collective action and ongoing commitment to transparency, the

CHAPTER 89: CLIMATE CHANGE LEGISLATION

Crude Awakening: The Modern Truths and Future of Oil

The Winds of Change: Analysing the Impact of Climate Change Legislation on the Oil Industry

Climate change legislation represents one of the most significant challenges and opportunities for the oil industry. As governments worldwide enact laws to reduce greenhouse gas emissions and transition to cleaner energy sources, the oil industry faces profound changes. This chapter explores the impact of climate change legislation on the oil industry, examining the strategies companies employ to adapt, the economic and environmental implications, and the broader context of the global energy transition.

Understanding Climate Change Legislation

Climate change legislation encompasses a range of policies and regulations aimed at reducing carbon emissions, promoting renewable energy, and enhancing energy efficiency. These laws are designed to mitigate the impacts of climate change and foster a sustainable energy future.

Key Components of Climate Change Legislation

- Emissions Reduction Targets: Setting legally binding targets for reducing greenhouse gas emissions.
- Carbon Pricing: Implementing carbon taxes or cap-and-trade systems to price carbon emissions and incentivize reductions.
- Renewable Energy Mandates: Requiring a certain percentage of energy to come from renewable sources.
- Energy Efficiency Standards: Establishing standards for energy efficiency in buildings, vehicles, and industrial processes.

Human Story: Imagine Samuel, a policy analyst working for a government agency, tasked with developing and implementing climate change legislation to meet national emissions reduction targets.

The Impact of Climate Change Legislation on the Oil Industry

Climate change legislation has far-reaching implications for the oil industry, affecting everything from production practices to market dynamics.

Regulatory Compliance and Costs

Compliance with climate change legislation imposes significant costs on the oil industry, including investments in cleaner technologies and operational changes.

- Carbon Pricing: Carbon taxes and cap-and-trade systems increase the cost of carbon emissions, incentivizing oil companies to reduce their carbon footprint.
- Technological Upgrades: Companies must invest in technologies such as carbon capture and storage (CCS) to comply with emissions reduction targets.
- Operational Changes: Meeting energy efficiency standards requires changes in production processes and infrastructure upgrades.

Human Story: Consider Maria, an operations manager at an oil refinery, overseeing the implementation of new technologies and processes to meet stringent emissions standards.

Expert Insight: Dr. Laura Green, an environmental economist, explains, "The cost of compliance with climate change legislation can be substantial, but it also drives innovation and efficiency improvements. Companies that invest in cleaner technologies can gain a competitive edge in the long run."

Market Shifts and Investment Strategies

Climate change legislation is reshaping market dynamics, influencing investment strategies and the competitive landscape of the energy sector.

- Shift to Renewables: Legislation promoting renewable energy creates opportunities for oil companies to diversify into renewable energy projects.
- Investment in Innovation: Companies are investing in research and development to create more sustainable energy solutions and reduce their reliance on fossil fuels.
- Stranded Assets: The risk of stranded assets, where investments in fossil fuel infrastructure become obsolete due to regulatory changes, is a growing concern.

Human Story: Picture John, an investor assessing the impact of climate change legislation on his portfolio, shifting investments from traditional oil companies to those that are actively diversifying into renewable energy.

Economic and Environmental Impacts

The economic and environmental impacts of climate change legislation are significant, influencing both the short-term profitability and long-term sustainability of the oil industry.

- Economic Adjustments: While the initial cost of compliance may be high, the long-term benefits of a more sustainable energy system include reduced environmental risks and new economic opportunities.
- Environmental Benefits: Reducing carbon emissions and promoting cleaner energy sources contribute to mitigating climate change and protecting ecosystems.

Human Story: Consider Emily, a local community member experiencing the benefits of reduced air pollution and improved environmental health as a result of stricter emissions standards.

Strategies for Adapting to Climate Change Legislation

Oil companies are employing various strategies to adapt to the challenges and opportunities presented by climate change legislation.

Diversification and Investment in Renewables

Many oil companies are diversifying their energy portfolios by investing in renewable energy projects and technologies.

- Renewable Energy Projects: Investing in wind, solar, and bioenergy projects to reduce dependence on fossil fuels.
- Energy Storage Solutions: Developing and deploying energy storage technologies to support the integration of renewable energy into the grid.

Human Story: Picture Raj, a project manager leading a new solar energy initiative for a major oil company, reflecting the industry's shift towards renewable energy.

Enhancing Operational Efficiency

Improving operational efficiency is a key strategy for

reducing emissions and complying with climate change legislation.

- Energy Efficiency Measures: Implementing energy efficiency measures in production processes, such as optimizing equipment and reducing energy waste.
- Carbon Capture and Storage: Investing in CCS technology to capture and store carbon emissions from industrial processes.

Human Story: Consider Ahmed, an engineer working on a CCS project at an oil refinery, aiming to significantly reduce the facility's carbon emissions.

Engaging in Policy Advocacy

Oil companies are actively engaging in policy advocacy to influence the development and implementation of climate change legislation.

- Collaborative Efforts: Participating in industry associations and multi-stakeholder initiatives to advocate for balanced and effective climate policies.
- Transparent Reporting: Enhancing transparency and reporting on climate-related risks and mitigation efforts to build trust with stakeholders.

Human Story: Picture Li, a corporate affairs executive at an oil company, working with policymakers and industry groups to shape climate change legislation that balances environmental goals with economic considerations.

Challenges and Opportunities

While climate change legislation presents significant challenges for the oil industry, it also offers opportunities for innovation, leadership, and growth.

Navigating Regulatory Uncertainty

The evolving nature of climate change legislation creates regulatory uncertainty, posing challenges for long-term planning and investment.

- Policy Volatility: Changes in political leadership and policy priorities can lead to shifts in climate change legislation, creating uncertainty for the industry.
- Risk Management: Developing robust risk management strategies to navigate regulatory uncertainty and ensure compliance with evolving standards.

Human Story: Consider Sarah, a risk manager at an oil company, developing strategies to navigate the uncertainties of climate change legislation and ensure long-term sustainability.

Driving Innovation and Competitive Advantage

The need to comply with climate change legislation is driving innovation in the oil industry, creating opportunities for competitive advantage.

- Technological Leadership: Companies that lead in developing and implementing sustainable technologies can gain a competitive edge in the transitioning energy market.
- Sustainable Practices: Adopting sustainable practices can enhance a company's reputation and attract investment from environmentally conscious investors.

Human Story: Picture Emily, a sustainability officer at an oil company, spearheading initiatives to reduce the company's carbon footprint and enhance its reputation as an industry leader in sustainability.

Thought-Provoking Questions: Navigating Climate Change Legislation

Reflecting on the impact of climate change legislation on

the oil industry raises several critical questions: How can oil companies balance the cost of compliance with the need for profitability? What role should innovation and technological advancement play in adapting to climate change legislation? And how can policymakers ensure that climate change legislation effectively reduces emissions while supporting economic growth?

Storytelling Techniques: Bringing Climate Change Legislation to Life

To illustrate the complexities of climate change legislation, envision the strategic discussions in boardrooms, the technological innovations in research labs, and the community impacts of reduced emissions. Highlight the voices of policymakers, industry leaders, engineers, and community members to provide a balanced and comprehensive narrative.

Actionable Insights: Strategies for Effective Adaptation

1. Investing in Innovation: Supporting research and development of cleaner technologies can help the industry meet regulatory standards and enhance efficiency.
2. Diversifying Energy Portfolios: Investing in renewable energy projects and technologies can reduce dependence on fossil fuels and create new opportunities for growth.
3. Enhancing Transparency: Implementing robust reporting practices on climate-related risks and mitigation efforts can build trust and accountability.
4. Engaging in Policy Advocacy: Actively participating in policy discussions and multi-stakeholder initiatives can help shape effective and balanced climate policies.

Conclusion: Navigating the Path to Sustainability

Climate change legislation is reshaping the oil industry, presenting both challenges and opportunities for companies to innovate, diversify, and lead in the transition to a sustainable energy future. This crude awakening highlights the importance of proactive adaptation and strategic planning in navigating the complexities of climate change legislation.

By understanding the impact of climate change legislation on the oil industry, we can better appreciate the importance of sustainable practices and technological innovation. The lessons learned from various adaptation strategies provide valuable insights for shaping policies and practices that promote environmental sustainability and economic growth. Through collective action and ongoing commitment to responsible practices, the oil industry can navigate the challenges of climate change legislation and contribute to a more stable and sustainable energy landscape.

CHAPTER 90: ENERGY SECURITY POLICIES

Crude Awakening: The Modern Truths and Future of Oil

Safeguarding the Future: Exploring National and International Energy Security Policies

Energy security is a critical concern for countries around the world. Ensuring a stable, reliable, and affordable supply of energy is vital for economic stability, national security, and public well-being. This chapter delves into the national and international energy security policies that shape the global energy landscape, examining their impacts, challenges, and the broader implications for the oil industry.

Understanding Energy Security

Energy security refers to the uninterrupted availability of energy sources at an affordable price. It encompasses various dimensions, including the reliability of energy supply, the affordability of energy, and the environmental sustainability of energy production and consumption.

Key Components of Energy Security

- Supply Reliability: Ensuring a continuous and stable supply of energy, regardless of disruptions.

- Affordability: Maintaining energy prices at levels that are accessible for consumers and businesses.
- Sustainability: Balancing energy needs with environmental protection and sustainable practices.

Human Story: Imagine Samuel, a policy advisor working to develop a comprehensive energy security strategy for his country, addressing the complexities of ensuring reliable, affordable, and sustainable energy.

National Energy Security Policies

Countries around the world adopt various strategies to enhance their energy security, reflecting their unique energy needs, resources, and geopolitical contexts.

The United States: Diversification and Independence

The United States has pursued a strategy of energy diversification and independence, focusing on reducing reliance on foreign oil and promoting a mix of energy sources.

- Shale Revolution: The development of shale oil and gas through hydraulic fracturing has significantly increased domestic production, reducing dependence on imported energy.
- Renewable Energy: Investment in renewable energy sources, such as wind and solar, is part of the strategy to diversify the energy mix and enhance sustainability.
- Strategic Petroleum Reserve (SPR): The SPR is a critical component of U.S. energy security, providing a buffer against supply disruptions.

Human Story: Consider Maria, an energy entrepreneur in Texas, benefiting from policies that support the development of renewable energy projects and contribute to national energy security.

European Union: Integration and Sustainability

The European Union (EU) focuses on energy integration and sustainability, aiming to create a unified energy market and promote clean energy.

- Energy Union: The EU's Energy Union strategy aims to ensure secure, sustainable, and affordable energy through market integration and policy coordination among member states.
- Renewable Energy Directive: The directive sets binding targets for renewable energy use, driving investment in clean energy technologies.
- Energy Efficiency: The EU emphasizes energy efficiency measures to reduce consumption and enhance energy security.

Human Story: Picture John, a policy analyst in Brussels, working on initiatives to improve energy efficiency and integrate renewable energy into the EU's energy system.

China: Strategic Planning and Diversification

China's energy security strategy involves long-term planning, diversification of energy sources, and securing international energy supplies.

- Five-Year Plans: China's Five-Year Plans outline strategic priorities for energy development, including increasing domestic production and expanding renewable energy.
- Belt and Road Initiative (BRI): The BRI includes investments in energy infrastructure abroad, securing supply routes and enhancing energy security.
- Coal and Renewables: While China continues to rely on coal, it is also investing heavily in renewable energy to diversify its energy mix.

Human Story: Consider Emily, a researcher in Beijing,

studying the impact of China's energy policies on its domestic energy landscape and international relations.

Expert Insight: Dr. Laura Green, an expert in energy policy, explains, "National energy security policies reflect the unique challenges and priorities of each country. Strategies like diversification, integration, and international cooperation are essential for ensuring a stable and sustainable energy future."

International Energy Security Policies

International cooperation and agreements play a crucial role in enhancing global energy security, addressing cross-border energy issues, and promoting stability in global energy markets.

International Energy Agency (IEA)

The IEA is an autonomous organization that works to ensure reliable, affordable, and clean energy for its member countries and beyond.

- Emergency Response Mechanism: The IEA coordinates collective responses to major disruptions in oil supply, such as releasing oil from strategic reserves.
- Energy Policy Recommendations: The IEA provides policy advice and recommendations to enhance energy security and promote sustainable energy practices.
- Data and Analysis: The IEA collects and disseminates energy data, providing valuable insights into global energy trends and security issues.

Human Story: Picture Raj, an analyst at the IEA, working on reports that assess global energy security and provide recommendations for member countries to enhance their energy resilience.

Organization of the Petroleum Exporting Countries (OPEC)

OPEC plays a significant role in global oil markets, influencing oil production and prices to ensure market stability and energy security for its member countries.

- Production Quotas: OPEC sets production quotas for member countries to manage oil supply and stabilize prices.
- Market Monitoring: OPEC monitors global oil markets and provides analysis to guide production decisions and ensure a balanced market.
- Dialogue and Cooperation: OPEC engages in dialogue with non-member oil producers and consumers to enhance cooperation and address energy security issues.

Human Story: Consider Ahmed, an economist at OPEC, analysing market data and contributing to discussions on production strategies to ensure stable oil supplies.

United Nations Framework Convention on Climate Change (UNFCCC)

The UNFCCC addresses the intersection of energy security and climate change, promoting international cooperation to reduce greenhouse gas emissions and transition to sustainable energy.

- Paris Agreement: The Paris Agreement commits countries to reducing emissions and enhancing adaptive capacities, impacting national energy security strategies.
- Nationally Determined Contributions (NDCs): Countries submit NDCs outlining their climate actions, including energy policies that contribute to global energy security and sustainability.
- Climate Finance: The UNFCCC facilitates climate finance to support developing countries in implementing sustainable energy projects.

Human Story: Picture Li, a negotiator at UNFCCC

conferences, working to align national energy security policies with global climate goals and secure funding for renewable energy projects.

Challenges and Opportunities

While energy security policies are essential for stability and resilience, they also face significant challenges and present opportunities for innovation and collaboration.

Geopolitical Tensions

Geopolitical tensions can threaten energy security, as conflicts and political instability disrupt energy supplies and markets.

- Supply Disruptions: Political conflicts in key oil-producing regions can lead to supply disruptions and market volatility.
- Resource Nationalism: Some countries may prioritize control over their energy resources, leading to restrictive policies and tensions with foreign investors.

Human Story: Consider Sarah, a geopolitical analyst, studying the impact of regional conflicts on global energy security and advising governments on mitigation strategies.

Technological Advancements

Technological advancements offer opportunities to enhance energy security through innovation and efficiency.

- Renewable Energy Technologies: Advancements in renewable energy technologies, such as solar and wind, provide cleaner and more sustainable energy sources.
- Energy Storage: Improved energy storage technologies enhance the reliability of renewable energy and support grid stability.

- Smart Grids: Smart grid technologies improve the efficiency and resilience of energy systems, enabling better management of supply and demand.

Human Story: Picture Emily, an engineer developing advanced energy storage solutions that enhance the reliability and sustainability of the energy grid.

The Future of Energy Security

The future of energy security will be shaped by evolving technologies, policy frameworks, and international cooperation. Ensuring a stable, affordable, and sustainable energy supply requires a multifaceted approach that addresses economic, environmental, and geopolitical dimensions.

Integrating Renewables and Sustainability

Integrating renewable energy sources into the energy mix is crucial for achieving long-term energy security and sustainability.

- Policy Support: Strong policy support and incentives are necessary to promote the adoption of renewable energy technologies.
- Infrastructure Development: Investing in infrastructure, such as transmission lines and storage facilities, is essential for integrating renewables into the energy grid.
- Public-Private Partnerships: Collaborations between governments and the private sector can drive innovation and investment in sustainable energy solutions.

Human Story: Consider Li, a policy advisor working on initiatives to support renewable energy development and create a more sustainable and resilient energy system.

Enhancing International Cooperation

International cooperation is vital for addressing global

energy security challenges and promoting stability in energy markets.

- Collaborative Frameworks: Strengthening collaborative frameworks, such as the IEA and UNFCCC, can enhance coordination and policy alignment.
- Information Sharing: Sharing information and best practices among countries can improve energy security and resilience.
- Joint Investments: Joint investments in energy infrastructure and technology development can create synergies and enhance global energy security.

Human Story: Picture Raj, a diplomat working on international energy agreements that foster cooperation and enhance global energy security.

Thought-Provoking Questions: Navigating Energy Security Policies

Reflecting on national and international energy security policies raises several critical questions: How can countries balance energy security with environmental sustainability? What role should technological innovation play in enhancing energy security? And how can international cooperation be strengthened to address global energy challenges?

Storytelling Techniques: Bringing Energy Security Policies to Life

To illustrate the complexities of energy security policies, envision the strategic discussions in government offices, the technological innovations in research labs, and the collaborative efforts in international forums. Highlight the voices of policymakers, engineers, investors, and community members to provide a balanced and comprehensive narrative.

Actionable Insights: Strategies for Enhancing Energy Security

1. Diversifying Energy Sources: Promoting a diverse energy mix that includes renewables and other sustainable sources can enhance energy security and reduce dependency on fossil fuels.
2. Investing in Technology: Supporting research and development of advanced energy technologies can improve efficiency and resilience.
3. Strengthening Cooperation: Enhancing international cooperation and collaborative frameworks can address cross-border energy issues and promote stability

PART X: INNOVATIONS AND FUTURE DIRECTIONS

CHAPTER 91: BIOFUELS AND OIL

Crude Awakening: The Modern Truths and Future of Oil

Green Alternatives: The Potential of Biofuels to Complement or Replace Oil

As the world grapples with the dual challenges of energy security and climate change, biofuels have emerged as a promising alternative to conventional fossil fuels. Derived from organic materials, biofuels offer the potential to reduce greenhouse gas emissions and provide a sustainable source of energy. This chapter explores the potential of biofuels to complement or replace oil, examining the benefits, challenges, and broader implications for the global energy landscape.

Understanding Biofuels

Biofuels are produced from biomass, which includes plant materials, agricultural residues, and other organic matter. They are categorized into different types based on their source materials and production processes.

Types of Biofuels

- First-Generation Biofuels: Derived from food crops such as corn, sugarcane, and soybeans. Examples include ethanol and biodiesel.
- Second-Generation Biofuels: Produced from non-food

biomass, such as agricultural residues, grasses, and woody plants. These include cellulosic ethanol and biomass-to-liquid fuels.
- Third-Generation Biofuels: Made from algae and other advanced biological processes, offering higher yields and lower environmental impacts.

Human Story: Imagine Samuel, a farmer in the Midwest, who grows corn for both food and ethanol production, contributing to the local economy and renewable energy efforts.

The Potential of Biofuels to Complement Oil

Biofuels have the potential to complement oil by providing an alternative source of liquid fuels for transportation and other applications.

Environmental Benefits

Biofuels can significantly reduce greenhouse gas emissions compared to conventional fossil fuels, contributing to climate change mitigation.

- Lower Carbon Emissions: Biofuels produce fewer carbon emissions because the carbon dioxide absorbed by plants during growth offsets the emissions released during combustion.
- Renewable Resource: Unlike fossil fuels, biofuels are derived from renewable resources that can be replenished over short periods.

Human Story: Consider Maria, an environmental activist advocating for increased use of biofuels to reduce the carbon footprint of the transportation sector and combat climate change.

Expert Insight: Dr. Laura Green, an environmental scientist, explains, "Biofuels offer a viable path to reducing greenhouse gas emissions, particularly in sectors where electrification is challenging. Their renewable nature and potential for carbon neutrality make them an attractive alternative to fossil fuels."

Energy Security and Economic Benefits

Biofuels can enhance energy security by diversifying energy sources and reducing dependence on imported oil.

- Domestic Production: Biofuels can be produced locally, reducing reliance on foreign oil, and enhancing national energy security.
- Economic Opportunities: The biofuel industry creates jobs in agriculture, manufacturing, and research, supporting local economies.

Human Story: Picture John, a worker at a biofuel refinery, benefiting from the economic opportunities created by the growing biofuel industry in his community.

The Challenges of Biofuels

Despite their potential, biofuels face several challenges that must be addressed to fully realize their benefits and ensure sustainable production.

Competition with Food Production

First-generation biofuels, derived from food crops, can compete with food production, and contribute to food price volatility.

- Food vs. Fuel Debate: The use of food crops for biofuel production can divert resources from food production, raising concerns about food security and prices.
- Sustainable Practices: Developing sustainable agricultural practices and advancing second- and third-generation biofuels can mitigate these concerns.

Human Story: Consider Emily, a food security advocate concerned about the impact of biofuel production on global food supplies and advocating for sustainable biofuel practices.

Environmental and Land Use Impacts

Biofuel production can have environmental impacts, including land use changes, water consumption, and biodiversity loss.

- Land Use Changes: Converting land for biofuel crops can lead to deforestation, habitat loss, and increased carbon emissions from land-use changes.
- Resource Intensity: Biofuel production can be resource-intensive, requiring significant amounts of water and fertilizers.

Human Story: Picture Raj, an environmental researcher studying the impact of large-scale biofuel plantations on local ecosystems and advocating for sustainable land management practices.

Expert Insight: Dr. Robert Lee, an expert in sustainable agriculture, notes, "While biofuels have the potential to reduce carbon emissions, it is essential to consider their broader environmental impacts. Sustainable land management and advancements in second- and third-generation biofuels are critical for minimizing these impacts."

Advancements and Innovations in Biofuels

Advancements in biofuel technologies and production methods are key to overcoming challenges and enhancing the sustainability and efficiency of biofuels.

Second- and Third-Generation Biofuels

Second- and third-generation biofuels offer significant advantages over first-generation biofuels, including reduced competition with food production and lower environmental impacts.

- Cellulosic Ethanol: Produced from non-food biomass, cellulosic ethanol offers a more sustainable alternative to traditional ethanol.
- Algae-Based Biofuels: Algae-based biofuels have high energy yields and can be grown on non-arable land, reducing competition with food production.

Human Story: Consider Ahmed, a researcher working on developing algae-based biofuels, aiming to create a sustainable and high-yield energy source that does not compete with food crops.

Technological Innovations

Technological innovations in biofuel production, such as advanced fermentation processes and genetic engineering, are improving efficiency and reducing costs.

- Advanced Fermentation: Innovations in fermentation technology are increasing the yield and efficiency of biofuel production from various biomass sources.
- Genetic Engineering: Genetic engineering of biofuel crops and microorganisms is enhancing their productivity and reducing environmental impacts.

Human Story: Picture Li, a bioengineer developing genetically modified microorganisms that can efficiently convert biomass into biofuels, pushing the boundaries of biofuel technology.

The Role of Policy and Regulation

Policy and regulatory frameworks play a crucial role in promoting the development and adoption of biofuels, addressing challenges, and ensuring sustainability.

Government Support and Incentives

Government policies and incentives can drive the growth of the biofuel industry by providing financial support, setting mandates, and promoting research and development.

- Renewable Fuel Standards (RFS): Mandates that require a certain percentage of transportation fuels to come from renewable sources, including biofuels.
- Subsidies and Grants: Financial incentives, such as subsidies and grants, support biofuel research, production, and infrastructure development.

Human Story: Consider Sarah, a policymaker advocating for stronger renewable fuel standards and increased government support for biofuel innovation and infrastructure.

Sustainability Standards

Implementing sustainability standards and certification schemes can ensure that biofuels are produced in an environmentally and socially responsible manner.

- Certification Schemes: Certification schemes, such as the Roundtable on Sustainable Biomaterials (RSB), set standards for sustainable biofuel production and provide third-party verification.
- Sustainable Practices: Encouraging sustainable agricultural practices and land management to minimize the environmental impact of biofuel production.

Human Story: Picture Emily, a sustainability officer at a biofuel company, working to ensure that the company's production practices meet international sustainability standards.

Thought-Provoking Questions: Navigating the Future of Biofuels

Reflecting on the potential of biofuels to complement or replace oil raises several critical questions: How can we balance the benefits of biofuels with the challenges of food security and environmental impacts? What role should government policies and incentives play in promoting sustainable biofuels? And how can technological innovations drive the future of biofuels?

Storytelling Techniques: Bringing Biofuels to Life

To illustrate the complexities of biofuels, envision the vibrant activity on biofuel farms, the cutting-edge research in laboratories, and the policy discussions in government offices. Highlight the voices of farmers, researchers, policymakers, and environmental advocates to provide a balanced and comprehensive narrative.

Actionable Insights: Strategies for Promoting Sustainable Biofuels

1. Investing in Innovation: Supporting research and development of second- and third-generation biofuels can enhance sustainability and reduce competition with food production.
2. Implementing Sustainability Standards: Adopting and enforcing sustainability standards can ensure environmentally and socially responsible biofuel production.
3. Enhancing Policy Support: Strengthening government policies and incentives can drive the growth of the biofuel industry and promote the transition to renewable energy.
4. Promoting Public Awareness: Educating the public about the benefits and challenges of biofuels can build support for sustainable energy initiatives.

Conclusion: Navigating the Path to a Sustainable Energy Future

Biofuels represent a promising alternative to conventional fossil fuels, offering the potential to reduce carbon emissions, enhance energy security, and support economic development. This crude awakening highlights the importance of sustainable practices, technological innovation, and supportive policies in realizing the full potential of biofuels.

By understanding the potential of biofuels to complement or replace oil, we can better appreciate the complexities and opportunities in the transition to a sustainable energy future. The lessons learned from various biofuel initiatives provide valuable insights for shaping policies and practices that promote environmental sustainability and energy security. Through collective action and ongoing commitment to innovation and sustainability, the biofuel industry can navigate the challenges ahead and contribute to a more stable and sustainable energy landscape.

CHAPTER 92: HYDROGEN ECONOMY

Crude Awakening: The Modern Truths and Future of Oil

A New Dawn: Exploring the Development of the Hydrogen Economy and Its Impact on Oil

The transition to a hydrogen economy represents a significant shift in the global energy landscape. Hydrogen, as a clean and versatile energy carrier, holds the potential to revolutionize various sectors, including transportation, industry, and power generation. This chapter delves into the development of the hydrogen economy, its potential to complement or replace oil, and the broader implications for the global energy framework.

Understanding the Hydrogen Economy

The hydrogen economy envisions a future where hydrogen serves as a primary energy carrier, offering a cleaner alternative to fossil fuels. Hydrogen can be produced from a variety of sources, stored, and transported in multiple forms, and utilized across diverse applications.

Key Components of the Hydrogen Economy

- Production: Hydrogen can be produced through various methods, including natural gas reforming (gray hydrogen),

water electrolysis using renewable energy (green hydrogen), and nuclear energy (yellow hydrogen).
- Storage and Distribution: Hydrogen can be stored as compressed gas, liquid hydrogen, or in chemical carriers, and transported through pipelines, tankers, or trucks.
- Utilization: Hydrogen can power fuel cells for transportation, serve as a feedstock in industrial processes, and generate electricity in power plants.

Human Story: Imagine Samuel, an engineer at a green hydrogen production facility, using solar and wind energy to produce hydrogen that powers local industries and vehicles.

The Potential of Hydrogen to Complement or Replace Oil

Hydrogen's versatility and environmental benefits position it as a potential complement or replacement for oil in various applications.

Environmental Benefits

Hydrogen is a clean energy carrier that produces no carbon emissions at the point of use, making it crucial for reducing greenhouse gas emissions and combating climate change.

- Zero Emissions: When used in fuel cells, hydrogen emits only water vapor, significantly reducing air pollution and greenhouse gas emissions.
- Renewable Production: Green hydrogen, produced using renewable energy, offers a sustainable alternative to fossil fuels.

Human Story: Consider Maria, an environmental activist advocating for the adoption of hydrogen technologies to reduce carbon emissions and promote cleaner air in urban areas.

Expert Insight: Dr. Laura Green, an environmental

scientist, explains, "Hydrogen's potential to produce zero emissions at the point of use makes it a key player in our transition to a low-carbon economy. Its versatility across various sectors enhances its appeal as a sustainable energy solution."

Energy Security and Economic Benefits

Hydrogen can enhance energy security by diversifying energy sources and reducing dependence on imported fossil fuels.

- Domestic Production: Hydrogen can be produced locally from various resources, enhancing energy security, and reducing reliance on foreign oil.
- Economic Opportunities: The hydrogen economy can create jobs in production, distribution, and infrastructure development, supporting local economies.

Human Story: Picture John, a worker at a hydrogen refuelling station, benefiting from the economic opportunities created by the growing hydrogen industry in his community.

Challenges of Developing the Hydrogen Economy

Despite its potential, the hydrogen economy faces significant challenges that must be addressed to achieve widespread adoption and integration.

High Production Costs

Producing green hydrogen, essential for a sustainable hydrogen economy, is currently more expensive than conventional fossil fuels.

- Cost of Electrolysis: Electrolysis, which splits water into hydrogen and oxygen using electricity, is energy-intensive and costly, particularly when using renewable energy sources.

- Infrastructure Investment: Significant investment is required to develop the infrastructure for hydrogen production, storage, and distribution.

Human Story: Consider Emily, a researcher working to develop more cost-effective electrolysis technologies, aiming to make green hydrogen competitive with conventional fossil fuels.

Storage and Distribution Challenges

Hydrogen's low energy density and the need for high-pressure storage present challenges for efficient and safe storage and distribution.

- Energy Density: Hydrogen has a lower energy density compared to liquid fuels like gasoline, requiring larger storage volumes or higher pressures.
- Safety Concerns: Hydrogen is highly flammable, necessitating stringent safety measures for storage and transportation.

Human Story: Picture Raj, an engineer designing advanced storage solutions to safely and efficiently store and transport hydrogen, addressing one of the key challenges of the hydrogen economy.

Expert Insight: Dr. Robert Lee, an expert in energy storage, notes, "While hydrogen offers significant benefits, its storage and distribution present technical challenges that require innovative solutions. Advances in materials science and engineering will be crucial for overcoming these hurdles."

Advancements and Innovations in Hydrogen Technology

Advancements in hydrogen technology are key to addressing challenges and accelerating the development of the hydrogen economy.

Green Hydrogen Production

Innovations in electrolysis and renewable energy integration are making green hydrogen more cost-competitive and sustainable.

- Advanced Electrolysis: Research into more efficient electrolysis processes and materials is reducing the cost and energy requirements of green hydrogen production.
- Renewable Integration: Integrating electrolysis with renewable energy sources, such as wind and solar, is enhancing the sustainability and scalability of green hydrogen production.

Human Story: Consider Ahmed, a scientist developing new catalysts for electrolysis, improving the efficiency of green hydrogen production, and reducing costs.

Hydrogen Fuel Cells

Advances in hydrogen fuel cell technology are expanding the applications of hydrogen in transportation and other sectors.

- Fuel Cell Vehicles: Hydrogen fuel cell vehicles (FCVs) offer a zero-emission alternative to conventional gasoline and diesel vehicles, with fast refuelling times and long driving ranges.
- Stationary Fuel Cells: Hydrogen fuel cells are being used for backup power and distributed generation, providing reliable and clean energy.

Human Story: Picture Li, a driver of a hydrogen fuel cell vehicle, enjoying the benefits of zero-emission transportation and fast refuelling at hydrogen stations.

The Role of Policy and Regulation

Government policies and regulations play a crucial role

in promoting the development and adoption of hydrogen technologies, addressing challenges, and ensuring sustainability.

Government Support and Incentives

Policies and incentives can drive the growth of the hydrogen economy by providing financial support, setting targets, and promoting research and development.

- Subsidies and Grants: Financial incentives, such as subsidies and grants, support hydrogen research, production, and infrastructure development.
- Renewable Hydrogen Standards: Setting standards and targets for renewable hydrogen production can drive investment and development.

Human Story: Consider Sarah, a policymaker advocating for stronger government support and incentives for hydrogen technology to accelerate the transition to a hydrogen economy.

International Cooperation

International cooperation is vital for addressing global challenges and promoting the development of a hydrogen economy.

- Collaborative Research: International collaboration on research and development can accelerate technological advancements and reduce costs.
- Harmonized Standards: Developing harmonized standards for hydrogen production, storage, and distribution can facilitate international trade and cooperation.

Human Story: Picture Emily, a researcher participating in an international collaboration to develop advanced hydrogen technologies and share best practices.

Thought-Provoking Questions: Navigating the Hydrogen Economy

Reflecting on the development of the hydrogen economy raises several critical questions: How can we overcome the high production costs and storage challenges of hydrogen? What role should government policies and incentives play in promoting hydrogen technologies? And how can international cooperation enhance the development of a global hydrogen economy?

Storytelling Techniques: Bringing the Hydrogen Economy to Life

To illustrate the complexities of the hydrogen economy, envision the innovative research in laboratories, the strategic discussions in government offices, and the practical applications of hydrogen in everyday life. Highlight the voices of engineers, scientists, policymakers, and community members to provide a balanced and comprehensive narrative.

Actionable Insights: Strategies for Promoting the Hydrogen Economy

1. Investing in Research and Development: Supporting research and development of advanced hydrogen technologies can reduce costs and improve efficiency.
2. Enhancing Policy Support: Strengthening government policies and incentives can drive the growth of the hydrogen economy and promote the transition to renewable energy.
3. Developing Infrastructure: Investing in infrastructure for hydrogen production, storage, and distribution is essential for widespread adoption.
4. Promoting International Collaboration: Enhancing international cooperation on hydrogen research and

development can accelerate progress and share best practices.

Conclusion: Navigating the Path to a Hydrogen Economy

The hydrogen economy represents a promising path towards a sustainable and low-carbon energy future. This crude awakening highlights the importance of technological innovation, supportive policies, and international cooperation in realizing the full potential of hydrogen.

By understanding the development of the hydrogen economy and its impact on oil, we can better appreciate the complexities and opportunities in the transition to sustainable energy. The lessons learned from various hydrogen initiatives provide valuable insights for shaping policies and practices that promote environmental sustainability and energy security. Through collective action and ongoing commitment to innovation and sustainability, the hydrogen economy can navigate the challenges ahead and contribute to a more stable and sustainable energy landscape.

CHAPTER 93: CARBON CAPTURE AND STORAGE

Crude Awakening: The Modern Truths and Future of Oil

Capturing Change: Analysing the Role of Carbon Capture and Storage in Mitigating Oil's Environmental Impact

As the world grapples with the urgent need to reduce greenhouse gas emissions, carbon capture and storage (CCS) has emerged as a pivotal technology in mitigating the environmental impact of fossil fuels, including oil. This chapter explores the development and potential of CCS, examining its effectiveness, challenges, and the broader implications for the oil industry and global climate goals.

Understanding Carbon Capture and Storage

Carbon capture and storage is a process that involves capturing carbon dioxide (CO_2) emissions from industrial processes and power generation, transporting it to a storage site, and injecting it deep underground for long-term isolation from the atmosphere.

Key Components of CCS

- Capture: CO_2 is captured from sources such as power plants and industrial facilities using various technologies, including pre-combustion, post-combustion, and oxy-fuel

combustion.
- Transport: The captured CO2 is then transported via pipelines, ships, or trucks to a storage site.
- Storage: CO2 is injected into geological formations such as depleted oil and gas fields, deep saline aquifers, or unmendable coal seams for long-term storage.

Human Story: Imagine Samuel, an engineer at a CCS facility, overseeing the capture of CO2 from a nearby refinery and ensuring its safe transport and storage underground.

The Potential of CCS to Mitigate Oil's Environmental Impact

CCS offers significant potential to reduce the carbon footprint of the oil industry, making it a crucial tool in the fight against climate change.

Reducing Carbon Emissions

CCS can capture up to 90% of CO2 emissions from industrial processes and power plants, significantly reducing greenhouse gas emissions.

- Industrial Applications: CCS can be applied to various industrial processes, including oil refining, cement production, and chemical manufacturing, where emissions are challenging to eliminate.
- Enhanced Oil Recovery (EOR): Captured CO2 can be used in enhanced oil recovery operations, where it is injected into oil reservoirs to increase oil extraction while simultaneously storing CO2.

Human Story: Consider Maria, an environmental scientist working on a project that uses captured CO2 for enhanced oil recovery, combining environmental benefits with economic gains.

Expert Insight: Dr. Laura Green, an environmental

scientist, explains, "Carbon capture and storage is a critical technology for achieving deep reductions in greenhouse gas emissions. Its ability to capture CO2 from hard-to-abate sectors makes it indispensable for meeting climate goals."

Supporting Net-Zero Goals

As countries and companies commit to net-zero emissions targets, CCS provides a viable pathway to achieving these ambitious goals by addressing emissions that are difficult to eliminate through other means.

- Transition Technology: CCS serves as a bridge technology, enabling the continued use of fossil fuels while transitioning to renewable energy sources.
- Complementing Renewables: By capturing emissions from fossil fuel-based power generation, CCS can complement renewable energy deployment and enhance grid stability.

Human Story: Picture John, a policy advisor advocating for the integration of CCS into national climate strategies to support net-zero targets and sustainable development.

Challenges of Implementing CCS

Despite its potential, CCS faces significant challenges that must be addressed to achieve widespread deployment and effectiveness.

High Costs and Economic Viability

The high costs associated with capturing, transporting, and storing CO2 pose a major barrier to the large-scale deployment of CCS.

- Capital and Operational Costs: The infrastructure required for CCS, including capture facilities, pipelines, and storage sites, involves substantial capital and operational costs.

- Economic Incentives: Government policies and economic incentives, such as carbon pricing and tax credits, are essential to make CCS economically viable.

Human Story: Consider Emily, an economist analysing the financial challenges of CCS projects and advocating for policy measures to support their economic feasibility.

Technical and Logistical Challenges

The technical complexity and logistical requirements of CCS present significant challenges, including ensuring the safety and integrity of CO_2 storage sites.

- Site Selection and Monitoring: Identifying suitable geological formations for CO_2 storage and implementing robust monitoring systems to detect leaks are critical for ensuring long-term storage integrity.
- Infrastructure Development: Developing the necessary infrastructure for CO_2 transport and storage requires careful planning and significant investment.

Human Story: Picture Raj, a geologist conducting assessments of potential CO_2 storage sites and developing monitoring protocols to ensure their safety and effectiveness.

Expert Insight: Dr. Robert Lee, an expert in carbon capture technologies, notes, "The technical and logistical challenges of CCS are significant, but advancements in technology and engineering are steadily improving its feasibility. Robust monitoring and regulatory frameworks are essential for building public trust and ensuring safety."

Advancements and Innovations in CCS Technology

Ongoing advancements and innovations in CCS technology are critical for addressing challenges and enhancing the efficiency and effectiveness of carbon capture and storage.

Advanced Capture Technologies

Innovations in capture technologies are improving the efficiency and reducing the costs of CO_2 capture from industrial processes and power generation.

- Solvent-Based Capture: New solvents and chemical processes are increasing the efficiency of CO_2 absorption and reducing energy requirements.
- Membrane Technologies: Advances in membrane technologies are providing more efficient and cost-effective options for separating CO_2 from other gases.

Human Story: Consider Ahmed, a chemical engineer developing advanced solvents for CO_2 capture, working to make the process more efficient and affordable.

Integrated CCS Systems

Integrating CCS with other technologies and processes can enhance its overall effectiveness and economic viability.

- CCUS (Carbon Capture, Utilization, and Storage): Integrating CO_2 utilization with storage can create additional value by using captured CO_2 in products such as fuels, chemicals, and building materials.
- Hybrid Systems: Combining CCS with renewable energy sources, such as using renewable electricity for CO_2 capture, can enhance sustainability and reduce costs.

Human Story: Picture Li, an engineer working on a project that integrates CCS with renewable energy, creating a hybrid system that maximizes environmental and economic benefits.

The Role of Policy and Regulation

Government policies and regulations play a crucial role in promoting the development and deployment of

CCS, addressing challenges, and ensuring environmental integrity.

Government Support and Incentives

Policies and incentives can drive the growth of CCS by providing financial support, setting emission reduction targets, and promoting research and development.

- Carbon Pricing: Implementing carbon pricing mechanisms, such as carbon taxes and cap-and-trade systems, can create economic incentives for CCS deployment.
- Tax Credits and Subsidies: Financial incentives, such as tax credits and subsidies, can reduce the financial burden on CCS projects and encourage investment.

Human Story: Consider Sarah, a policymaker advocating for stronger government support and incentives for CCS to accelerate its adoption and impact.

Regulatory Frameworks

Robust regulatory frameworks are essential for ensuring the safety, effectiveness, and public acceptance of CCS projects.

- Storage Regulations: Establishing regulations for site selection, monitoring, and reporting can ensure the safety and integrity of CO_2 storage sites.
- Public Engagement: Engaging with communities and stakeholders to build trust and address concerns about CCS projects is crucial for their successful implementation.

Human Story: Picture Emily, a community liaison officer working to engage local communities and address their concerns about a new CCS project in their area.

Thought-Provoking Questions: Navigating the Future of CCS

Reflecting on the role of carbon capture and storage in mitigating oil's environmental impact raises several critical questions: How can we overcome the high costs and technical challenges of CCS? What role should government policies and incentives play in promoting CCS technologies? And how can we ensure the long-term safety and integrity of CO2 storage sites?

Storytelling Techniques: Bringing CCS to Life

To illustrate the complexities of CCS, envision the innovative research in laboratories, the strategic discussions in government offices, and the practical applications of CCS in industrial settings. Highlight the voices of engineers, scientists, policymakers, and community members to provide a balanced and comprehensive narrative.

Actionable Insights: Strategies for Promoting CCS

1. Investing in Research and Development: Supporting research and development of advanced CCS technologies can reduce costs and improve efficiency.
2. Enhancing Policy Support: Strengthening government policies and incentives can drive the growth of CCS and promote its integration into climate strategies.
3. Developing Infrastructure: Investing in infrastructure for CO2 transport and storage is essential for widespread adoption of CCS.
4. Ensuring Public Engagement: Engaging with communities and stakeholders can build trust and support for CCS projects.

Conclusion: Navigating the Path to a Sustainable Energy Future

Carbon capture and storage is a critical technology for mitigating the environmental impact of oil and achieving deep reductions in greenhouse gas emissions. This crude awakening highlights the importance of technological innovation, supportive policies, and public engagement in realizing the full potential of CCS.

By understanding the role of CCS in mitigating oil's environmental impact, we can better appreciate the complexities and opportunities in the transition to a sustainable energy future. The lessons learned from various CCS initiatives provide valuable insights for shaping policies and practices that promote environmental sustainability and energy security. Through collective action and ongoing commitment to innovation and sustainability, CCS can navigate the challenges ahead and contribute to a more stable and sustainable energy landscape.

CHAPTER 94: SYNTHETIC FUELS

Crude Awakening: The Modern Truths and Future of Oil

Crafting the Future: The Development of Synthetic Fuels and Their Potential to Replace Conventional Oil

As the world seeks sustainable alternatives to conventional fossil fuels, synthetic fuels have emerged as a promising option. Synthetic fuels, or synfuels, are produced from non-petroleum sources and offer the potential to significantly reduce greenhouse gas emissions while providing a reliable energy source. This chapter explores the development of synthetic fuels, their potential to replace conventional oil, and the broader implications for the global energy landscape.

Understanding Synthetic Fuels

Synthetic fuels are liquid or gaseous fuels produced from feedstocks other than crude oil. These feedstocks can include natural gas, coal, biomass, and even carbon dioxide captured from the atmosphere.

Key Types of Synthetic Fuels

- Fischer-Tropsch Fuels: Produced through the Fischer-Tropsch process, which converts carbon monoxide and hydrogen into liquid hydrocarbons. Feedstocks include natural gas, coal, and biomass.

- Biofuels: Produced from biological materials such as plant oils, agricultural residues, and algae. Examples include biodiesel and bioethanol.
- Electrofuels: Also known as e-fuels, these are produced using electricity, typically from renewable sources, to convert water and carbon dioxide into fuels like synthetic methane and methanol.

Human Story: Imagine Samuel, a chemical engineer working at a cutting-edge synthetic fuel facility, where innovative technologies transform carbon dioxide into synthetic gasoline.

The Potential of Synthetic Fuels to Replace Conventional Oil

Synthetic fuels offer several advantages that could allow them to replace conventional oil in various applications, contributing to a cleaner and more sustainable energy future.

Environmental Benefits

Synthetic fuels can significantly reduce greenhouse gas emissions compared to conventional fossil fuels, making them a key component of climate change mitigation strategies.

- Carbon Neutrality: Some synthetic fuels, particularly those produced from biomass or captured carbon dioxide, can be nearly carbon-neutral, as the CO_2 emitted during combustion is offset by the CO_2 absorbed during production.
- Reduced Emissions: Synthetic fuels generally produce fewer pollutants, such as sulphur oxides and particulate matter, compared to conventional petroleum fuels.

Human Story: Consider Maria, an environmental advocate who promotes the use of synthetic fuels to reduce air pollution and greenhouse gas emissions in her city.

Expert Insight: Dr. Laura Green, an environmental scientist, explains, "Synthetic fuels have the potential to play a significant role in our transition to a low-carbon economy. Their ability to utilize captured carbon dioxide and renewable energy sources makes them an attractive alternative to traditional fossil fuels."

Energy Security and Economic Benefits

Synthetic fuels can enhance energy security by diversifying energy sources and reducing dependence on imported oil.

- Domestic Production: Synthetic fuels can be produced domestically from a variety of feedstocks, reducing reliance on foreign oil, and enhancing national energy security.
- Economic Opportunities: The development of a synthetic fuel industry can create jobs in production, research, and infrastructure development, supporting local economies.

Human Story: Picture John, a worker at a synthetic fuel plant, benefiting from the economic opportunities created by the growing synthetic fuel industry in his community.

Challenges of Developing Synthetic Fuels

Despite their potential, synthetic fuels face several challenges that must be addressed to achieve widespread adoption and effectiveness.

High Production Costs

The production of synthetic fuels is currently more expensive than conventional fossil fuels, posing a significant barrier to their widespread adoption.

- Feedstock and Process Costs: The cost of feedstocks

and the energy-intensive processes required to produce synthetic fuels contribute to their higher production costs.
- Economies of Scale: Achieving cost reductions will require scaling up production and improving process efficiencies.

Human Story: Consider Emily, a researcher working to develop more cost-effective methods for producing synthetic fuels, aiming to make them competitive with conventional fuels.

Technological and Logistical Challenges

The technical complexity and logistical requirements of synthetic fuel production and distribution present significant challenges.

- Infrastructure Development: Developing the necessary infrastructure for large-scale synthetic fuel production and distribution requires significant investment and planning.
- Feedstock Availability: Ensuring a reliable and sustainable supply of feedstocks, such as biomass or captured CO_2, is critical for the long-term viability of synthetic fuels.

Human Story: Picture Raj, an engineer designing advanced systems for the efficient production and distribution of synthetic fuels, addressing one of the key challenges in the industry.

Expert Insight: Dr. Robert Lee, an expert in fuel technologies, notes, "While synthetic fuels offer significant benefits, their production and distribution present technical challenges that require innovative solutions. Advances in process engineering and feedstock management will be crucial for their success."

Advancements and Innovations in Synthetic Fuel Technology

Ongoing advancements and innovations in synthetic fuel

technology are critical for addressing challenges and enhancing the efficiency and effectiveness of synthetic fuel production.

Fischer-Tropsch Process Improvements

Innovations in the Fischer-Tropsch process, which converts carbon monoxide and hydrogen into liquid hydrocarbons, are improving the efficiency and reducing the costs of synthetic fuel production.

- Catalyst Development: Advances in catalyst design and materials are increasing the efficiency of the Fischer-Tropsch process and reducing energy consumption.
- Process Optimization: Improving process conditions and integrating renewable energy sources can enhance the sustainability and economic viability of Fischer-Tropsch fuels.

Human Story: Consider Ahmed, a chemical engineer developing new catalysts for the Fischer-Tropsch process, working to make synthetic fuels more efficient and affordable.

Biofuel Advancements

Advancements in biofuel production technologies are expanding the range of feedstocks and improving the sustainability of biofuels.

- Second-Generation Biofuels: Produced from non-food biomass, such as agricultural residues and waste materials, these biofuels offer a more sustainable alternative to first-generation biofuels.
- Algae-Based Biofuels: Algae can produce high yields of biofuels without competing with food crops for land, offering a promising feedstock for sustainable biofuel production.

Human Story: Picture Li, a researcher working on algae-based biofuels, developing innovative methods to harness the potential of algae for sustainable fuel production.

The Role of Policy and Regulation

Government policies and regulations play a crucial role in promoting the development and adoption of synthetic fuels, addressing challenges, and ensuring sustainability.

Government Support and Incentives

Policies and incentives can drive the growth of the synthetic fuel industry by providing financial support, setting targets, and promoting research and development.

- Renewable Fuel Standards (RFS): Mandates that require a certain percentage of transportation fuels to come from renewable sources, including synthetic fuels.
- Subsidies and Grants: Financial incentives, such as subsidies and grants, support synthetic fuel research, production, and infrastructure development.

Human Story: Consider Sarah, a policymaker advocating for stronger government support and incentives for synthetic fuels to accelerate their adoption and impact.

Sustainability Standards

Implementing sustainability standards and certification schemes can ensure that synthetic fuels are produced in an environmentally and socially responsible manner.

- Certification Schemes: Certification schemes, such as the Roundtable on Sustainable Biomaterials (RSB), set standards for sustainable synthetic fuel production and provide third-party verification.
- Sustainable Practices: Encouraging sustainable agricultural practices and land management to minimize

the environmental impact of biofuel production.

Human Story: Picture Emily, a sustainability officer at a synthetic fuel company, working to ensure that the company's production practices meet international sustainability standards.

Thought-Provoking Questions: Navigating the Future of Synthetic Fuels

Reflecting on the potential of synthetic fuels to replace conventional oil raises several critical questions: How can we overcome the high production costs and technical challenges of synthetic fuels? What role should government policies and incentives play in promoting synthetic fuels? And how can we ensure the long-term sustainability and environmental benefits of synthetic fuel production?

Storytelling Techniques: Bringing Synthetic Fuels to Life

To illustrate the complexities of synthetic fuels, envision the innovative research in laboratories, the strategic discussions in government offices, and the practical applications of synthetic fuels in everyday life. Highlight the voices of engineers, scientists, policymakers, and community members to provide a balanced and comprehensive narrative.

Actionable Insights: Strategies for Promoting Synthetic Fuels

1. Investing in Research and Development: Supporting research and development of advanced synthetic fuel technologies can reduce costs and improve efficiency.
2. Enhancing Policy Support: Strengthening government policies and incentives can drive the growth of the synthetic fuel industry and promote the transition to renewable energy.

3. Developing Infrastructure: Investing in infrastructure for synthetic fuel production, storage, and distribution is essential for widespread adoption.

4. Implementing Sustainability Standards: Adopting and enforcing sustainability standards can ensure environmentally and socially responsible synthetic fuel production.

Conclusion: Navigating the Path to a Sustainable Energy Future

Synthetic fuels represent a promising alternative to conventional fossil fuels, offering the potential to reduce carbon emissions, enhance energy security, and support economic development. This crude awakening highlights the importance of technological innovation, supportive policies, and sustainable practices in realizing the full potential of synthetic fuels.

By understanding the development of synthetic fuels and their potential to replace conventional oil, we can better appreciate the complexities and opportunities in the transition to a sustainable energy future. The lessons learned from various synthetic fuel initiatives provide valuable insights for shaping policies and practices that promote environmental sustainability and energy security. Through collective action and ongoing commitment to innovation and sustainability, the synthetic fuel industry can navigate the challenges ahead and contribute to a more stable and sustainable energy landscape.

CHAPTER 95: ADVANCED REFINING TECHNIQUES

Crude Awakening: The Modern Truths and Future of Oil

Refining the Future: Exploring Advanced Refining Techniques That Reduce Environmental Impact

The oil refining industry is undergoing a transformative shift towards more sustainable and environmentally friendly practices. Advanced refining techniques are emerging as critical tools in reducing the environmental impact of oil processing, improving efficiency, and meeting stringent regulatory standards. This chapter delves into these new techniques, their benefits, challenges, and the broader implications for the oil industry and global sustainability efforts.

Understanding Oil Refining

Oil refining is the process of converting crude oil into useful products such as gasoline, diesel, jet fuel, and petrochemicals. Traditional refining methods, while effective, often result in significant environmental pollution, including greenhouse gas emissions, water contamination, and hazardous waste.

Key Processes in Traditional Refining

- Distillation: Separates crude oil into different fractions based on boiling points.
- Cracking: Breaks down large hydrocarbon molecules into smaller, more valuable ones.
- Reforming: Alters the molecular structure of hydrocarbons to increase the quality of gasoline.
- Hydrotreating: Removes sulphur and other impurities to produce cleaner fuels.

Human Story: Imagine Samuel, a refinery operator who has worked with traditional methods for years but is now training on new, advanced refining techniques aimed at reducing the plant's environmental footprint.

The Need for Advanced Refining Techniques

As environmental regulations tighten and societal demands for cleaner energy grow, the oil refining industry must innovate to reduce its ecological impact. Advanced refining techniques offer promising solutions to these challenges.

Environmental Benefits

Advanced refining techniques can significantly reduce emissions, waste, and resource consumption, contributing to a more sustainable oil industry.

- Lower Emissions: Techniques such as hydro processing and gasification can reduce carbon dioxide and other harmful emissions.
- Waste Minimization: Processes like solvent de-asphalting and advanced catalytic cracking produce fewer by-products and hazardous waste.
- Energy Efficiency: Improved thermal and process efficiency reduces the overall energy consumption of

refineries.

Human Story: Consider Maria, an environmental scientist who collaborates with refineries to implement advanced techniques that lower emissions and improve efficiency, ultimately benefiting the local community.

Expert Insight: Dr. Laura Green, an environmental engineer, explains, "Advanced refining techniques are essential for the future of the oil industry. They not only help in meeting regulatory requirements but also play a crucial role in reducing the overall environmental impact of oil production and processing."

Key Advanced Refining Techniques

Several advanced refining techniques are being developed and implemented to make the refining process cleaner and more efficient.

Hydro processing

Hydro processing combines hydrotreating and hydrocracking to produce cleaner fuels by removing impurities and breaking down larger molecules.

- Hydrotreating: Uses hydrogen to remove sulphur, nitrogen, and other impurities, resulting in lower sulphur emissions.
- Hydrocracking: Breaks down heavy hydrocarbons into lighter, more valuable products, improving fuel quality and yield.

Human Story: Picture John, a refinery engineer overseeing the installation of new hydro processing units, which are expected to drastically reduce sulphur emissions and improve fuel quality.

Gasification

Gasification converts heavy hydrocarbons and waste materials into synthesis gas (syngas), which can be used to produce clean fuels and chemicals.

- Feedstock Flexibility: Gasification can process a wide range of feedstocks, including heavy oils, residues, and biomass.
- Cleaner Products: The process produces fewer pollutants and higher-quality fuels compared to traditional methods.

Human Story: Consider Emily, a researcher at a pilot gasification plant, working on optimizing the process to handle various feedstocks and produce cleaner fuels.

Solvent De-Asphalting

Solvent de-asphalting separates heavy fractions of crude oil using solvents, producing de-asphalted oil and asphaltenes with fewer contaminants.

- Enhanced Efficiency: This technique improves the efficiency of downstream processes and reduces waste.
- Better Quality Products: The resulting products are cleaner and more valuable, with fewer impurities.

Human Story: Picture Raj, a chemical engineer implementing solvent de-asphalting at his refinery, reducing waste, and improving the quality of the refined products.

Expert Insight: Dr. Robert Lee, an expert in chemical engineering, notes, "Techniques like solvent de-asphalting and gasification represent significant advancements in refining technology. They allow refineries to process a wider range of feedstocks more efficiently and with less environmental impact."

Challenges in Implementing Advanced Techniques

Despite their benefits, advanced refining techniques face several challenges that must be addressed to achieve widespread adoption.

High Initial Costs

The installation and operation of advanced refining technologies can be expensive, posing a significant barrier for many refineries.

- Capital Investment: The high capital costs associated with advanced equipment and infrastructure can deter investment.
- Operational Costs: Maintaining and operating advanced refining systems may require specialized skills and additional resources.

Human Story: Consider Sarah, a refinery manager tasked with justifying the investment in new technologies to stakeholders who are concerned about the high initial costs.

Technical and Logistical Challenges

Implementing advanced refining techniques involves overcoming various technical and logistical hurdles, including integration with existing systems and ensuring reliability.

- System Integration: Integrating new technologies with existing refining processes can be complex and requires careful planning and execution.
- Reliability and Maintenance: Advanced systems must be reliable and easy to maintain to ensure continuous operation and cost-effectiveness.

Human Story: Picture Ahmed, a project manager responsible for coordinating the integration of new hydro processing units into an existing refinery, facing technical challenges, and ensuring minimal disruption.

The Role of Policy and Regulation

Government policies and regulations play a crucial role in promoting the adoption of advanced refining techniques, addressing challenges, and ensuring environmental compliance.

Government Support and Incentives

Policies and incentives can drive the growth of advanced refining technologies by providing financial support and setting stringent environmental standards.

- Regulatory Standards: Implementing strict emissions and waste reduction standards encourages refineries to adopt cleaner technologies.
- Financial Incentives: Subsidies, tax credits, and grants can offset the high initial costs of installing advanced refining equipment.

Human Story: Consider Emily, a policy advocate working to promote government support and incentives for refineries that adopt advanced, environmentally friendly technologies.

Industry Collaboration

Collaboration within the industry can accelerate the development and deployment of advanced refining techniques.

- Knowledge Sharing: Sharing best practices and technological innovations can help refineries improve their operations and reduce environmental impact.
- Joint Ventures: Collaborative projects and joint ventures can pool resources and expertise, making advanced technologies more accessible.

Human Story: Picture Li, a sustainability officer at a major oil company, leading industry-wide initiatives to promote the adoption of advanced refining techniques.

Thought-Provoking Questions: Navigating Advanced Refining Techniques

Reflecting on the development of advanced refining techniques raises several critical questions: How can refineries balance the high initial costs with long-term environmental and economic benefits? What role should government policies and incentives play in promoting these technologies? And how can industry collaboration enhance the adoption and effectiveness of advanced refining techniques?

Storytelling Techniques: Bringing Advanced Refining to Life

To illustrate the complexities of advanced refining techniques, envision the innovative research in laboratories, the strategic discussions in boardrooms, and the practical applications in refineries. Highlight the voices of engineers, scientists, policymakers, and community members to provide a balanced and comprehensive narrative.

Actionable Insights: Strategies for Promoting Advanced Refining Techniques

1. Investing in Research and Development: Supporting research and development of advanced refining technologies can reduce costs and improve efficiency.
2. Enhancing Policy Support: Strengthening government policies and incentives can drive the adoption of advanced refining techniques and promote environmental compliance.
3. Developing Infrastructure: Investing in infrastructure

for advanced refining processes is essential for widespread adoption.

4. Fostering Industry Collaboration: Promoting collaboration and knowledge sharing within the industry can accelerate the deployment and effectiveness of advanced refining technologies.

Conclusion: Navigating the Path to Sustainable Refining

Advanced refining techniques represent a promising path towards a more sustainable and environmentally friendly oil industry. This crude awakening highlights the importance of technological innovation, supportive policies, and industry collaboration in realizing the full potential of advanced refining techniques.

By understanding the development and benefits of advanced refining techniques, we can better appreciate the complexities and opportunities in the transition to sustainable oil refining. The lessons learned from various initiatives provide valuable insights for shaping policies and practices that promote environmental sustainability and energy security. Through collective action and ongoing commitment to innovation and sustainability, the refining industry can navigate the challenges ahead and contribute to a more stable and sustainable energy landscape.

CHAPTER 96: DIGITAL TRANSFORMATION IN OIL

Crude Awakening: The Modern Truths and Future of Oil

Digitally Driven: The Impact of Digital Technologies on Oil Production and Distribution

In the era of rapid technological advancement, the oil industry is undergoing a significant transformation driven by digital technologies. From exploration and production to distribution and retail, digital innovations are revolutionizing the way oil companies operate, improving efficiency, reducing costs, and enhancing environmental sustainability. This chapter explores the profound impact of digital transformation on the oil industry, examining key technologies, benefits, challenges, and the broader implications for the global energy landscape.

Understanding Digital Transformation in Oil

Digital transformation in the oil industry involves the integration of advanced digital technologies into various aspects of the oil production and distribution process. These technologies include big data analytics, the Internet of Things (IoT), artificial intelligence (AI), and blockchain, among others.

Key Technologies Driving Digital Transformation

- Big Data Analytics: Analysing vast amounts of data to optimize operations, predict equipment failures, and enhance decision-making.
- Internet of Things (IoT): Connecting sensors and devices to monitor and control operations in real-time, improving efficiency and safety.
- Artificial Intelligence (AI): Using machine learning and AI algorithms to analyse data, automate processes, and optimize performance.
- Blockchain: Providing secure and transparent transaction records, improving supply chain management, and reducing fraud.

Human Story: Imagine Samuel, a data scientist working for an oil company, leveraging big data analytics to predict equipment failures and optimize production processes, resulting in significant cost savings and increased efficiency.

The Impact of Digital Technologies on Oil Production

Digital technologies are transforming oil production by enhancing exploration, drilling, and extraction processes, leading to improved efficiency, safety, and sustainability.

Enhanced Exploration and Drilling

Advanced digital tools are revolutionizing the way oil companies explore and drill for oil, making these processes more accurate, efficient, and cost-effective.

- Seismic Imaging and Data Analysis: Using advanced seismic imaging and big data analytics to create detailed subsurface models, improving the accuracy of oil and gas exploration.

- Automated Drilling Rigs: Implementing automated drilling rigs equipped with IoT sensors and AI algorithms to optimize drilling operations and reduce downtime.

Human Story: Consider Maria, a geologist using advanced seismic imaging and data analysis to identify promising drilling locations, significantly increasing the success rate of exploration projects.

Expert Insight: Dr. Laura Green, a petroleum engineer, explains, "Digital technologies are transforming exploration and drilling by providing more accurate data and automating complex processes. These advancements are not only improving efficiency but also reducing environmental impact and operational risks."

Optimized Production and Maintenance

Digital technologies enable oil companies to optimize production processes and implement predictive maintenance strategies, enhancing operational efficiency and reducing costs.

- Predictive Maintenance: Using IoT sensors and AI to monitor equipment health and predict failures before they occur, reducing downtime and maintenance costs.
- Production Optimization: Leveraging big data analytics to optimize production parameters, maximize output, and minimize waste.

Human Story: Picture John, a maintenance engineer using predictive maintenance tools to monitor the health of critical equipment, preventing costly breakdowns, and ensuring smooth operations.

The Impact of Digital Technologies on Oil Distribution

Digital transformation extends beyond production to the distribution and retail aspects of the oil industry, enhancing supply chain management, logistics, and customer engagement.

Improved Supply Chain Management

Digital technologies are revolutionizing supply chain management by providing greater visibility, transparency, and efficiency.

- Blockchain for Transparency: Implementing blockchain technology to create secure, transparent records of transactions, improving traceability and reducing fraud.
- IoT for Real-Time Tracking: Using IoT sensors to monitor the location and condition of oil shipments in real-time, ensuring timely deliveries and reducing losses.

Human Story: Consider Emily, a logistics manager using blockchain and IoT technologies to track oil shipments in real-time, ensuring timely deliveries and enhancing supply chain transparency.

Enhanced Customer Engagement

Digital technologies are transforming the way oil companies engage with customers, providing personalized experiences and improving service quality.

- AI-Powered Customer Service: Using AI chatbots and virtual assistants to handle customer inquiries and provide personalized support.
- Digital Platforms: Developing mobile apps and online platforms to enable customers to track shipments, manage orders, and access services conveniently.

Human Story: Picture Raj, a customer service manager implementing AI-powered chatbots to provide customers with instant support and personalized service, enhancing customer satisfaction.

Challenges of Digital Transformation

Despite the numerous benefits, the digital transformation of the oil industry faces several challenges that must be addressed to achieve successful implementation and adoption.

High Implementation Costs

The high costs associated with implementing and maintaining advanced digital technologies pose a significant barrier for many oil companies.

- Initial Investment: The capital investment required for digital infrastructure, software, and training can be substantial.
- Ongoing Maintenance: Maintaining and updating digital systems requires ongoing investment and specialized skills.

Human Story: Consider Sarah, a financial manager tasked with justifying the investment in digital technologies to stakeholders concerned about the high initial and ongoing costs.

Cybersecurity Risks

The increased reliance on digital technologies exposes oil companies to cybersecurity risks, including data breaches and cyberattacks.

- Data Protection: Ensuring the security of sensitive data and protecting systems from cyber threats is critical.
- Cybersecurity Measures: Implementing robust cybersecurity measures and protocols to safeguard digital infrastructure and prevent unauthorized access.

Human Story: Picture Ahmed, a cybersecurity specialist working to protect his company's digital systems from cyber threats, ensuring the security and integrity of critical data.

The Role of Policy and Regulation

Government policies and regulations play a crucial role in promoting digital transformation in the oil industry, addressing challenges, and ensuring compliance with standards.

Government Support and Incentives

Policies and incentives can drive the digital transformation of the oil industry by providing financial support and promoting technological innovation.

- Innovation Grants: Offering grants and subsidies to support research and development of advanced digital technologies in the oil industry.
- Tax Incentives: Providing tax incentives for companies investing in digital infrastructure and cybersecurity measures.

Human Story: Consider Emily, a policy advocate working to promote government support and incentives for digital transformation in the oil industry, helping companies overcome financial barriers.

Regulatory Standards

Establishing regulatory standards for digital technologies can ensure their safe and effective implementation and promote industry-wide best practices.

- Cybersecurity Regulations: Implementing regulations to ensure robust cybersecurity measures are in place, protecting sensitive data and critical infrastructure.
- Data Privacy Standards: Establishing standards for data privacy and protection to ensure the secure handling of personal and operational data.

Human Story: Picture Li, a regulatory compliance officer

ensuring that his company meets all regulatory standards for digital technologies and data protection, maintaining high levels of security and compliance.

Thought-Provoking Questions: Navigating Digital Transformation

Reflecting on the impact of digital technologies on oil production and distribution raises several critical questions: How can oil companies balance the high costs of digital transformation with long-term benefits? What role should government policies and incentives play in promoting digital technologies? And how can the industry address cybersecurity risks and ensure data protection?

Storytelling Techniques: Bringing Digital Transformation to Life

To illustrate the complexities of digital transformation, envision the innovative research in tech labs, the strategic discussions in boardrooms, and the practical applications in oil fields and refineries. Highlight the voices of data scientists, engineers, policymakers, and community members to provide a balanced and comprehensive narrative.

Actionable Insights: Strategies for Promoting Digital Transformation

1. Investing in Research and Development: Supporting research and development of advanced digital technologies can reduce costs and improve efficiency.
2. Enhancing Policy Support: Strengthening government policies and incentives can drive the digital transformation of the oil industry and promote technological innovation.
3. Implementing Robust Cybersecurity Measures: Ensuring robust cybersecurity measures are in place to protect sensitive data and critical infrastructure is essential for

successful digital transformation.

4. Fostering Industry Collaboration: Promoting collaboration and knowledge sharing within the industry can accelerate the adoption and effectiveness of digital technologies.

Conclusion: Navigating the Path to a Digital Future

Digital transformation is revolutionizing the oil industry, offering significant benefits in terms of efficiency, cost savings, and environmental sustainability. This crude awakening highlights the importance of technological innovation, supportive policies, and robust cybersecurity measures in realizing the full potential of digital technologies.

By understanding the impact of digital transformation on oil production and distribution, we can better appreciate the complexities and opportunities in the transition to a digitally driven energy future. The lessons learned from various digital initiatives provide valuable insights for shaping policies and practices that promote technological innovation and energy security. Through collective action and ongoing commitment to innovation and sustainability, the oil industry can navigate the challenges ahead and contribute to a more stable and sustainable energy landscape.

CHAPTER 97: INNOVATIONS IN DRILLING TECHNOLOGY

Crude Awakening: The Modern Truths and Future of Oil

Drilling Deep: Examining Recent Innovations in Drilling Technology and Their Impact

The quest for oil has driven technological advancements for decades, and recent innovations in drilling technology are transforming the industry. These advancements not only enhance efficiency and reduce costs but also minimize environmental impact and improve safety. This chapter delves into the latest innovations in drilling technology, their benefits, challenges, and the broader implications for the oil industry and global energy landscape.

Understanding Drilling Technology

Drilling technology encompasses the methods and equipment used to explore for and extract oil from beneath the earth's surface. Traditional drilling methods, while effective, often involve significant environmental risks, high costs, and operational challenges.

Key Components of Traditional Drilling

- Rotary Drilling: The most common method, involving a rotating drill bit to bore through the earth.
- Directional Drilling: Allows the drill bit to be steered in various directions, enabling access to reserves that are not directly beneath the drilling rig.
- Hydraulic Fracturing (Fracking): Involves injecting high-pressure fluid into rock formations to create fractures, allowing oil and gas to flow more freely.

Human Story: Imagine Samuel, a seasoned drilling engineer who has seen the evolution of drilling technology over the years, now working with cutting-edge tools that are revolutionizing the industry.

The Need for Innovations in Drilling Technology

As the oil industry faces increasing pressure to improve efficiency, reduce costs, and minimize environmental impact, innovations in drilling technology are essential. These advancements address various challenges and offer significant benefits.

Environmental Benefits

Recent innovations in drilling technology are designed to minimize the environmental impact of oil extraction, making the industry more sustainable.

- Reduced Footprint: Advanced drilling techniques can reduce the physical footprint of drilling operations, preserving more of the natural landscape.
- Lower Emissions: New technologies can reduce greenhouse gas emissions associated with drilling operations.
- Improved Waste Management: Innovations in waste management techniques help minimize the environmental impact of drilling by reducing waste and improving disposal methods.

Human Story: Consider Maria, an environmental scientist working with oil companies to implement new drilling technologies that significantly reduce their environmental footprint.

Expert Insight: Dr. Laura Green, an environmental engineer, explains, "The advancements in drilling technology are crucial for making oil extraction more sustainable. These technologies not only improve efficiency but also play a vital role in reducing the environmental impact of drilling operations."

Key Innovations in Drilling Technology

Several recent innovations are revolutionizing the way oil is extracted, improving efficiency, safety, and sustainability.

Horizontal Drilling

Horizontal drilling allows the drill bit to be steered horizontally after reaching the desired depth, enabling access to a larger area from a single drilling site.

- Increased Productivity: Horizontal drilling can access multiple reservoirs from a single well, significantly increasing productivity.
- Reduced Surface Impact: By drilling multiple horizontal wells from a single site, the surface impact is minimized, preserving the natural landscape.

Human Story: Picture John, a drilling engineer overseeing a horizontal drilling operation that accesses multiple oil reservoirs from a single site, reducing the environmental impact and increasing efficiency.

Automated Drilling Systems

Automated drilling systems leverage advanced software

and robotics to control drilling operations with minimal human intervention.

- Enhanced Precision: Automation improves the precision of drilling operations, reducing the risk of errors and accidents.
- Increased Efficiency: Automated systems can operate continuously without the need for breaks, significantly increasing drilling efficiency.
- Improved Safety: By minimizing human involvement in hazardous tasks, automated drilling systems enhance safety on drilling sites.

Human Story: Consider Emily, a technician monitoring an automated drilling rig, ensuring the operation runs smoothly and safely with minimal human intervention.

Expert Insight: Dr. Robert Lee, an expert in drilling technology, notes, "Automation in drilling is transforming the industry by increasing precision and efficiency while reducing risks. These systems represent the future of safe and sustainable oil extraction."

Managed Pressure Drilling (MPD)

Managed pressure drilling (MPD) is an advanced technique that precisely controls the pressure within the wellbore, enhancing safety and efficiency.

- Improved Control: MPD allows for precise control of wellbore pressure, reducing the risk of blowouts and other pressure-related issues.
- Enhanced Safety: By maintaining optimal pressure, MPD minimizes the risks associated with drilling, improving overall safety.
- Cost Savings: The increased control and efficiency provided by MPD can lead to significant cost savings in drilling operations.

Human Story: Picture Raj; a drilling supervisor using MPD technology to safely and efficiently manage the pressure within a complex well, ensuring the operation's success and safety.

Challenges in Implementing Advanced Drilling Technologies

Despite their numerous benefits, advanced drilling technologies face several challenges that must be addressed to achieve widespread adoption and effectiveness.

High Initial Costs

The high costs associated with implementing and maintaining advanced drilling technologies pose a significant barrier for many oil companies.

- Capital Investment: The initial investment required for advanced drilling equipment and infrastructure can be substantial.
- Operational Costs: Maintaining and operating advanced systems may require specialized skills and additional resources.

Human Story: Consider Sarah, a financial manager tasked with justifying the investment in new drilling technologies to stakeholders who are concerned about the high initial and ongoing costs.

Technical and Logistical Challenges

Implementing advanced drilling technologies involves overcoming various technical and logistical hurdles, including integration with existing systems and ensuring reliability.

- System Integration: Integrating new technologies with

existing drilling processes can be complex and requires careful planning and execution.
- Reliability and Maintenance: Advanced systems must be reliable and easy to maintain to ensure continuous operation and cost-effectiveness.

Human Story: Picture Ahmed, a project manager responsible for coordinating the integration of new drilling technologies into existing operations, facing technical challenges, and ensuring minimal disruption.

The Role of Policy and Regulation

Government policies and regulations play a crucial role in promoting the adoption of advanced drilling technologies, addressing challenges, and ensuring environmental compliance.

Government Support and Incentives

Policies and incentives can drive the growth of advanced drilling technologies by providing financial support and promoting technological innovation.

- Regulatory Standards: Implementing strict environmental and safety standards encourages oil companies to adopt cleaner and safer drilling technologies.
- Financial Incentives: Subsidies, tax credits, and grants can offset the high initial costs of installing advanced drilling equipment.

Human Story: Consider Emily, a policy advocate working to promote government support and incentives for oil companies that adopt advanced, environmentally friendly drilling technologies.

Industry Collaboration

Collaboration within the industry can accelerate the development and deployment of advanced drilling technologies.

- Knowledge Sharing: Sharing best practices and technological innovations can help oil companies improve their operations and reduce environmental impact.
- Joint Ventures: Collaborative projects and joint ventures can pool resources and expertise, making advanced technologies more accessible.

Human Story: Picture Li, a sustainability officer at a major oil company, leading industry-wide initiatives to promote the adoption of advanced drilling technologies.

Thought-Provoking Questions: Navigating Innovations in Drilling

Reflecting on the impact of recent innovations in drilling technology raises several critical questions: How can oil companies balance the high initial costs with long-term environmental and economic benefits? What role should government policies and incentives play in promoting these technologies? And how can industry collaboration enhance the adoption and effectiveness of advanced drilling technologies?

Storytelling Techniques: Bringing Drilling Innovations to Life

To illustrate the complexities of drilling innovations, envision the innovative research in laboratories, the strategic discussions in boardrooms, and the practical applications in drilling fields. Highlight the voices of engineers, scientists, policymakers, and community members to provide a balanced and comprehensive narrative.

Actionable Insights: Strategies for Promoting Advanced Drilling Technologies

1. Investing in Research and Development: Supporting research and development of advanced drilling technologies can reduce costs and improve efficiency.
2. Enhancing Policy Support: Strengthening government policies and incentives can drive the adoption of advanced drilling techniques and promote environmental compliance.
3. Developing Infrastructure: Investing in infrastructure for advanced drilling processes is essential for widespread adoption.
4. Fostering Industry Collaboration: Promoting collaboration and knowledge sharing within the industry can accelerate the deployment and effectiveness of advanced drilling technologies.

Conclusion: Navigating the Path to Sustainable Drilling

Innovations in drilling technology represent a promising path towards a more efficient, safe, and environmentally friendly oil industry. This crude awakening highlights the importance of technological innovation, supportive policies, and industry collaboration in realizing the full potential of advanced drilling techniques.

By understanding the development and benefits of advanced drilling technologies, we can better appreciate the complexities and opportunities in the transition to sustainable oil extraction. The lessons learned from various initiatives provide valuable insights for shaping policies and practices that promote environmental sustainability and energy security. Through collective action and ongoing commitment to innovation and sustainability, the drilling industry can navigate the challenges ahead and contribute to a more stable and sustainable energy landscape.

CHAPTER 98: OIL AND ARTIFICIAL INTELLIGENCE

Crude Awakening: The Modern Truths and Future of Oil

Smart Oil: Analysing How Artificial Intelligence is Transforming the Oil Industry

The oil industry, long seen as a bastion of traditional engineering and manual processes, is now embracing the digital age. Artificial Intelligence (AI) is at the forefront of this transformation, bringing unprecedented changes in efficiency, safety, and sustainability. This chapter explores how AI is reshaping the oil industry, its benefits, challenges, and the broader implications for the future of energy.

Understanding Artificial Intelligence in the Oil Industry

Artificial Intelligence encompasses a range of technologies that enable machines to perform tasks that typically require human intelligence, such as learning, reasoning, problem-solving, and decision-making. In the oil industry, AI is being integrated into various stages of the value chain, from exploration and production to distribution and retail.

Key AI Technologies Transforming the Oil Industry

- Machine Learning: Algorithms that analyse large datasets

to identify patterns and make predictions, improving decision-making and operational efficiency.
- Predictive Analytics: Using historical data to predict future events, such as equipment failures or production trends, allowing for proactive management.
- Robotics and Automation: AI-driven robots and automated systems that perform complex tasks with high precision and minimal human intervention.
- Natural Language Processing (NLP): Enabling machines to understand and respond to human language, improving communication and data analysis.

Human Story: Imagine Samuel, an AI specialist working for a major oil company, developing machine learning models to predict equipment failures and optimize drilling operations, resulting in significant cost savings and increased efficiency.

The Impact of AI on Oil Exploration and Production

AI technologies are revolutionizing exploration and production processes, making them more accurate, efficient, and cost-effective.

Enhanced Exploration and Drilling

Advanced AI tools are transforming the way oil companies explore and drill for oil, leading to more efficient and successful operations.

- Seismic Data Analysis: AI algorithms analyse vast amounts of seismic data to create detailed subsurface models, improving the accuracy of oil and gas exploration.
- Automated Drilling Systems: AI-driven automation optimizes drilling operations, reducing downtime and operational risks.

Human Story: Consider Maria, a geologist using AI-powered seismic data analysis tools to identify promising drilling locations, significantly increasing the success rate of exploration projects.

Expert Insight: Dr. Laura Green, a petroleum engineer, explains, "AI technologies are transforming exploration and drilling by providing more accurate data and automating complex processes. These advancements are not only improving efficiency but also reducing environmental impact and operational risks."

Optimized Production and Maintenance

AI enables oil companies to optimize production processes and implement predictive maintenance strategies, enhancing operational efficiency and reducing costs.

- Predictive Maintenance: AI-powered sensors monitor equipment health and predict failures before they occur, reducing downtime and maintenance costs.
- Production Optimization: Machine learning algorithms analyse production data to optimize parameters, maximize output, and minimize waste.

Human Story: Picture John, a maintenance engineer using AI-driven predictive maintenance tools to monitor the health of critical equipment, preventing costly breakdowns, and ensuring smooth operations.

The Impact of AI on Oil Distribution and Retail

AI is not only transforming production but also revolutionizing the distribution and retail aspects of the oil industry, enhancing supply chain management, logistics, and customer engagement.

Improved Supply Chain Management

AI technologies are enhancing supply chain management by providing greater visibility, transparency, and efficiency.

- Demand Forecasting: AI analyses historical data and market trends to forecast demand accurately, optimizing inventory levels and reducing waste.
- Logistics Optimization: Machine learning algorithms optimize routing and scheduling for transportation, ensuring timely deliveries and reducing costs.

Human Story: Consider Emily, a logistics manager using AI tools to optimize the supply chain, ensuring timely deliveries, and reducing operational costs.

Enhanced Customer Engagement

AI is transforming the way oil companies engage with customers, providing personalized experiences and improving service quality.

- AI-Powered Customer Service: AI chatbots and virtual assistants handle customer inquiries and provide personalized support.
- Digital Platforms: AI-driven digital platforms enable customers to manage orders, track shipments, and access services conveniently.

Human Story: Picture Raj, a customer service manager implementing AI-powered chatbots to provide customers with instant support and personalized service, enhancing customer satisfaction.

Challenges of Implementing AI in the Oil Industry

Despite the numerous benefits, the implementation of AI in the oil industry faces several challenges that must be addressed to achieve successful adoption and effectiveness.

High Implementation Costs

The high costs associated with developing and integrating AI technologies pose a significant barrier for many oil companies.

- Capital Investment: The initial investment required for AI infrastructure, software, and training can be substantial.
- Ongoing Maintenance: Maintaining and updating AI systems requires ongoing investment and specialized skills.

Human Story: Consider Sarah, a financial manager tasked with justifying the investment in AI technologies to stakeholders who are concerned about the high initial and ongoing costs.

Data Security and Privacy Concerns

The increased reliance on digital technologies exposes oil companies to data security and privacy risks, including cyberattacks and data breaches.

- Data Protection: Ensuring the security of sensitive data and protecting systems from cyber threats is critical.
- Regulatory Compliance: Adhering to data privacy regulations and standards to ensure the secure handling of personal and operational data.

Human Story: Picture Ahmed, a cybersecurity specialist working to protect his company's AI systems from cyber threats, ensuring the security and integrity of critical data.

The Role of Policy and Regulation

Government policies and regulations play a crucial role in promoting the adoption of AI in the oil industry, addressing challenges, and ensuring compliance with standards.

Government Support and Incentives

Policies and incentives can drive the digital transformation of the oil industry by providing financial support and promoting technological innovation.

- Innovation Grants: Offering grants and subsidies to support research and development of AI technologies in the oil industry.
- Tax Incentives: Providing tax incentives for companies investing in AI infrastructure and cybersecurity measures.

Human Story: Consider Emily, a policy advocate working to promote government support and incentives for AI adoption in the oil industry, helping companies overcome financial barriers.

Regulatory Standards

Establishing regulatory standards for AI technologies can ensure their safe and effective implementation and promote industry-wide best practices.

- Cybersecurity Regulations: Implementing regulations to ensure robust cybersecurity measures are in place, protecting sensitive data and critical infrastructure.
- Data Privacy Standards: Establishing standards for data privacy and protection to ensure the secure handling of personal and operational data.

Human Story: Picture Li, a regulatory compliance officer ensuring that his company meets all regulatory standards for AI technologies and data protection, maintaining high levels of security and compliance.

Thought-Provoking Questions: Navigating AI in the Oil Industry

Reflecting on the impact of AI on the oil industry raises several critical questions: How can oil companies balance the high costs of AI implementation with long-term benefits? What role should government policies and incentives play in promoting AI technologies? And how can the industry address data security and privacy concerns?

Storytelling Techniques: Bringing AI Transformations to Life

To illustrate the complexities of AI in the oil industry, envision the innovative research in tech labs, the strategic discussions in boardrooms, and the practical applications in oil fields and refineries. Highlight the voices of data scientists, engineers, policymakers, and community members to provide a balanced and comprehensive narrative.

Actionable Insights: Strategies for Promoting AI in the Oil Industry

1. Investing in Research and Development: Supporting research and development of AI technologies can reduce costs and improve efficiency.
2. Enhancing Policy Support: Strengthening government policies and incentives can drive the adoption of AI in the oil industry and promote technological innovation.
3. Implementing Robust Cybersecurity Measures: Ensuring robust cybersecurity measures are in place to protect sensitive data and critical infrastructure is essential for successful AI adoption.
4. Fostering Industry Collaboration: Promoting collaboration and knowledge sharing within the industry can accelerate the adoption and effectiveness of AI technologies.

Conclusion: Navigating the Path to an AI-Driven Future

Artificial Intelligence is revolutionizing the oil industry, offering significant benefits in terms of efficiency, cost savings, and environmental sustainability. This crude awakening highlights the importance of technological innovation, supportive policies, and robust cybersecurity measures in realizing the full potential of AI.

By understanding the impact of AI on oil production and distribution, we can better appreciate the complexities and opportunities in the transition to a digitally driven energy future. The lessons learned from various AI initiatives provide valuable insights for shaping policies and practices that promote technological innovation and energy security. Through collective action and ongoing commitment to innovation and sustainability, the oil industry can navigate the challenges ahead and contribute to a more stable and sustainable energy landscape.

CHAPTER 99: FUTURE OF OIL TRANSPORTATION

Crude Awakening: The Modern Truths and Future of Oil

On the Move: Exploring Innovations in the Transportation of Oil

The transportation of oil, from extraction sites to refineries and markets, is a critical component of the global energy supply chain. Innovations in oil transportation are not only improving efficiency and safety but also addressing environmental concerns and enhancing sustainability. This chapter delves into the latest advancements in oil transportation, their benefits, challenges, and the broader implications for the oil industry and global energy landscape.

Understanding Oil Transportation

Oil transportation involves moving crude oil and refined products from production sites to refineries and end-users. Traditional methods include pipelines, tankers, railways, and trucks, each with its own set of challenges and advantages.

Key Methods of Oil Transportation

- Pipelines: The most efficient and cost-effective method for

transporting large volumes of oil over long distances.
- Tankers: Ships that transport oil across oceans, essential for international trade.
- Railways: Used for transporting oil in regions lacking pipeline infrastructure.
- Trucks: Ideal for short-distance and flexible oil transportation.

Human Story: Imagine Samuel, a logistics coordinator who has managed the complex task of transporting oil through various methods, now witnessing a transformation in the industry with new technologies.

The Need for Innovations in Oil Transportation

The oil industry faces increasing pressure to enhance the efficiency, safety, and sustainability of oil transportation. Innovations are essential to address these challenges and meet the demands of a changing energy landscape.

Environmental Benefits

Recent innovations in oil transportation aim to minimize environmental impact, reduce emissions, and enhance sustainability.

- Reduced Emissions: New technologies and methods can significantly reduce greenhouse gas emissions associated with oil transportation.
- Spill Prevention: Advanced monitoring and detection systems help prevent and mitigate oil spills, protecting ecosystems.
- Energy Efficiency: Innovations enhance the energy efficiency of transportation methods, reducing the overall carbon footprint.

Human Story: Consider Maria, an environmental scientist working with oil companies to implement new transportation technologies that minimize environmental impact and enhance safety.

Expert Insight: Dr. Laura Green, an environmental engineer, explains, "Innovations in oil transportation are crucial for making the industry more sustainable. These technologies not only improve efficiency but also play a vital role in reducing the environmental impact and enhancing safety."

Key Innovations in Oil Transportation

Several recent innovations are revolutionizing the transportation of oil, improving efficiency, safety, and sustainability.

Advanced Pipeline Technologies

Innovations in pipeline technology are making oil transportation more efficient, safer, and environmentally friendly.

- Smart Pipelines: Equipped with sensors and IoT technology, smart pipelines monitor flow, detect leaks, and optimize performance in real-time.
- Advanced Materials: New materials, such as composite and corrosion-resistant alloys, enhance the durability and safety of pipelines.
- Automated Monitoring Systems: AI and machine learning algorithms analyse data from sensors to predict maintenance needs and prevent failures.

Human Story: Picture John, a pipeline engineer overseeing the implementation of smart pipeline technology, significantly reducing the risk of leaks, and enhancing operational efficiency.

Innovations in Oil Tankers

Technological advancements are transforming the design and operation of oil tankers, making them more efficient and environmentally friendly.

- Double-Hull Designs: Modern tankers are built with double hulls to prevent spills in case of a collision or grounding.
- LNG-Powered Ships: Tankers powered by liquefied natural gas (LNG) produce fewer emissions compared to traditional fuel oil.
- Ballast Water Treatment Systems: Advanced systems treat ballast water to prevent the spread of invasive species, protecting marine ecosystems.

Human Story: Consider Emily, a captain of an LNG-powered oil tanker, navigating international waters with advanced systems that enhance safety and environmental performance.

Expert Insight: Dr. Robert Lee, an expert in maritime engineering, notes, "The advancements in tanker technology are significant. LNG-powered ships and double-hull designs are reducing emissions and preventing spills, making maritime oil transportation safer and more sustainable."

Rail and Truck Innovations

Innovations in rail and truck transportation are enhancing the flexibility and safety of oil logistics, particularly in regions lacking pipeline infrastructure.

- Automated Rail Systems: AI-driven systems optimize routing, monitor conditions, and ensure the safe transport of oil by rail.
- Hybrid and Electric Trucks: The adoption of hybrid and electric trucks reduces emissions and improves fuel efficiency in short-distance oil transportation.
- Enhanced Tank Car Designs: Modern tank cars are designed with improved safety features to prevent leaks

and withstand impacts.

Human Story: Picture Raj, a logistics manager utilizing automated rail systems and hybrid trucks to transport oil efficiently and safely, even in challenging terrains.

Challenges in Implementing Advanced Transportation Technologies

Despite their numerous benefits, advanced transportation technologies face several challenges that must be addressed to achieve widespread adoption and effectiveness.

High Initial Costs

The high costs associated with implementing and maintaining advanced transportation technologies pose a significant barrier for many oil companies.

- Capital Investment: The initial investment required for advanced infrastructure, vehicles, and systems can be substantial.
- Operational Costs: Maintaining and operating advanced transportation systems may require specialized skills and additional resources.

Human Story: Consider Sarah, a financial manager tasked with justifying the investment in new transportation technologies to stakeholders who are concerned about the high initial and ongoing costs.

Technical and Logistical Challenges

Implementing advanced transportation technologies involves overcoming various technical and logistical hurdles, including integration with existing systems and ensuring reliability.

- System Integration: Integrating new technologies with

existing transportation networks can be complex and requires careful planning and execution.
- Reliability and Maintenance: Advanced systems must be reliable and easy to maintain to ensure continuous operation and cost-effectiveness.

Human Story: Picture Ahmed, a project manager responsible for coordinating the integration of new transportation technologies into existing logistics operations, facing technical challenges, and ensuring minimal disruption.

The Role of Policy and Regulation

Government policies and regulations play a crucial role in promoting the adoption of advanced transportation technologies, addressing challenges, and ensuring environmental compliance.

Government Support and Incentives

Policies and incentives can drive the growth of advanced transportation technologies by providing financial support and promoting technological innovation.

- Regulatory Standards: Implementing strict environmental and safety standards encourages oil companies to adopt cleaner and safer transportation technologies.
- Financial Incentives: Subsidies, tax credits, and grants can offset the high initial costs of installing advanced transportation systems.

Human Story: Consider Emily, a policy advocate working to promote government support and incentives for oil companies that adopt advanced, environmentally friendly transportation technologies.

Industry Collaboration

Collaboration within the industry can accelerate the development and deployment of advanced transportation technologies.

- Knowledge Sharing: Sharing best practices and technological innovations can help oil companies improve their operations and reduce environmental impact.
- Joint Ventures: Collaborative projects and joint ventures can pool resources and expertise, making advanced technologies more accessible.

Human Story: Picture Li, a sustainability officer at a major oil company, leading industry-wide initiatives to promote the adoption of advanced transportation technologies.

Thought-Provoking Questions: Navigating Innovations in Oil Transportation

Reflecting on the impact of recent innovations in oil transportation raises several critical questions: How can oil companies balance the high initial costs with long-term environmental and economic benefits? What role should government policies and incentives play in promoting these technologies? And how can industry collaboration enhance the adoption and effectiveness of advanced transportation technologies?

Storytelling Techniques: Bringing Transportation Innovations to Life

To illustrate the complexities of transportation innovations, envision the innovative research in tech labs, the strategic discussions in boardrooms, and the practical applications in the field. Highlight the voices of engineers, scientists, policymakers, and community members to provide a balanced and comprehensive narrative.

Actionable Insights: Strategies for Promoting Advanced

Transportation Technologies

1. Investing in Research and Development: Supporting research and development of advanced transportation technologies can reduce costs and improve efficiency.
2. Enhancing Policy Support: Strengthening government policies and incentives can drive the adoption of advanced transportation techniques and promote environmental compliance.
3. Developing Infrastructure: Investing in infrastructure for advanced transportation processes is essential for widespread adoption.
4. Fostering Industry Collaboration: Promoting collaboration and knowledge sharing within the industry can accelerate the deployment and effectiveness of advanced transportation technologies.

Conclusion: Navigating the Path to Sustainable Oil Transportation

Innovations in oil transportation represent a promising path towards a more efficient, safe, and environmentally friendly oil industry. This crude awakening highlights the importance of technological innovation, supportive policies, and industry collaboration in realizing the full potential of advanced transportation techniques.

By understanding the development and benefits of advanced transportation technologies, we can better appreciate the complexities and opportunities in the transition to sustainable oil logistics. The lessons learned from various initiatives provide valuable insights for shaping policies and practices that promote environmental sustainability and energy security. Through collective action and ongoing commitment to innovation and sustainability, the oil transportation industry can navigate the challenges ahead and contribute to a more stable and sustainable energy landscape.

CHAPTER 100: THE ROLE OF OIL IN A SUSTAINABLE FUTURE

Crude Awakening: The Modern Truths and Future of Oil

Balancing Act: Discussing the Potential Role of Oil in a Sustainable Energy Future

As the world grapples with climate change and the urgent need for a sustainable energy future, the role of oil remains a complex and contentious issue. While renewable energy sources are crucial for reducing greenhouse gas emissions, oil continues to play a significant role in the global energy mix. This chapter explores the potential role of oil in a sustainable energy future, examining the balance between reducing reliance on fossil fuels and leveraging advancements in technology to minimize environmental impact.

Understanding the Current Energy Landscape

The global energy landscape is characterized by a mix of fossil fuels, renewables, and nuclear energy. Oil, despite its environmental drawbacks, remains a dominant source of energy due to its high energy density, versatility, and established infrastructure.

Key Facts About Oil in the Energy Mix

- Energy Density: Oil has a high energy density, making it an efficient source of energy for transportation, heating, and industrial processes.
- Versatility: Oil is used in various forms, including gasoline, diesel, jet fuel, and petrochemicals.
- Infrastructure: A vast global infrastructure supports the extraction, refining, transportation, and distribution of oil, contributing to its continued dominance.

Human Story: Imagine Samuel, a veteran in the oil industry who has witnessed its evolution over decades. He now ponders the future of oil in a world increasingly focused on sustainability.

The Challenges of Oil in a Sustainable Future

The primary challenge of including oil in a sustainable future is its significant contribution to greenhouse gas emissions and environmental degradation. Addressing these challenges requires innovative solutions and a balanced approach.

Environmental Impact

The extraction, refining, and combustion of oil result in considerable greenhouse gas emissions, air pollution, and ecological damage.

- Greenhouse Gas Emissions: Oil combustion releases large amounts of carbon dioxide (CO_2), contributing to global warming.
- Air Pollution: Emissions from oil-based fuels include pollutants such as sulphur dioxide (SO_2), nitrogen oxides (NOx), and particulate matter, impacting air quality and public health.
- Ecological Damage: Oil spills and leaks can cause severe

harm to marine and terrestrial ecosystems.

Human Story: Consider Maria, an environmental activist who has dedicated her life to raising awareness about the environmental impacts of oil and advocating for cleaner alternatives.

Expert Insight: Dr. Laura Green, an environmental scientist, explains, "The environmental impact of oil is undeniable, but technological advancements and policy measures can help mitigate these effects. It's about finding the right balance."

Innovations and Technologies for Sustainable Oil Use

Advancements in technology offer pathways to reduce the environmental impact of oil and integrate it into a more sustainable energy future.

Carbon Capture and Storage (CCS)

CCS technology captures CO_2 emissions from industrial processes and power plants, storing it underground to prevent it from entering the atmosphere.

- Emission Reduction: CCS can capture up to 90% of CO_2 emissions, significantly reducing the carbon footprint of oil-based energy.
- Enhanced Oil Recovery (EOR): Captured CO_2 can be used in EOR operations, increasing oil extraction efficiency while sequestering carbon.

Human Story: Picture John, an engineer working at a CCS facility, overseeing the capture and storage of CO_2 from a nearby refinery, contributing to both energy production and emission reduction.

Biofuels and Synthetic Fuels

Biofuels and synthetic fuels offer renewable alternatives

that can be blended with or replace traditional oil-based fuels, reducing overall emissions.

- Biofuels: Produced from biological materials, biofuels can be blended with conventional fuels to reduce carbon intensity.
- Synthetic Fuels: Made from captured CO_2 and renewable hydrogen, synthetic fuels can provide a carbon-neutral alternative to traditional fossil fuels.

Human Story: Consider Emily, a researcher developing advanced biofuels that can be used in existing engines and infrastructure, providing a bridge to a more sustainable energy system.

Enhanced Efficiency and Reduced Emissions

Technological advancements in refining and combustion can improve the efficiency of oil use and reduce emissions.

- Refining Innovations: Advanced refining techniques can produce cleaner fuels with lower sulphur content and higher energy efficiency.
- Combustion Technologies: Innovations in engine design and combustion technology can reduce fuel consumption and emissions in vehicles and industrial processes.

Human Story: Picture Raj, an automotive engineer working on the next generation of engines that are more efficient and produce fewer emissions, contributing to cleaner transportation.

The Role of Policy and Regulation

Government policies and regulations are crucial in shaping the future role of oil in a sustainable energy landscape. Effective policies can drive innovation, reduce emissions, and promote the transition to cleaner energy sources.

Regulatory Measures

Implementing strict environmental regulations can ensure that oil production and use are as sustainable as possible.

- Emission Standards: Setting stringent emission standards for vehicles, refineries, and power plants to reduce pollution and greenhouse gas emissions.
- Renewable Energy Targets: Mandating the integration of renewable energy sources into the energy mix to gradually reduce reliance on fossil fuels.

Human Story: Consider Sarah, a policy advisor working to develop and implement regulations that encourage the use of cleaner technologies and renewable energy sources.

Economic Incentives

Providing economic incentives can accelerate the adoption of sustainable practices and technologies in the oil industry.

- Subsidies and Tax Credits: Offering financial incentives for companies investing in renewable energy, carbon capture, and advanced refining technologies.
- Carbon Pricing: Implementing carbon pricing mechanisms, such as carbon taxes or cap-and-trade systems, to incentivize emission reductions.

Human Story: Picture Ahmed, a financial analyst advocating for carbon pricing to drive investment in sustainable technologies and reduce the carbon footprint of the oil industry.

Thought-Provoking Questions: Navigating Oil's Role in a Sustainable Future

Reflecting on the potential role of oil in a sustainable future raises several critical questions: How can we balance the continued use of oil with the urgent need to reduce emissions? What role should government policies and economic incentives play in promoting sustainable practices? And how can technological innovations help mitigate the environmental impact of oil?

Storytelling Techniques: Bringing Sustainable Oil Practices to Life

To illustrate the complexities of integrating oil into a sustainable future, envision the innovative research in laboratories, the strategic discussions in government offices, and the practical applications in refineries and fields. Highlight the voices of engineers, scientists, policymakers, and community members to provide a balanced and comprehensive narrative.

Actionable Insights: Strategies for Promoting Sustainable Oil Use

1. Investing in Research and Development: Supporting research and development of sustainable technologies can reduce the environmental impact of oil and improve efficiency.
2. Enhancing Policy Support: Strengthening government policies and incentives can drive the adoption of cleaner technologies and promote environmental compliance.
3. Developing Infrastructure: Investing in infrastructure for carbon capture, biofuels, and renewable energy integration is essential for a sustainable energy future.
4. Fostering Industry Collaboration: Promoting collaboration and knowledge sharing within the industry can accelerate the adoption and effectiveness of sustainable practices.

Conclusion: Navigating the Path to a Sustainable Energy Future

The role of oil in a sustainable future is complex and multifaceted. While the transition to renewable energy sources is essential, oil can still play a role if managed responsibly and innovatively. This crude awakening highlights the importance of technological innovation, supportive policies, and industry collaboration in realizing a sustainable energy landscape.

By understanding the potential role of oil in a sustainable future, we can better appreciate the complexities and opportunities in the transition to cleaner energy. The lessons learned from various initiatives provide valuable insights for shaping policies and practices that promote environmental sustainability and energy security. Through collective action and ongoing commitment to innovation and sustainability, the oil industry can navigate the challenges ahead and contribute to a more stable and sustainable energy landscape.

www.ingramcontent.com/pod-product-compliance
Lightning Source LLC
Chambersburg PA
CBHW071443220526
45472CB00003B/638